Chinese Research Perspectives on the Environment, Volume 1

Chinese Research Perspectives: Environment

VOLUME 1

The titles published in this series are listed at brill.com/crp

Chinese Research Perspectives on the Environment, Volume 1

Urban Challenges, Public Participation, and Natural Disasters

Edited by

Yang Dongping
Friends of Nature

BRILL

LEIDEN • BOSTON
2013

This book is the result of a copublication agreement between Social Sciences Academic Press and Koninklijke Brill NV. These articles were selected and translated into English from the original 《中国环境发展报告 (2011)》 (*Zhongguo huanjing fazhan baogao 2011*) and《中国环境发展报告 (2012)》 (*Zhongguo huanjing fazhan baogao 2012*) with the financial support of the Chinese Fund for the Humanities and Social Sciences.

This publication has been typeset in the multilingual "Brill" typeface. With over 5,100 characters covering Latin, IPA, Greek, and Cyrillic, this typeface is especially suitable for use in the humanities. For more information, please see www.brill.com/brill-typeface.

ISSN 2212-7496
ISBN 978-90-04-24953-0 (hardback)
ISBN 978-90-04-24954-7 (e-book)

This book is printed on acid-free paper.

MIX
Paper from responsible sources
FSC® C109576
FSC www.fsc.org

Printed by Printforce, the Netherlands

CONTENTS

PART II

ENVIRONMENTAL POLLUTION AND ECOLOGICAL PROTECTION

PART III

URBAN ENVIRONMENTAL ISSUES AND LIVABILITY

PART VIII

APPENDIX

LIST OF FIGURES

LIST OF TABLES

ACKNOWLEDGEMENTS

Chinese Research Perspectives on the Environment, Volume 1 is a collection of articles selected from The China Environment Yearbooks published in Chinese separately in 2011 and 2012.

The translation of this edition could not have been accomplished without the team work of a group of volunteers. We are indebted to Zhou Yong who translated the two long and informative General Reports plus several other chapters and Cai Jindong who translated the Annual Indexes with numerous tables and figures. We are grateful to all the other volunteers who translated the remaining chapters. Almost all of them have participated in the translation project of the previous *China Environment Yearbooks*. Their dedication to Friends of Nature and environmental protection is truly admirable. They are Chai Liang, Chen Minghui, Cheng Juebin, Cheng Yinan, Gu Qing, Li Zhaohui, Sheng Changying, Song Yi, Sun Fei, Wan Feng, Wang Lili, Wang Xiuqiong, Zhai Zheng, Zhao Fang, and Zhou Wei.

Erika Scull did the final editing for the whole book, and joined Professors Judith Shapiro of American University and Guobin Yang of the University of Pennsylvania on the International Advisory Board. We are deeply grateful to this dedicated group for their support and wisdom.

Finally, we would like to thank all the individuals and organizations who have offered their warm support to the Green Book translation project. We hope that you will continue to give your attention to the project and become more actively involved in environmental protection, a shared cause for all citizens who cherish nature and have a strong sense of responsibility for our common planet.

The Green Book Project of Friends of Nature
Social Sciences Academic Press, Beijing
Brill, Leiden

LIST OF CONTRIBUTORS

Throughout this volume, Chinese names are ordered according to standard practice in China, with surnames preceding given names. To add clarity this ordering, surnames are in capital letters in the following list of contributors.

Chief Editor

YANG Dongping (杨东平), president of Friends of Nature, is a professor at the Beijing Institute of Technology.

Contributors

BO Xiaobo (薄晓波) is a Ph.D. candidate for Environment and Resources Protection Law at China University of Political Science and Law, and a volunteer at the Center for Legal Assistance to Pollution Victims.

CHENG Yuanyuan (陈媛媛) is a reporter with China Environment News.

DU Huanzheng (杜欢政) is the deputy director of the Center for Studies on China's Circular Economy and Environment, professor and dean of the Circular Economy and Technology Institute for the Yangtze Delta, Zhejiang, and Jiaoxing Institute.

FAN Xiao (范晓) is a senior engineer at the Sichuan Bureau of Geology and Mineral Resources.

FENG Jia (冯嘉) is a Ph.D. student of Environment and Resources Protection Law at China University of Political Science and Law. She is also a volunteer at the Center for Legal Assistance to Pollution Victims and a visiting scholar at the Law School of Columbia University in New York.

FRIENDS OF NATURE (FON) is the first officially registered environmental non-governmental organization in China. It is based in Beijing with chapters in Nanjing, Guangdong, Shenzhen, and other major Chinese cities.

GU Chaolin (顾朝林) is a professor at the School of Architecture, Tsinghua University.

GU Xunzhong (顾训中) is the director of the Shanghai Jingguang NPO Service Center, a member of Friends of Nature's Shanghai Chapter, and a director on the Board of Friends of Nature.

HU Jing (胡静) is a J.D. and associate professor at the School of Civil, Commercial, and Economic Laws of the China University of Political Science and Law, with research focused on environmental law.

HU Kanping (胡勘平) is a member of Friends of Nature, director of Research and Communications at the Chinese Ecological Civilization Research and Promotion Association, and a professor at the Zhejiang Agricultural and Forestry University.

HU Mengjie (胡梦婕) is an M.S. student of environmental engineering at the Circular Economy and Technology Institute for the Yangtze Delta, Zhejiang.

LI Bo (李波) is the former executive director of Friends of Nature and currently a senior advisor there. He is also a research fellow at the India China Institute of The New School in New York.

LI Dun (李楯) is a professor and head of the Experts Network at the Tsinghua University Center for the Study of Contemporary China and executive director of the Social Policy Research Institute, School of Public Policy and Management at Tsinghua University. Professor Li is also an advisor with FON.

LO Sze Ping (卢思骋) is a director on the Board of Friends of Nature and founder of Forward Works, an organization whose mission is to promote social progress and ecological conservation.

LÜ Zhongmei (吕忠梅) is an S.J.D. and doctoral tutor at the Zhongnan University of Economics and Law, and the president of the Environmental Law Research Center of the same university. She is also the president of the Hubei University of Economics and of the Environmental Resources Law Institute of Zhongnan University of Economics and Law. She is a delegate to the Tenth and Eleventh National People's Congress.

MAO Da (毛达) is a Ph.D. candidate for Environmental History at Beijing Normal University. He is a consultant for the Work Group of Municipal Solid Waste and also a member of Friends of Nature.

QIE Jianrong (郄建荣) is a member of Friends of Nature and a senior journalist with *Legal Daily*, who has been awarded one of the Best Journalists Reporting Environmental Issues for three consecutive years by Friends of Nature.

Erika Scull (International Advisory Board) has an M.A. from American University's School of International Service in Global Environmental Politics and is a former Boren Scholarship recipient for her work on China's environment.

Judith Shapiro (International Advisory Board) is the director of the Natural Resources and Sustainable Development M.A. program at American University's School of International Service in Washington, D.C. She has published extensively on modern and contemporary China, including *Mao's War against Nature* (Cambridge, 2001), and *China's Environmental Challenges* (Polity, 2012).

SHEN Xiaohui (沈孝辉), a member and a council member of Friends of Nature, is a member on the Chinese National Committee for Man and the Biosphere, and senior engineer of the State Forestry Administration. He has done extensive research on protected areas, forests, wetlands, deserts, and wild animals in China and has been an activist in environmental protection.

SHI Tie (石铁) is an M.A. student at the School of Civil, Commercial, and Economic Laws at the China University of Political Science and Law, with research focused on environmental law.

WANG Jian (王建) is a water environmentalist and a member of Friends of Nature.

WANG Jingjing (王晶晶) is the deputy director of the Institute of Environmental and Public Affairs.

WANG Shekun (王社坤) is a J.D. and instructor of law at Peking University, and a researcher at the Resources, Energy & Environmental Law Institute, Peking University.

WANG Tao (汪韬) is a reporter with *Southern Weekly*.

YAN Weiqi (颜维琦) is a journalist with *Guangming Daily*.

YANG Changjiang (杨长江) is a journalist with the *China Inspection and Quarantine Times*. As a member of Friends of Nature, he has shown deep concern for problems concerning municipal refuse.

YANG Chuanmin (杨传敏) is a senior journalist reporting environmental issues for the *Southern Metropolis Daily*.

Guobin Yang (International Advisory Board) is an associate professor of Communication and Sociology in the Annenberg School for Communication and Department of Sociology at the University of Pennsylvania. He has published widely on environmental NGOs and activism in China, and is the author of *The Power of the Internet in China: Citizen Activism Online*.

YANG Weihe (杨纬和) is head of a project on waste management at the Huanyou Science and Technology Research Center.

YI Yimin (易懿敏) is project officer at Moving Mountains, where she researches the social and environmental impact of economic development in China and overseas. Yi is a member of Friends of Nature.

YU Yin (于音) has worked and lived in Thailand for many years. Her work focuses on studying Chinese development influence in the Mekong region. Currently she is based in Beijing, working with Greenovation Hub, a grassroots environmental NGO.

ZENG Xiangbin (曾祥斌) is a lawyer at the Bejing Yinke Law Firm, Wuhan Office, leader of the Wuhan Chapter of Friends of Nature, and a member of the legal team in the public-interest litigation against the chromium slag pollution in Yunnan.

ZHAO Zhangyuan (赵章元) is the former director of the Center for Research on Lake Environments and director of the Center for Research on Offshore Environments of the Chinese Research Academy of Environmental Sciences, member of the council of the Chinese Geophysical Society, and an expert frequently invited by the Environmental Engineering Assessment Center affiliated to the former State Environmental Protection Administration.

ZHANG Boju (张伯驹) is the director of the Department of Research and Investigation of Friends of Nature and a member of the executive committee of Green Society Environmental Action Network. He was also a UNEP Youth Advisor.

ZHENG Yisheng (郑易生), a member and consultant of Friends of Nature, is a research fellow of the Institute of Quantitative and Technical Economics of the Chinese Academy of Social Science.

ZHOU Lei (周雷) is a doctor of anthropology and a post-doc fellow at the School of Social and Behavioral Sciences at Nanjing University.

ZHU Yucheng (朱渝铖) is an M.A. student of Industrial Economics at the Circular Economy and Technology Institute for the Yangtze Delta, Zhejiang.

INTRODUCTION

Judith Shapiro[1]

China's environmental future is, arguably, the future of the planet. China has an enormous footprint as a consumer of resources and producer of pollution, a special position as the world's manufacturing hub in a globalized marketplace running up against the limits of resources, and a newfound role as a superpower and leading voice of developing countries in the global debate about how to protect our dwindling resources. Critical to China's environmental future is the role of its public citizens' groups, which are increasingly mature, well-organized, sophisticated in the use of social media and other tools to leverage influence in environmental decision-making, and globally networked.

This series, the *Chinese Research Perspectives on the Environment* (formerly the *China Environment Yearbook*), is a product of one of the leading and most respected Chinese non-governmental organizations, Friends of Nature. Each year, through a compendium of perspectives from Chinese civil society, the organization seeks to shape public debate, provide information, and offer a counterpoint to the voices of government research institutions and government policy makers that shape China's environmental future. They do so in a politically perilous landscape, yet they have managed to expand the space for public participation during the last few decades to a truly remarkable degree.

A clear case can be made that the national government—or some parts of it—is inching toward a greater alliance with the green movement, as part of an overall trend toward an expanded civil society and rule-by-law. Top government officials have recognized that cleaning up the environment requires a combination of clear laws, transparent standards and practices, development of institutions, defined rights, a democratic process, encouragement of public participation, and engagement with the media.

Nonetheless, Chinese civil society faces severe limitations in its ability to organize, communicate, and conduct activities. Yet even as there are

[1] Parts of this essay have been adapted from *China's Environmental Challenges* (Polity Books) and are reprinted with permission.

tremendous changes with respect to the emphasis on sustainable development from the central leadership, as well as new demands for pollution controls and food safety from influential middle class consumers, the grassroots are feeling their power and becoming active, often forming a partnership with the upper reaches of the bureaucracy in order to pressure corrupt developers, self-interested local officials, lower-and middle-level bureaucrats, and polluting factory operators.

It is important not to over-generalize about Chinese civil society, for China is a large, diverse, and unevenly developed country, with unequal distribution of resources and local empowerment. The eastern seaboard and urban areas have different characteristics and dynamics than rural areas and interior western provinces. But even in remote areas such as the Nu River (怒江) region near Myanmar and the Upper Yangtze near the Tibetan Autonomous Region, there are signs of an increasingly vibrant civil society; environmental activists and filmmakers have helped to educate and inspire local people to oppose a series of dams that would displace them.

In recent years, as evidenced by this volume, there has been a remarkable shift in the strategies and activities of Chinese environmental NGOs (or ENGOs), which remain constrained by the need to register with the government and haunted by the threat of closure. Understanding the acceptable limits on their activities is a subtle game, with the degree of freedom available in the country at any given moment shifting. Freedoms contract during major Party congresses, for example, and during important anniversaries. Often, the demarcation line is not visible until it has been crossed. Nonetheless, ENGOs are expanding their range from traditional programs of environmental education and non-confrontational activities like trash pick-up, tree planting, and advocating for the protection of endangered species, and are now assuming advocacy roles similar to those adopted by ENGOs in other parts of the world. For example, they have started to pressure the government to take greater responsibility for China's role in climate change and served as advocates for communities victimized by pollution. They are exploring tactics of "symbolic politics," as pioneered worldwide by groups like Greenpeace, putting up banners in public places and mounting actions designed to garner media attention. They are taking the lead to pressure the government for greater transparency in disclosing air and water pollution statistics, while taking on powerful corporate interests such as the multinational corporations Monsanto (for genetically modified foods) and Hewlett Packard (for failing to remove toxic components from computers). The adoption of these strategies represents a maturation of Chinese activism and an indication of increased confidence and willingness to take on politically difficult issues. It may also indicate that activists feel safer. The central government's stated commitments to environmental protection provide them with tacit permission, as these official pronouncements open up avenues for "accountability politics." Citizens now have ammunition to hold government agencies to their promises to clean up pollution, improve public safety, and put China on the road to sustainable development.

Internet organization has also become a powerful tool for civil society organizing outside formal group structures, such as when citizens mobilized themselves via cell phone messages to rally against a paraxylene (PX) chemical plant planned for the coastal resort city of Xiamen (厦门) in Fujian Province and later in Dalian (大连), Liaoning Province. The citizens of Shanghai opposed the building of a magnetic levitation rail line through their neighborhood and organized themselves to "take a walk" on certain days; the phrase is so innocuous and the words in it so common that it cannot be filtered and banned, and participants cannot be arrested because they can claim that they just happened to be in the area. Such protests can be understood as a wave of "not in my back yardism" (NIMBYism) on the part of a newly empowered Chinese middle class. Unfortunately, in most cases, as elsewhere in the world, NIMBYism often results in the removal of the offending project to a more vulnerable, less well organized community.

Another very significant tool for Chinese ENGOs is their alliance with courageous investigative journalists and newspaper editors, who in China have long played a role as advocates for the disenfranchised by speaking truth to power and providing a conduit for information from the grassroots to top leaders who might be in a position to provide justice. The path-breaking news magazine, *Southern Weekend* (南方周末) repeatedly publishes environmental stories at considerable risk to its editors and writers, but despite firings and demotions, the publication's prestige and moral authority have allowed it to remain open. Blogs, the Internet, and instant messaging have also opened up space for the conveyance of environmental information.

Another strategy that Chinese NGOs have recently adopted with enthusiasm is that of survey research, a form of information politics. The E-waste Civil Action Network conducted a study revealing that 60% of Beijing's e-waste is being sold to trash collectors, who pay more than government recycling stations, and that the waste continues to end up in Guangdong Province where cities like Guiyu (贵屿) are hubs of the underground market. Groups have formed domestic and international alliances to conduct extensive investigations into the polluting practices of multinational corporations, such as when Friends of Nature, the Institute of Public and Environmental Affairs, Green Beagle, EnviroFriends, and Nanjing Green Stone researched Apple Inc.'s supply chain and exposed serious problems in 27 factories. Over the course of seven months, they interviewed villagers affected by chemical plants that supply Apple components, took water samples that revealed serious pollution levels, and produced an extensive report that they made available electronically, accusing Apple of failing to monitor its supply chain. This exposé came to global attention, and in alliance with United States-based NGOs like the Natural Resources Defense Council, China's Institute for Public and Environmental Affairs led by water activist Ma Jun (马军) held discussions in California with Apple executives. Chinese groups have continued to pressure Apple to become more transparent and to take responsibility for its suppliers, forcing Apple to undertake regular talks with environmental groups. This campaign demonstrates increased sophistication in networking and the use of the media to disseminate findings, willingness to take risks by conducting undercover investigations, fearlessness in taking on a powerful multinational corporation whose business is important for China, ability to connect with ordinary villagers who trust such groups to be their voices for justice, and capacity to form transboundary alliances with other members of global civil society to put pressure on a major transnational corporation.

A significant trend in domestic environmental politics is the rise in frequency and scope of environmental litigation and citizens' growing awareness of the courts as a venue to resolve environmental disputes. Although there are still tremendous weaknesses in the Chinese legal system with regard to implementation and enforcement, and the number of such lawsuits is difficult to quantify, the increase in the use of such tactics is a signal of rising confidence in the rule of law and a growing sense that citizens have a right to a clean environment that the state has an obligation to protect. Although the use of the Chinese courts on behalf of pollution victims is far from new, the strategy appears to have achieved broader acceptance, with even ordinary farmers aware of the possibility of bringing their cases to court. This marks a new tool compared with the Chinese culture of dispute resolution. Traditionally, wronged citizens seek to bring a petition to a higher authority by traveling to the front gate of an agency and trying to pass a letter to an influential person inside, or to display a large poster on the sidewalk outside detailing their woes, or to use personal influence or bribery to try to gain redress. The numbers of legal petitioners have grown astronomically as pollution has worsened throughout the country, and more than 40 new specialized courts or tribunals dedicated to hearing environmental lawsuits are now hearing cases, many of them brought by public interest plaintiffs including NGOs, private citizens, and environmental protection bureaus.

In this volume, which encompasses two years, we see a new focus on the environmental impact of natural disasters, from mudslides due to torrential rains to pollution spills of oil and chromium. We see a record of the struggle to preserve a nature reserve from a dam, the Xiaonanhai (小南海), which ENGOs fought hard but ultimately lost. We learn about the influence of foreign debates over urban "livability" and consumption patterns, and the intense concern over the urban waste crisis. There are developments in the ability of ENGOs to pressure the government to enforce its environmental laws, as in the essay over the Ministry of Environmental Protection's failure to sue over a mining disaster, as well as new pressures for information disclosure, particularly with respect to the fine particle matter air pollution struggle in Beijing, which has now become well known internationally and is a success story for citizens' right to know. Finally, we see here an increasing sensitivity to international criticism over China's environmental behavior overseas in places like Africa and Southeast Asia. Domestic ENGOs are taking up the call to ask Chinese investors to consider their responsibility not to shift environmental harm beyond China's borders.

In only a generation and a half, China has undergone a transition that now touches every conceivable facet of the country, including the government, the people, civil society, and even intangible cultural characteristics such as identity and tradition. China's rapid economic and much slower political "openings" have drastically changed the landscape, both physically and metaphorically, in ways that could not have been imagined a few decades ago.

One fact is clear: China's environment is beset by numerous interconnected and countervailing forces and pressures. Environmental challenges are posed by many aspects of China's present situation, including its astronomical growth, the evolving role of the state, and the emergence of civil society. At the same time, progress toward a sustainable future is hampered by an East-West divide, unequal access to resources, corruption, lax enforcement of regulations, and unresolved and often unacknowledged ethnic conflicts. Even the progress made by middle class and green activists can be illusory if the environmental problems that are tackled in one region are simply shifted to another.

Emerging civil society faces questions about its own ability and intent, as it intersects in unusual ways with China's changing demographics and economy. A newly empowered middle class is beginning to demand more of the government with respect to food safety and pollution standards. Chinese cities have become cleaner as community groups and loosely organized citizens confront government planners and businesspeople. Some measure of justice has been meted out to corrupt industry barons and officials. At the same time, at least some of these demands also stem from broader popular desires for more consumption and affluence, both of which accompany a lifestyle that contributes to environmental degradation. So, while ENGOs press the government for action, they are also doing some soul searching. Can the environmental movement advocate an alternate path and still draw support from a middle class that at times is focused on wealth and conspicuous consumption? Successful protests against polluters may shut down a factory, only to cause it to relocate in a more marginal, less powerful community. Severe environmental problems are occurring in ethnic minority communities on the periphery of the country, where a grab for resources is thinly disguised as a development program. In these hinterlands, poor people will bear the heaviest burden of climate change. Yet these communities suffer racial and class prejudices. This will be an evolving area for China's burgeoning environmental movement as it seeks to balance the desires of its urban support base and wider ideals of environmental justice.

Complex challenges within China overlap and intersect in ways that rule out quick fixes. And while there are clear reasons for hope, time is running short. There is huge pressure for China to continue on its current path of rapid growth coupled with environmental degradation, both within and beyond borders. The ability to shift course is constrained by historical and cultural baggage that puts the emphasis on global recognition and conspicuous consumption.

There are encouraging signs that a shift is already under way, with greater confidence in the legal system, increased public participation and information transparency, and broad recognition that development that chokes and harms is undesirable. Significant forces are working tirelessly for greater attention to and emphasis on sustainability, including within the Chinese state system. The international community's recognition of their efforts is critical: China and the world are so intertwined that what happens in China not only impacts all of us, but what all of us do has an impact in China.

VOLUME OVERVIEW

GENERAL REPORT

ENVIRONMENTAL DEGRADATION AND THE WATER CRISIS IN RURAL CHINA: AN INTRODUCTION

Li Dun

Abstract: Although according to government reports environmental conditions in China are improving, the public has hardly been reassured after renewed problems with environmental detection indices and environmental incidents. There were many uncertainties about both China's development and the global situation in 2011. By studying structural and regulatory issues, we will gain a better understanding of the institutional obstacles to the country's sustainable development.

Keywords: pollution, ecosystem deterioration, water, Chinese characteristics, significance of actions

The public seems to have gotten used to disasters in recent years, regardless of whether they are droughts, floods, bridge/road collapses, casualties, and property losses in China or floods as far as away as Australia and South America, tornadoes in the Unites States that came earlier than usual, and temperatures as low as –60°C in Russia. It seems as if God could really be blamed for all these tragedies, were it not for the news commentators who, from time to time, remind us to think about human beings' responsibility.[1] Nonetheless, we are now inevitably faced with the relationship between natural disasters and human responsibility after the Great Japanese Earthquake and the resulting tsunami and nuclear accidents.

2011 marks the tenth anniversary of China's WTO membership. Investors, businesses, consumers, and producers worldwide are all contributing to the rapid growth of the Chinese economy. Nonetheless, there are concerns that the well-established core concepts of being human-oriented and sustainable, the judgment that peace, development, and cooperation are the current themes of the world, and the strategic idea that development will only be achieved by changing the industrial structure and the growth/consumption patterns and building an ecological civilization, have all been set aside by the subprime mortgage crisis, the EU debt crisis, the tensions over the South China Sea, and the unrest in the Middle East and North Africa. What will the prospects of China and the rest of the world be

[1] Source: "News Weekly," CCTV, July 16, 2011.

if we continue to stick to the existing growth pattern that opposes being human-oriented and sustainable? That is our concern in this essay.

I. *New Figures Seen in a New Way*

Environmental and ecosystem deterioration has become less severe in China, according to figures from the Chinese government's *Report on the State of the Environment in China* (中国环境状况公报).[2] Nonetheless, there have been actions and arguments in which monitoring and detection methods are challenged, in addition to frequent media reports on environmental incidents. How should we look at those figures?

A. *Water Pollution*

The previous environmental reports published by the Chinese government show that the poorest overall water quality of the seven largest drainage systems in China occurred in 2002, when water at Levels IV, V, and over V represented 70.9% of the whole. Later, that percentage gradually decreased to 45% in 2008, 42.7% in 2009 and 40.1% in 2010. Nonetheless, the Yellow and Liao (辽河) Rivers remain moderately polluted and the Hai River (海河) remains heavily polluted.

As for the state-controlled lakes and reservoirs, pollution became the most severe in 2008, as water at Levels IV, V, and over V represented 78.6% (the percentage was also above 70% in each of the previous four years) before the percentage decreased to 76.9% in 2009 and increased to 77% in 2010. The water in Tai (太湖), Dian (滇池), Dalai (达赉湖), Baiyangdian (白洋淀), and Dongting (洞庭湖) Lakes was above Level V; the water in the Chao (巢湖), Hongze (洪泽湖), Poyang (鄱阳湖), and Nansi (南四湖) Lakes was at Level V.

China is faced with a concurrence of problems such as groundwater pollution, excessive water exploitation, reduced water levels, subsidence, and large areas of groundwater depression cones. The 2011 version of the environmental report shows that 57.2% of the 4,110 monitored points in 182 Chinese cities had poor or very poor groundwater quality. Cities where water quality is getting worse are mostly in north, northeastern, and northwestern China. Compliant water represented only 76.5% of the

---

[2] The *Report on the State of the Environment in China* is published by the Ministry of Environmental Protection in June every year. All the following figures with unidentified sources are from this report published in previous years.

water sampled in 113 Chinese cities and at 395 centralized drinking-water sources, according to this report.

Seawater at Levels I and II represented 62.7% of the total in 2010, down 10.2 percentage points from 2009; seawater at Level III represented 14.1%, up 8.1 percentage points from 2009; and seawater at Levels IV and above IV represented 23.2%, up 2.1 percentage points from 2009. Water quality in the Bohai Sea was poor (or moderately polluted); water quality in the East China Sea was very poor (or heavily polluted); water quality in Liaodong Bay was poor; and water quality in Bohai Bay, the estuary of the Yangtze River, Hangzhou Bay as well as the estuaries of the Min (闽江) and Pearl Rivers was very poor.

A Greenpeace report published in 2011 reveals that international clothing brands have caused water pollution in China. The report states that numerous toxic and harmful substances (such as heavy metals and chemicals) are above China's water pollution standard and these substances pose direct and severe threats to human health and the ecosystem, as they are carcinogenic and may interfere with humans' and animals' endocrine systems; it is difficult for such substances to be naturally degraded in their surroundings and they may accumulate in bodies; and they may be carried through ocean currents, atmospheric deposition, and food chains to places even as far as the polar regions. This report reveals a problem that is facing not only China, but also the entire world.

The Greenpeace report pointed out that numerous global brands opt to open factories in China in order to reduce costs (because China's laws have lower environmental requirements than do the laws of developed countries); local wastewater treatment plants are unable to remove the majority of toxic and harmful substances from wastewater; considerable amounts of persistent toxic and harmful substances are discharged into rivers, lakes, and seas; alkylphenol and perfluorinated compounds (PFCs), which are persistent and able to interfere with the endocrine system, have been found in fish in the Yangtze River; polyoxyethylene nonylphenyl ethers and nonylphenol, resulting from the decomposition of the former in water, can all be found from within the wastewater generated by the daily washing of clothes. The substances outside China's standard, even at tiny amounts, may cause serious harms to humans.[3]

[3] Greenpeace, "Toxic Fashion—A Survey on Water Pollution by Global Clothes Brands in China; Toxic Fashion 2: Toxins in Clothes—a Survey on Residual Toxic and Harmful Substances from Global Clothes Brands," 2011.

Toxic and harmful substances including heavy metals such as copper, lead, mercury, chromium, and cadmium as well as chemicals such as nonyphenol/octylphenol, trichloromethane, dioxin, dibutyl phthalate (DBP), formaldehyde, azo compounds and perfluorooctane sulfonates (PFOS) have been found from within the wastewater discharged by China's textile and garment industries, according to an earlier Greenpeace report.[4]

Media reports say: "Oysters at the estuary of the Pearl River contain copper 740 times as high as the national standard and that is not the only such case along the 18,000-kilometer coastline of China. Seafood is being polluted with heavy metals, red tides, and organic compounds, according to many ocean and research reports." "About 1.08 million tons of pollutants flowed from the Pearl River into the South China Sea in 2010, leading to excessive heavy metals in some kinds of shellfish, according to the *Environmental Report on the South China Sea off Guangdong Province 2010* (2010年广东省海洋环境质量公报) published by the Administration of Ocean and Fisheries of Guangdong Province in April 2011." "Heavy metal pollution has become a problem throughout China's marine environment." "Shellfish toxins such as paralytic shellfish poisoning (PSP) and diarrheic shellfish poisoning (DSP) produced in red tides may pose direct threats to human beings through shellfish species prone to heavy metal pollution."[5]

An oil spill occurred at the Bohai Penglai 19-3 Oilfield, China's largest offshore oil and gas field, on June 4, 2011. The disaster was not under control by the end of August.[6] About 700 barrels (or 115 cubic meters) of oil had spilled and 2,600 barrels (or 416.45 cubic meters) of mineral oil-based drilling mud had spilled and had become sediment on the sea bed by the end of the year, according to a report published by the Xinhua News Agency. ConocoPhillips China, a wholly-owned subsidiary of U.S.-based ConocoPhillips, was blamed for the oil spill. China National Offshore Oil Corporation (CNOOC) holds 51% of the interest in the project that caused the accident.

The public was outraged for three reasons: 1) the accident went unreported long after it occurred; 2) ConocoPhillips showed a bad attitude; and 3) the oil spill involved CNOOC. Eleven environmental NGOs such as the Green Beagle Environment Institute (Green Beagle), the Institute

---

4 Greenpeace, "Pollution by Fashion—an Environmental Survey on Two Textile Industry Towns in China," 2010.

5 Lü Minghe, "Seafood vs. Pollution," *Southern Weekly*, October 13, 2011.

6 Source: "News Weekly," CCTV, August 20, 2011.

of Public & Environmental Affairs (IPE) and the Friends of Nature (FON) jointly wrote letters to CNOOC and ConocoPhillips asking them to disclose the details of the accident and the oil spill clean-up and to apologize to both the victims and the public. Later on, victims of the pollution sued the polluters for compensation, but none of them saw good results. ConocoPhillips claimed that no evidence showed that the oil spill in the Bohai Sea had caused environmental pollution. It agreed to set a compensation fund for the oil spill to settle "reasonable claims" afterward, but still said that, "The current result shows that the oil spill has tiny persistent effects on the environment."[7]

Given the public attention, CNOOC announced in January 2012 that ConocoPhillips would provide 10 billion RMB as pollution compensation. There is no further information on how compensation will be made available, but the public still questions what the compensation amount will be and how it will be allocated.

A natural gas leak occurred in CNOOC's pipelines in Zhuhai (珠海), Guangdong Province in December 2011. No oil spill was found except for thinly scattered oil stains, according to the South China Sea Branch of the State Oceanic Administration (SOA).[8]

B. *Air Pollution and Acid Rain*

The percentage of urban air quality at or better than Level 2 (which is compliant) increased from 33.8% in 2000 to 72.8% in 2008, 76.8% in 2009, and 81.7% in 2010. The percentage of urban air quality worse than Level 3 was higher than 33% in 2000 and 2001, before decreasing annually to 1.4% in 2008 and 1.3% in 2009. It rose to 1.8% in 2010.

Although the government has announced that the number of days when blue skies are seen in cities is increasing, volunteers in Beijing such as Lu Weiwei (卢为薇) and Fan Tao (范涛) (the co-authors of the *Daily Records of Blue Skies* (蓝天日记)) took photos of blue skies on a daily basis. Some citizens and environmental NGOs in other cities measured the local air quality, because the national monitoring standard covers only particles

[7] Hu Junchao and Wang Yu, "ConocoPhillips Denies That It Had Claimed There Was No Pollution, Still With No Comment on Compensation," Xinhuanet, http://news.xinhuanet.com/2011-12/21/c_111269507.htm (accessed December 21, 2011).

[8] Zhou Wenchao, "No Oil Spill Found in the Area Where a Natural Gas Leak Occurred to CNOOC Pipelines, Only Bits of Oil Stains," China National Radio, http://www.cnr.cn/china/gdgg/201112/t20111223_508964977.shtml (accessed December 23, 2011).

with diameters no smaller than 10 microns.[9] It fails to cover particles with diameters as small as 2.5 microns that can never be removed from human bodies once they enter the lungs. Such fine particles contain many toxic and harmful substances, which may cause tracheitis, cardiovascular diseases, and even cancers. Among citizens who measured air quality were tycoons and opinion leaders such as Pan Shiyi (潘石屹) and Zhang Yue (张跃). Problems revealed by their measurements have been confirmed by the results of multi-year regional monitoring and analysis performed by several organizations in cities such as Beijing, Guangzhou, Lanzhou, Wuhan, Chongqing, and Nanjing as well as measurements that the U.S. embassy in Beijing collects on its premises. Those who refuted these measurements or who said that they were unscientific or unable to represent the overall situation included Du Shaozhong (杜少中), deputy director of the Beijing Municipal Environmental Protection Bureau (BJEPB), and some Chinese academicians. Take the $PM_{2.5}$ measurements collected by the U.S. embassy from October 18 through October 25, 2011 versus the air quality measurements published by the BJEPB in the same period for example. The measurements collected by the U.S. embassy indicated that the local air quality rating was Unhealthy, Very Unhealthy, or Hazardous on days when the lowest rating released by the BJEPB was Slightly Polluted. Compare relevant local measurements with the U.S. standard and we see that the former exceeds the latter by several times. Multiple media organizations reported in late 2011 that $PM_{2.5}$ would likely be included into China's *Ambient Air Quality Standard* (环境空气质量标准). Nonetheless, there is no information on when it will be included or whether it will be included into government assessment. Local governments in places such as Shanghai and Beijing announced in early 2012 that they would include $PM_{2.5}$ into the local air quality indices and regularly release the information.

Another report says that when a sandstorm contains pollutants from coal burning, the amounts of toxic and harmful substances such as mercury, chromium, selenium, lead, arsenic, boron, antimony, molybdenum, cadmium, thallium, fluorin, and sulfur in total suspended particulate matter (TSP) and respirable particles can be several or even scores of times greater than usual. Spreading with sandstorms, such toxic substances are harmful to humans, livestock, and industrial/agricultural production and

9 Feng Jie and Lü Zongshu, "I Measure Air for My Homeland," *Southern Weekly*, October 27, 2011.

may affect up to 18.75% of China's land area and up to 20.77% of the Chinese population. These sandstorms affect northwestern and north China as well as the Yellow and Yangtze Rivers and can reach the provinces of Hubei, Hunan, Anhui, Zhejiang, and Fujian, as well as Hong Kong and Taiwan. They may even reach the Korean Peninsula and Japan.[10]

Industrial waste gas emission is the primary source of various air pollutants, according to the *Report on Identifying Air Pollution Sources in China* (中国大气污染源定位报告) published by the IPE on December 15, 2011. Specifically, the power, cement manufacturing, chemical, iron and steel, and coking industries have the largest numbers of major waste-gas-emitting companies. Among them, there are 1,178 power companies, or 34% of all the major waste-gas-emitting companies monitored by competent authorities.

Among cities (including counties) where acid rain is monitored, cities where acid rain occurred represented 52.8% of the total in 2008, 52.9% in 2009 and 50.4% in 2010. Cities where the acid rain occurrence was at least 75% represented 11.5% of the total in 2008, 10.9% in 2009, and 11% in 2010.

C. *Soil Erosion and Soil Pollution*

The area of soil erosion represented 37.2% of China's total land area at the end of 2010. This figure was made up of water-eroded areas, which represented 16.8% of the total land area, and the wind-eroded area, which represented 20.4% of the total land area.

The *Report on the State of the Environment in China 2010* said: 1) lots of soil and agricultural-product samples and data had been collected; 2) the national soil pollution state survey database and sample library had been created; and 3) general and special reports on a nationwide soil pollution state survey had been written. Nonetheless, it did not disclose details and specific data, including the ones about sites polluted by heavy metals, oil and chemical products, and polychlorinated biphenyl (PCB) as well as the state of farmland in wastewater-irrigated areas.

In addition, media reports have revealed a soil pollution problem with cities and industrial areas (or brownfield sites) that are polluted by heavy metals, scrapped electronics, organic petrochemical pollutants, and/or persistent organic pollutants. The reports say: the sites are polluted up

---

[10] Greenpeace, "The True Cost of Coal—Coal Dust Storms: Toxic Wind," 2011.

to 15 meters underground; the pollutants may remain toxic for up to a hundred years, they may cause cancer; and they may threaten more people through groundwater. Although the former State Environmental Protection Administration (SEPA) and the Ministry of Land and Resources (MLR) had completed a nationwide three-year soil pollution survey long before, no information on brownfield sites has been disclosed. Documents such as the *Guidelines on Risk Assessment on Polluted Sites* (污染场地风险评估导则) issued by the Ministry of Environmental Protection (MEP) in 2009 and the *Circular on Enhancing Management over Redeveloped Brownfield Sites* (关于加强工业企业再开发利用环境场地管理通知) issued by the State Council in 2011 to all the ministries, commissions, and provinces for comments have yet to be made public, but lots of brownfield sites have been used for housing developments, including affordable housing.[11]

D. *Solid Waste Pollution*

The amount of industrial solid waste generated is increasing year by year, as the amount reached 2.41 billion tons in 2010. However, the amount of discharged industrial solid wastes decreased to 4.98 million tons in 2010. The amounts of industrial solid wastes that are reused, stored, and treated are all increasing, as they reached 1.62 billion tons, 239.18 million tons, and 572.64 million tons respectively in 2010.

According to the *Report on the State of the Environment in China*, 237 hazardous and medical waste treatment facilities, 31 radioactive waste repositories, and 4 dioxin monitoring centers had been built by the end of 2010, when solid waste management center projects had been implemented by the central government and half of the provincial governments. The MEP issued policy documents such as the *Guide to Planning the Disposal of Scrapped Electrical Appliances and Electronics* (废弃电器电子产品处理发展规划编制指南). Meanwhile, chromium slag treatment facilities were still being built in both Tianjin and Jilin Province.

The 2011 version of the above-mentioned report did not mention the amount of urban domestic garbage and the garbage treatment methods that citizens worry about as well as a possibly huge amount of construction waste.

[11] Bao Xiaodong and Zhang Xinyuan, "Are We Building Lots of Houses on Toxic Land?" *Southern Weekly*, December 22, 2011.

E. *Radiation and Other Types of Pollution*

The 2010 version of the *Report on the State of the Environment in China* said the numbers of nuclear facilities and projects that involve the use of nuclear technologies continued to increase in 2010 with good overall environmental quality in terms of radiation. It also stated the ambient ionizing radiation remains at a steady level and the ambient electromagnetic radiation is good. The report said that the tritium concentrations in seawater near the discharge outlets of Dayawan (大亚湾核电厂), Ling'ao (岭澳核电厂), and Tianwan (田湾核电厂) Nuclear Power Plants are higher than the background values before these plants were put into operation, but the resulting additional doses that the public is exposed to are much lower than the limits specified by the state. The mining, smelting, and processing of some associated radioactive minerals such as uranium ore from Baiyun'ebo (白云鄂博) have certain effects on the surroundings of the company. The Chinese government pushed for the disposal of legacy radiation sources while enhancing the construction of urban radioactive waste repositories across municipalities and provinces in 2010. Meanwhile, 29 provinces (including autonomous regions and municipalities directly under the central government) have completed their repositories. In addition, the Chinese government has conducted ambient radiation monitoring capacity assessment on regions such as Jiangsu, Zhejiang, Beijing, Sichuan, and Guangdong.

A 9-magnitude earthquake and a resulting tsunami hit Japan on March 11, 2011. The earthquake and the tsunami led to nuclear leaks at the Fukushima Daiichi Nuclear Power Plant, where all the protective mechanisms were damaged or failed. Explosions occurred at the plant and resulted in excessive radiation.

The 252,000 square kilometers open sea within 800 kilometers east of Fukushima had received significant radioactive pollution by August 2011, as radionuclides such as Cs-137, Sr-90, and Cs-134, which are undetectable in normal seawater, were detected, according to the monitoring data from China's SOA. Among them, the highest concentrations of Cs-137 and Sr-90 exceeded the background values measured in China's seas by 300 and 10 times respectively. The highest concentrations of Cs-137 and Cs-134 exceeded China's seawater quality standards. These radioactive substances will have significant long-term effects on marine species, the marine ecosystem, and even human health since they all have a half-life of about 30 years and can accumulate in animals and plants and be passed on through the food chain. As a result, rice produced in Fukushima

Prefecture was prohibited from being sold because excessive radioactive cesium was detected in it, before the same metal was detected in milk powder produced there.[12, 13]

A nuclear crisis resulted from this incident in Japan, a technology superpower always known for its awareness of potential dangers and seriousness at work. The crisis shattered Japan's confidence in the safety of nuclear energy. Russia recommended at the G8 summit in May that the *Convention on Nuclear Safety* be rewritten to avoid any worldwide catastrophic consequence.[14] In June, the IAEA Ministerial Conference on Nuclear Safety discussed the issues of enhancing nuclear safety standards as well as establishing a nuclear emergency preparedness and response mechanism and a global nuclear safety framework.[15]

In China, panicked consumers raced to stock up salt despite the fact that governmental officials said: 1) The nuclear accident in Japan has no effects on us; 2) Our nuclear power plants are absolutely safe; and 3) The police will arrest anyone who starts or spreads rumors. It suggested that it was impossible to reassure the public with words from governmental officials, not to mention action by the police. On the one hand, "technology" has its inherent dangers, as modern technologies that are able to change nature are always double-edged swords. Nuclear technologies, for example, can destroy human life if they are used in wars. Can we make sure that they are definitely harmless if they are used for peace? Nuclear leaks and explosions as well as attacks by terrorists or criminals are possible as long as there are nuclear power plants. The probabilities of such accidents or incidents may be only one in one hundred thousand or one in one million, but it is difficult to achieve 100% safety and security.

More importantly, do we need so many man-made things? Take nuclear energy for example again. The general answer is: We need more electricity. But Chinese President Hu Jintao proposed in the report delivered at

---

12 Wen Xue, "Excessive Radioactive Cesium Found in Rice Produced in Fukushima," the Tokyo Channel of the People's Daily Online, http://japan.people.com.cn/35467/7648140.html (accessed November 17, 2011).

13 "Meiji Holdings Co. Will Recall 400,000 Cans of Milk Powder after Radioactive Cesium Found in Powder," China News, http://www.chinanews.com/gj/2011/12-06/3512840.shtml (accessed December 7, 2011).

14 Wei Zhong, "The G8 Summit Focuses on Nuclear Safety, and Japan Wants to Host a Summit on Nuclear Safety," China News, http://www.chinanews.com/gj/2011/05-27/3073427.shtml (accessed May 27, 2011).

15 Liu Gang, "The International Conference on Nuclear Safety Closes with no Significant Result," Xinhuanet (Vienna), http://news.xinhuanet.com/world/2011-06/24/c_121582081.htm (accessed June 24, 2011).

the 17th National Congress of the Communist Party of China (CPC) that China should change its growth and consumption patterns. Economic growth should center on public welfare, and consumption should be at a moderate level.

What we need may be "adjustments" rather than "more of something." Nuclear radiation may cause pollution, of course, but reminds us to make further adjustments to our thoughts, to the international landscape, to economic development, to energy policies and future energy development and applications, and to our standard of living and consumption patterns.

F. *The Ecosystem: Reduced Forests, Grassland, and Biodiversity*

The forest coverage is 20.36% across China, 35.67% in its eastern part, 33.3% in its central part and 17.05% in its western part, which represents 71% of its land area.

A Greenpeace report says that the total area of natural tropical forests at least 300 meters above sea level decreased by 24.68% in the central mountainous region of Hainan Province in 2010 compared with 2001. The reasons were their being replaced by artificial forests for pulp and rubber making and the violation of laws by large companies such as Asia Pulp & Paper Company (金光纸业) (APP).[16]

Grassland represents 41.7% of China's land area, with a livestock carrying capacity of about 240.13 million sheep equivalents. China's grassland ecosystem as a whole is deteriorating.[17] The amount of meat and milk produced per hundred hectare of grassland in China is only 4% that of the United States in the same climate zone and 1% that of New Zealand.[18] China's stockbreeding industry is regulated by economic policies in a country where modern industries and traditional agriculture coexist. The industry is faced with increasing cost-of-living pressure on the producer side and a growing market demand on the consumer side. It is supported by purchasing pasture from outside the grassland system and exhausting the carrying capacity and groundwater resources within that system. Aimed at selling products to the outside, such a stockbreeding industry

---

[16] Greenpeace, "The Disappearing Tropical Rainforests—Survey Report on Changes in Hainan's Natural Tropical Forests 2001–2010," 2011.

[17] The Ministry of Agriculture, *Monitoring Report on the Grassland across China*, 2008.

[18] Wang Xiaoyi, a report made at the Graduate School of the Chinese Academy of Social Sciences, 2008.

is referred to by one researcher as "an economy constantly in the red."[19] One sociologist has pointed out that: "Nomadism is actually not an outdated way of production. Instead, it adapts to changes in the grassland, as nomads move from place to place to find fresh pasture for their animals, thereby protecting the grassland."[20]

China's main ecosystems do not provide good service functions. Biodiversity degradation has not been effectively contained. Genetic resources are being lost. Alien species are posing severe threats to local ecosystems. Major ecosystem problems remain outstanding because of the effects of natural factors such as the global warming, as well as human activities such as land reclamation, overgrazing, and damage to land reserved for ecosystem restoration in recent years.

As for the climate, 2010 was the fourteenth year with temperatures on the high side since 1997. Temperatures were 1–2°C higher than normal in the Qinghai-Tibet Plateau, southern and eastern Xinjiang Uygur Autonomous Region, central Gansu Province, most parts of Ningxia Hui Autonomous Region, Yunnan Province, and southern Sichuan Province. The nationwide average number of hot days was 11.1 days, or 4.1 days more than normal. It was the highest year since 1961. The numbers of hot days were at least 15 days more than normal in southern Jiangxi and Hunan Provinces, southwestern Fujian Province, and northern Hainan Province.

There are different versions of China's carbon emission level. One article stated that China ranked 92nd by cumulative emission per capita from 1904 through 2004.[21] Another said that according to rankings made by a U.K.-based risk assessment organization in 2010, China ranked 26th among 183 countries by carbon emission per capita.[22]

Since China is a big country with a huge population, there should be differences in requirements between China's energy-consuming region (eastern China), energy-consuming groups (medium- and high-income groups), energy-consuming industries, and between western China, the

[19] Da Lintai, *Pasture Areas and the Market*, the Social Sciences Academic Press (China), 2010.

[20] Wang Xiaoyi, "Rural Areas Reduced to an Appendage and Environmental Degradation," *Xuehai*, No. 2/2010.

[21] Chang Hong, "China Ranks 92nd by Carbon Emission Per Capita," People's Daily Online, http://cppcc.people.com.cn/GB/45579/10652306.html (accessed December 25, 2009).

[22] Tan Liya, "The United Kingdom Releases Global Rankings by Per-Capita Carbon Emission Index," *Global Times*, November 17, 2010.

agricultural population, and farming. Assessments should be made by eastern, central, and western China (each part is larger than a lot of countries), and by region, group, or industry (some regions, groups, or industries represent significant shares of global totals). China should have more than one type of measure when it comes to carbon emission reduction.

II. *Water—The Foundation for Survival*

The Central Committee of the CPC and the State Council jointly issued the *Decision on Accelerating Water Conservation System Reform and Development* (关于加快水利改革发展的决定) on January 29, 2011. This was the CPC's first ever decision on water issues made in the form of a No. 1 document since the People's Republic of China was founded in 1949. The CPC Central Committee held a working meeting on water conservation in Beijing in July 2011.

What are the problems with water in China? What are the effects of water-relevant problems? How should we address such problems in China? With different interests and values, groups have differing answers to these questions.

A. *Water Problems: Survival, Ecosystem Safety, and Economic & National Security*

The water crisis affects people's well-being, the ecosystem, economic development, and national security.

As for the people's well-being, pollution is the primary water problem. The most severe pollution problem is that there are persistent organic pollutants in water that cannot be naturally degraded, such as heavy metals, oil products, and other toxic and harmful substances. The other severe problems include water bodies' poor abilities to dissolve oxygen, high concentrations of E. coli, and exposure to schistosomic eggs.

Water body pollution poses direct threats to human life and health, as it may cause cancers, diseases (liver, kidney, stomach, intestine, and vascular), and harm to the nervous and reproductive systems. Moreover, it has adverse effects on the development of fetuses, infants, and children, and in worse cases, causes physical handicaps, reduced intelligence, and mental retardation.

The Ministry of Health (MOH) and the Ministry of Water Resources (MWR) have said that more than 300 million people have no access to clean water in China, where about 45.6% of the 46 major cities are faced with poor water quality.[23] They have said that the public has been paying attention to the cleanness and quality of tap and bottled water since China modified its drinking-water standard.

Severe water shortages are the second biggest water problem. In recent years, there have been frequent media reports that large numbers of humans and domestic animals in China's rural areas are short of water because of droughts. Additionally, Chinese cities are also faced with the increasing water shortage. Among the 669 cities in China, 400 of them are undersupplied with water and 110 others are severely short of water. Among the 32 ultra-large cities each with a population of one million or more, 30 have long been faced with water shortages. Among the 14 open coastal cities, 9 are severely short of water, according to statistics from the MWR.

Water pollution interacts with the water shortage. Water is a renewable resource. Groundwater vaporizes, drops to the ground, and then infiltrates the soil or flows into rivers, lakes, and seas in the same way as does surface water after it flows out and is drawn by humans. It is from such a natural circulation that human, animals, and plants receive water indispensable for their survival and the ecosystem is likewise maintained. On the one hand, excessive hydropower projects and water diversion have caused rivers to dry up.[24] Occupying wetlands and riverbeds and turning lakes into farmland leads to a sharp decrease of water areas. Excessive impermeable surfaces in cities, industrial areas, and along highways blocks natural water circulation. Climate change brings about reduced precipitation and melting glaciers. All these have led to less available water. On the other hand, water consumption has been increasing as a result of the construction of a large number of residential buildings; changes in lifestyle such as tap water being introduced into more kitchens and the wide use of washers, toilets, and showers at home; and water consumption for urban landscaping and green belts (non-local plants are introduced and require irrigation), for luxury purposes such as golfing, skiing, and private swimming pools, and for industrial and mining activities.[25]

The natural circulation of water has also been changed. Excrement, which would otherwise be fertilizer for traditional agriculture, now enters

[23] http://zhidao.baidu.com/question/319976015.html%20%202011-10-2.
[24] Source: "Economic 30 Minutes," CCTV, August 2–3, 2011.
[25] Source: "Economic 30 Minutes," CCTV, August 5–6, 2011.

sewage systems and becomes waste that has to be treated at a high cost. Rainwater and snowmelt, which would otherwise infiltrate the soil, can only flow into sewers because of blockage by concrete ground surfaces, reducing the amount of water available to agriculture and ecosystems and increasing the amount of wastewater to be treated. The use of chemical fertilizers that replace excrement ends up adding to water pollution. Water that is used for plantation aquaculture and golfing purposes becomes wastewater that cannot be reused without treatment. A considerable part of the water used for industrial and mining activities ultimately becomes wastewater that contains persistent organic pollutants, and that is therefore extremely difficult to treat. Pollutants in the air enter water through rain and snow. Pollutants in water enter farmland and grassland, before entering agricultural, livestock, and aquatic products, thereby entering a new circulation through the food chain. In China, a considerable number of wastewater treatment facilities are used for nothing but allowing local governments to pass inspections conducted by higher-level governments. To make things worse, studies show that the currently available wastewater treatment technologies are unable to degrade or clear persistent organic pollutants, as they can only relocate them at best. As a result, such pollutants' harms to humans, animals, plants, and ecosystems will be persistent.[26]

Water problems are becoming increasingly important for the country on a macro level, as they relate to food security and may affect economic and national security. The CPC Central Committee has therefore decided to "implement the most stringent water resources management system and accelerate defining three red lines: water resources development and utilization controls, water utilization efficiency controls, and restrictions on discharging wastewater into water function areas," which suggests the severity of water problems in China.[27]

B. *The Structural and Institutional Reasons behind Water Problems*

The amount of freshwater resources per capita in China is 31.7% of the global average. The spatiotemporal distribution of such resources is

[26] Greenpeace, "Toxic Fashion—a Survey on Water Pollution by Global Clothes Brands in China," 2011.

[27] Zhang Jinlong, "The CPC Central Committee Holds a Working Meeting on Water Conservancy in Beijing," Xinhuanet, http://news.xinhuanet.com/politics/2011-07/09/c_121645412.htm (accessed July 9, 2011).

extremely uneven in the country.[28] The argument that water problems result from a large population (or a large land area) with a relatively small amount of water has been widely accepted in China. Ironically, there are quite a few countries that have a much higher population density and smaller amounts of water resources and farmland per capita than does China, but are better developed and able to provide better living conditions for their people than China.

Generally speaking, ecosystems have been destroyed by man's way of living, wants, and even unrealistic expectations, in addition to the worldwide inequality and poverty. As far as China is concerned, it has a different story from that of developed and other developing countries. Firstly, as a result of a lack of synchronization between industrialization and urbanization, more than 200 million Chinese farmers have no choice but to sign contracts on a per-household basis and work on small areas of farmland that they do not own. They are unable to adopt advanced irrigation, farming, and aquaculture technologies, to jointly implement rural water conservancy projects and maintain local ecosystems in an autonomous manner, or to communicate to the decision makers their demands for protecting their rights and asking the government to subsidize rural water conservancy projects and agricultural operations. Having been confined to the places where their permanent residence is registered, Chinese farmers have had to rely on overly intensivefarming, grazing, fishing, and hunting over the past decades as a result of an increasing population. This, to a certain extent, has worsened soil erosion and led to the massive consumption of water and other resources in an inefficient manner. In addition, young farmers have had to leave the countryside and find jobs in cities over the past couple of decades. The existing rural water conservancy works have long been left unmaintained. The groundwater level is going down. All these factors have worsened the problems caused by a combination of water shortage, lack of money to buy water, and the wastage of water for agricultural purposes.

Secondly, it is under the excuses of "development" and "modernization" as well as the differences between China's old and new systems that the central and local governments as well as domestic and foreign large companies are contributing to the exploitation of water, land, and other resources across China. Under the influence of consumerism, people have a growing number of material wants (including excessive ones) and

---

[28] Xie Fuzhan, the *International Statistical Yearbook 2008*, China Statistics Press.

generally overlook the limited availability of resources, bringing about a lifestyle characterized by high water consumption.

After the CPC Central Committee's working meeting on water conservancy in 2011, China Central Television (CCTV), as China's leading media organization, made two programs titled "Secrets behind the race to build hydropower stations on rivers across China" and "Hydropower stations built on a fault zone." The shows commented on activities where local governments attempted to maximize certain groups' short-term and local interests at the sacrifice of the public interest.[29] As an example, organizations involved in the Xiaonanhai Hydropower Project (小南海水电工程), which had been surrounded by controversy for several years, were allowed in late 2011 to change the boundary of a nature reserve once again, depriving rare fish unique to the Yangtze River's upper reaches of the last space for their survival. To increase the financial benefits from Xiaonanhai and another hydropower project, the project participants have repeatedly changed the boundary of the national reserve that was founded in 1996 and received the State Council's approval for a higher protection level in 2000. It turned out that both the fishery/environmental administrations and experts involved in the assessment on the nature reserve actually agreed to reduce the nature reserve's area twice, which occurred in April 2005 and November 2011. They did all this without considering the long-term interest of the nation.

Another major event is the controversial Poyang Lake Hydro-Complex Project (鄱阳湖水利枢纽工程) initiated as a result of unusually low water levels for several consecutive years. It is believed by some that this project will not only cause numerous ecosystem problems, but also threaten water safety and ecosystems in municipalities and provinces along the lower reaches of the Yangtze River. Moreover, it is intended to use the Three Gorges Dam approach to solve problems caused by the Three Gorges Project. The Comprehensive Poyang Lake Area Plan has yet to be approved, but the Poyang Lake Hydro-Complex Project is still being pushed.

All these are occurring in a scenario where public engagement and interest balancing are absent in making decisions on public affairs and where arguments on decisions that will likely cause irrevocable consequences of resources utilization are absent, along with lack of information

---

[29] Source: "Economic 30 Minutes," CCTV, August 2–3, 2011.

disclosure. As a result, a majority of the people have no access to relevant information; nobody is willing or able to take a long-term view.

It is under such a scenario that China is faced with water problems including:

Water is managed by the CPC plus the administration system, as the People's Congress and court systems play an insignificant role. Moreover, both experts and media organizations have weak voices, and public engagement has little influence on water management;

Water management is decentralized in an administration system that requires centralization. In a single administrative region, in particular, water for agricultural purposes, water for cities and wastewater treatment are separately managed. River sections or water areas of a single river or lake in different administrative regions are also separately managed;

The government has a weak capacity of overall balancing in the administration system. When it comes to making decisions on water issues, there is lack of institutional support for balancing and coordination between the authority in charge of economic development and other authorities in charge of water resources, land, environmental protection, forestry (and also animal and plant species protection), and cultural relics protection;

The government lacks systematic abilities to make decisions with the overall public interest (including their long-term interest) and willingness in mind. It even lacks abilities to make decisions on the basis of its overall interest and willingness;

Water management by the government agency in charge of water resources is fundamentally within the scope of project-based water management, as the concept of ecological balance has no influence on real-world activities. The idea that "Man should, and is able to, conquer nature" still has influence. The idea that decision making in an open society requires participation by different groups under the background of diversified interests has not been established, as the idea of "the competent authority naturally represents the people's overall interest" that came into being in the period of a planned economy remains prevalent. The concepts of social justice and public interest have never been established, as project companies always pursue high return on investment and local governments only consider the demand and GDP growth from projects. "Sustainability" is nothing but a theoretical concept. Plans aimed at catching up with developed countries, national-power competition plans based on the concept of "confrontation between enemy states," and high-demand plans under the influence of consumerism have caused an overestimation of the demand for water and hydropower in China, where sciences and technologies are confused with each other and little attention is paid to humanities. In addition, China lacks both an understanding of the concept of "system" and research on complexities. As for values, China has yet to learn how to properly deal with nature. All these factors have led to the prevalence of the man-, government-, capital-, demand-, and technology-centric theories.

C. *Rule of Law, Good Governance, and Basin Management—Alternative Solutions to the Water Crisis*

The current solutions to ecological problems such as the water crisis should include the government acting as the leader; developing a unified plan; and making unified arrangements for reasonable water resources development, optimized allocation, comprehensive conservation, effective protection, and scientific management.

Another alternative solution is to adopt a new management mode under the principles including "Be human-oriented," "The people's well-being comes first," and "Man coexists with nature in harmony." This mode is comprised of: integrated management of basins, rural and urban areas, and water/land/resources as well as protection (including maintenance, restoration, and optimization) and development (including use and reuse); information disclosure; collaboration among government agencies; and public engagement.

Since resources necessary for human survival, including water and land, are generally given, water as a renewable resource should be regarded as something for public use in most cases. The overall consideration should then focus on sustainable, fair use that allows for the interests of different groups and their offspring. High efficiencies are realized through market-oriented operations; the government should endeavor to maintain fair use. Laws should be developed to define the government's power and responsibility. Efforts should be made to change the practice in which the government tends to shirk responsibility and make adjustments through pricing and in which companies interact with governments and tend to rely on administrative actions for development.

It should be made clear that only the People's Congress is entitled to review and decide on projects that will change water bodies and land in an irrevocable manner. In addition, it is necessary to clarify courts' responsibility for judging cases involving disputes on resources, the environment, and the ecosystem so that they are obliged to handle such cases.

The government's competent administrative body in the basin should set up a basin management committee as an organization that coordinates interests and makes decisions. Efforts should be made to change the old mechanism in which different functional departments and local governments within the government system work separately. A mechanism in which multiple government agencies collaborate with each other should be created. Efforts should be made to improve government agencies' capacity for collaboration.

A mechanism in which the local people's congress, villagers' and residents' autonomous organizations, NGOs, and companies participate in basin management, should be created. It is necessary to clarify base management-relevant governments' and companies' responsibility in terms of information disclosure.

Water should be properly allocated for domestic, ecological, and production purposes under the principle of being sustainable, fair, and effective, provided that the health of river, lake, and wetland ecosystems and the normal functions of rivers and other water bodies are maintained. Sufficient ground space should be reserved for water by promoting water utilization projects for farmland, watercourse improvement, and ecosystem rehabilitation. Advanced technologies should be used to increase the water use efficiency and enhance water reuse.

Both using water for luxury purposes and drawing deep groundwater that is a non-renewable resource must be prohibited.

Efforts should be made to reduce water pollution from sources and to protect glaciers and snow-capped mountains as the water sources of the Yangtze, Yellow, and Lancang (澜沧江) Rivers.

It is necessary to research and improve seawater desalination and other seawater utilization technologies.

The State Council should coordinate basin management and address disputes between basins. Courts including the supreme one should handle all lawsuits relevant to administrative legislation on water, land, and other resources as well as administrative decision-making. And it is necessary to create a public-interest litigation system relevant to water, land, and other resources.

It is necessary to formulate the *Basin Law* (流域法) and revise regulations to facilitate the performance of the above-mentioned tasks.

In fact, the solution to severe ecological problems—changing the industrial structure and the growth/consumption patterns and rehabilitating the ecosystem—has long been put forward in a series of official documents such as the report delivered at the CPC's 17th National Congress.

Nonetheless, the existing structure and interest pattern make it difficult to implement this solution.

Moreover, it was proposed at the CPC Central Committee's working meeting on water conservancy in July 2011 that the principles included "The people's well-being comes first" (坚持民生优先) and "Create harmony between man and water" (坚持人水和谐). It was also proposed at the meeting that "we should endeavor to push aquatic ecosystem protection and aquatic environment rehabilitation, give high priorities to protection and natural restoration, and maintain healthy river/lake ecosystems, thereby improving rural and urban living environments."

Water problems in China can be alleviated as long as the above-mentioned objectives are achieved.

III. *Action Is Critical*

What have NGOs and civil society done? Some environmental NGOs have carried out activities that only benefit themselves, since environmental protection has become a profession. Some NGOs, including environmental ones, focus their efforts on implementing projects for their existence and applying for project funds. Some foreign NGOs that have offices and projects in China have recognized China's specialness, as they are inclined to do technical environmental protection and work with the government. Environmental NGOs have actually played a very limited role in objecting to pollution, preventing and controlling pollution, and rehabilitating ecosystems.

Valuable and effective actions in 2011 included:

1) The Green Choice Alliance (GCA) continuously revealed illegal pollution by Apple's China-based suppliers. The information was downloaded by the public more than 140,000 times, making Apple, which had refused to change its policy, change its mind in September 2011 and start a dialogue with several environmental NGOs. Apple said they were willing to investigate the suppliers and urge them to make improvements.
2) Greenpeace revealed that suppliers of more than ten globally renowned brands such as Puma, Nike, Adidas, H&M, C&A, Lining, Youngor, Lacoste, Calvin Klein, and Metersbonwe emitted various toxic and harmful substances during production in China. There were about 4,000 media reports on this published in the Chinese mainland, along

with reports published by some influential foreign media organizations. By September 16, 2011, 157 of the 970 surveyed major media organizations in 126 countries had published reports on this. As a result, Puma, Nike, Adidas, H&M, C&A, and Lining promised they would eliminate all the toxic and harmful substances from within their supply chains and products while disclosing environmental information. Some other brands are in negotiations.

3) FON, the Green Volunteer League of Chongqing, and the Qujing (曲靖市) Municipal Environmental Protection Bureau launched environmental public-interest litigation against the Yunnan-based Luliang Chemical Industry Company (陆良化工实业有限公司) and Heping Technologies (和平科技有限公司) for chromium slag contamination. The Intermediate People's Court of Qujing heard the case in October 2011. Greenpeace conducted surveys and sampling at pollution sites in China.
4) Shan Shui Conservation Center contributed to the government's decision to add "respect the local culture" and "let farmers and herdsmen play a leading role in local ecosystem protection" into its policies for the National Comprehensive Ecological Protection Experiment Zone in the Three Rivers Source area (三江源国家生态保护综合实验区).
5) Multiple environmental NGOs successfully prevented a highway that runs along the Bohai Bay in Panjin (盘锦), Liaoning Province from running through a local wetland.
6) Some environmental NGOs called on the government to stop building the Xiaonanhai Hydropower Station.
7) Dalian citizens launched a street movement and made the municipal government decide to move a local paraxylene (PX) maker.
8) Some urban citizens and the Beijing-based Green Beagle measured respirable particles in the air on their own initiative.
9) Environmental organizations in multiple places and some urban citizens launched or participated in classified garbage collection.
10) Citizens in Nanjing, Jiangsu Province acted to protect parasol trees.
11) Twenty-one low carbon ways of living were promoted in Beijing.
12) Guangzhou citizens called on the local government to stop the subway redecoration and "Light Up the City" projects.

Actions by NGOs and civil society are far from able to solve environmental and ecological problems that are occurring at an increasingly fast rate. When it comes to pollution prevention/control and ecological rehabilitation, the desired mainstream belief that requires acceptance by a sufficient majority of the people is far from being in place. A lot of Chinese environmental organizations lack a vision, a mission, core values, and work objectives that are clear and attractive enough to deserve dedication and participation by their members and volunteers; localized and effective action tactics; environmental and ecological ideas, ethics, and theories that they can provide to the global civil society to direct future human behavior; and leaders in non-governmental environmental and ecosystem protection campaigns who are recognized by the public in the process of a civil society emerging.

While tackling environmental problems, we should not overlook: the ignorance caused by a shortage of information, lack of initiative, as well as the indifference resulting from one's not seeing immediate harms to oneself; the peremptoriness of advantaged groups involved in environmental pollution and ecological damages; and the lack of consciousness and the corruption of experts.

A considerable amount of environmental pollution and ecological damage is irreversible, and even the rehabilitation part requires high costs. Are we losing opportunities because we are delaying at a time when numerous problems such as environmental damage across China can only be solved through further reforms?

PART ONE

SPECIAL FOCUS: NATURAL AND UNNATURAL DISASTERS

HUMAN FACTORS IN NATURAL DISASTERS: DEBRIS FLOWS, DROUGHTS, AND FLOODS

Shen Xiaohui

Abstract: After the severe drought that hit Southwest China, the floods across the country, and the debris flow in Zhouqu County, Gansu Province, some government authorities failed to reflect on the disasters amid questions by the public. They seemed to have hoped they could evade responsibility. Ironically, some of them contradicted themselves with pseudoscience and pseudo-environmentalism. We have already paid too much in our relationship with nature. Every disaster is a warning from nature, and we must first learn from previous disasters. This article links human factors with several natural disasters across China in 2010.

Keywords: debris flow, drought, flood, virgin forest, man-made forest, hydraulic work, river ecosystem

I. *The Relationship between the Debris Flow in Zhouqu and Economic Development*

Environmental damage can be seen all over the country. A direct financial loss of 600–800 million RMB is caused by geological disasters such as landslides, collapses, debris flows, and soil erosion every year.

A debris flow is severe erosion, including soil erosion, which occurs instantaneously and at high speeds. A heavy rainfall swept Zhouqu County (舟曲县), Gansu Province on the evening of August 7, 2010. The resulting debris flow ruined half the county seat. According to official statistics, 1,481 people were killed and 284 others missing after the disaster. Like the earthquakes that hit Sichuan Province in 2008 and Yushu County (玉树), Qinghai Province in 2010, this disaster shocked the country.

After the debris flow occurred, government authorities attributed it to the combined effects of large-scale natural factors such as geology, topography, hydrology, and meteorology. Specifically, the blame was laid on: wobbly, cracked, and weathered rocks and steep terrain; mountain rocks loosened by the Sichuan Earthquake; a lasting drought that led to more cracked rocks on the mountains; and a heavy rainfall. It can be said that all four natural factors are beyond human control. But were there any human factors that contributed to this huge disaster?

Zhouqu used to be a beautiful town surrounded by forests, rivers, and mountains. It was a place where all the mountains were covered by trees, according to the *Zhouqu County Records* (舟曲县志). Nonetheless, great changes have occurred to its mountains since the 1950s.

Local environmental changes began from deforestation that lasted five decades. Wood was transported down the Bailong River (白龙江) to other places before roads were built in the 1970s. Hundreds of rafts floating down the river were seen at that time. Up to 250,000 cubic meters of wood was produced each year. About 126,500 hectares of forests disappeared in the 38 years from 1952—when the Zhouqu Forest Industry Bureau (舟曲林业局) was founded—until 1990. A severe debris flow blocked the Bailong River with 34,000,000 cubic meters of debris in 1981. This is when the local people began to realize the consequences of deforestation. Beginning in 1985, the number of felled trees decreased, while the number of planted ones rose, but it was too late. Forest coverage has fallen sharply from 67% to 20%. Almost all the rocks on the mountains along the Bailong River into Zhouqu County are bare of vegetation. Hurting the ecosystem is easy, but rehabilitating it is difficult in such a place with steep mountains and severe soil erosion. Reforestation efforts have seen little fruit since they began more than two decades ago. According to the local people, the planted trees have low growth rates and almost all of them are small.

Next, there was hydropower development. The forest industry used to contribute 95% of the local revenue—known as wood-based revenue—before 1998. Commercial felling was terminated when the natural forest protection project began. The local source of revenue was then severed. It was in 2001 that Zhouqu County found another source of revenue—building hydropower plants on the Bailong River and its branches. There have been 55 hydropower plants approved in the county over the past decade, including 41 plants that have been built or are being built. The Gongba River (拱坝河) alone houses more than ten hydropower plants, which are only about ten kilometers away from each other. Building a hydropower plant requires blasting rocks, which leads to loosened rocks and damage to the vegetation. Landslides and soil erosion follow soon after. According to one study, 322.8 hectares of land have been occupied by local hydropower plants, built and being built, with a forecasted soil erosion of 749,000 tons.[1] Economic development based on resource over-

1 Du Yueying, "Human Causes for the Debris Flow in Zhouqu County," *China Economic Times*, August 10, 2010.

use comes with huge pitfalls—potential geological disasters. According to Fan Xiao (范晓), the chief engineer on the Geological Survey Team at the Sichuan Provincial Geology and Mineral Resources Bureau, hydropower works damage the rocks around their reservoirs. This is because of the large-scale excavation that is done to build side slopes, as well as road and tunnel building after the reservoirs begin storing water. So far, the total dumped debris from the local hydropower works has reached 38,348,000 cubic meters. It is dumped into valleys and watercourses, providing loose materials for the formation of landslides and debris flows.

Road building has also been a source of environmental destruction in Zhouqu County. Building roads in mountainous areas also requires blasting, which further loosens the mountains' rocks. The results are collapses, landslides, and other geological disasters. Roads to 22 townships in Zhouqu have been built over the past decade, bringing a significantly improved transportation infrastructure, but at a high cost. Take Provincial Road No. 313, for example. There are 13 lakes along the 17 kilometer road from Lianghekou (两河口) to the county seat. Such a density is closely tied to making slopes for road building, according to a survey by the Zhouqu County Water Resources Bureau.

The gold mines in Zhouqu County are small and rich, making them another magnet attracting investment. Twenty-nine companies had been granted concessions for mineral exploitation by the end of 2009. After years of gold mining, almost all the trees on the mountains along local rivers have been felled. Slopes are covered by grayish black soil, while grayish black mud and gold dredge can be seen floating on the rivers.

The area of farmland in Zhouqu County has increased to about 14,353 hectares from 7,087 hectares over the past four decades. Slopes with gradients less than 40 degrees have all been reclaimed for farms. Soil erosion affects 42% of the county, or 124,500 hectares. Farmlands on slopes and gullies are the main sources of soil erosion. Every kilogram of grain produced on such farmland means the loss of 40–60 kilograms of soil, according to some studies. This results in the stony desertification of such farmland.

Today, there are already 86 potential sources of debris flows across Zhouqu County. Potential sources of landslides can be found in 116 villages in 22 townships of the county. Zhouqu has a resource-exhausting development pattern ranging from wood- and hydropower-based revenue to mining-based revenue at the sacrifice of the environment. Such an unsustainable pattern has turned a picturesque county into a place with frequent geological disasters that is no longer suitable for human inhabitation. Zhouqu is the epitome of environmental damage in China:

improper development has hurt the ecosystem and, hence, increased natural disaster risks.

II. *The Relationship between Droughts/Floods and Destroying Virgin Forests*

A. *Historical Review of Virgin Forest Destruction*

Virgin forests in the Yellow River basin have endured the most severe destruction in China, which now has a muddy river and heartland hit by frequent floods and droughts. In the second half of the 20th century, 135 large state-owned forest companies focused their operations on virgin forests in other river basins including the upper reaches of the Yangtze River (the Southwest China forest area) and the Songhua River basin (the Northeast China and Inner Mongolia forest areas which include the Daxing'anling (大兴安岭), Xiaoxing'anling (小兴安岭), and Changbai (长白山) mountains). Nearly all the exploitable resources disappeared after forty years of over-felling. As a result, the Yangtze River has become like another Yellow River with a deteriorating ecosystem.

Changes in the Songhua River over the last hundred years are another typical case. The main water source in the upper reaches of the river is the well-known Changbai Mountain Forest Farm. The local virgin forest was well maintained before the 19th century thanks to the Qing Dynasty's ban on felling. As a result, the Songhua River always had a steady water level free from any sharp changes caused by unexpected weather, and there were nearly no floods or droughts. In contrast, frequent floods have swept the river basin since the late 20th century. The forest farms managed by all 18 forest industry bureaus around the water source were flooded in 2010. This implies that it is no longer possible for the river to stay at a steady water level due to the lower water storage capacity from decades of over-felling and the later replacement of virgin forests with man-made forests, farmland, and emerging cities. The unprecedented floods throughout the Songhua River basin have been caused by unprecedented destruction of the virgin forests around the Changbai Mountains.

One study shows that every cubic meter of wood felled along a river will lead to 400 kilograms of soil and sand flowing down the river each year. China has produced more than six billion cubic meters of wood over the past six decades. The total wood cubage is actually closer to ten billion cubic meters if the wood used as fuel and for other purposes, as well as forest resource consumption caused by over-felling is considered. In other words, the annual amount of resulting soil and sand has already reached four billion tons.

B. *Problems with Man-Made Forests*

China has created the world's largest man-made forests with a total area of 61.67 million hectares. Forest coverage has been increasing fast in a lot of areas where mountains look green again. In fact, these forests are providing considerable financial benefits, but they fall far behind expectations in terms of quality and ecological benefits.

Consider the controversial man-made forests consisting of eucalyptus and rubber trees. What role did they play in the most recent drought that swept Southwest China? The five southwestern provinces in China have an abundance of forests and water resources, but they were hit by a severe drought that lasted from the autumn of 2009 through the spring of 2010. Local forests failed to play their desired role against the drought. Instead, numerous trees were killed by the drought. Some experts claim that a eucalyptus forest is like a water pump, while a rubber tree forest is actually a "green desert." Both of these types of man-made forests contributed to the drought in Southwest China. Nonetheless, some local officials in charge of the forest industry claimed that "there is no scientific evidence that the planted eucalyptus and rubber trees are the reason for the drought."[2] Below is our analysis of this issue.

1. *Does a Eucalyptus Forest Act as a Water Pump?*

This statement defending man-made eucalyptus forests fails to distinguish between "causing the drought" from "worsening the drought." It is known that droughts are primarily caused by the climatic factor of global warming and that they are worsened by more complex factors, including replacing virgin and natural forests with man-made ones consisting of a single tree species. As the most representative tree species of Australia, eucalyptus matures in 4–7 years while it takes at least 20 years for pine trees to do so. One study has shown that a pine tree needs to consume 1,000 liters of water to generate one kilogram of dry mass, while a eucalyptus needs only 510 liters to do so. Despite high water utilization, eucalyptuses consume more water within a unit of time than do other trees since they grow fast

---

[2] Li Shenglin, "The Chief Scientist at the Chinese Academy of Forestry Says There Is No Scientific Ground for Eucalyptuses as the Cause for Droughts," www.southcn.com, April 23, 2010.

and generate large amounts of dry mass. "The plantation of extensive forests of eucalyptus in any deforested catchment will substantially decrease water yield from that catchment," reads the Food and Agriculture Organization's 1985 report called *The Ecological Effects of Eucalyptus*. In 2005, a study on planted eucalyptus forests in Argentina also demonstrated that large areas of such forests will not only lead to water loss from soil and dried-up rivers, but also affect the groundwater quality. China has made no in-depth study on the environmental impact of such forests, making it easier to massively plant eucalyptuses and, hence, worsen the related problems.

Asia Pulp & Paper Group (APP) signed land leases with local governments in provinces such as Hainan, Guangxi, and Guangdong at the beginning of the 21st century, when it had large areas of natural forests felled to plant eucalyptuses there. APP planned to plant 1.83 million hectares of eucalyptus in places such as Lancang (澜沧) and Simao (思茅) in Yunnan Province and more than 133,333 hectares of eucalyptus had been planted there by the end of 2008. The total area of eucalyptus forests reached about 533,333 hectares in Guangxi Zhuang Autonomous Region amid suspicions that Yunnan's ecosystem would be destroyed by replacing natural forests with eucalyptus ones. Today, eucalyptus has been planted in more than 600 counties in 17 southern municipalities with a total area of about 2.67 million hectares.

In 2010, Greenpeace investigators went to eight reservoirs and their water sources in counties such as Long'an (隆安), Napo (那坡), and Pingguo (平果) where water shortages were severe. They saw that water shortages were dire around seven of the reservoirs and their water sources where there were large areas of eucalyptuses. As a result, it was very difficult for local farmers to obtain water for drinking and farmland irrigation. The land was humid before the eucalyptuses were planted and has now become drier, according to local villagers. Unsurprisingly, the counties of Pubei (浦北), Baise (百色), and Teng (藤), which were covered by the forest ownership reform launched by Guangxi's forest industry authority with a focus on eucalyptus plantations, were all among the main areas hit by the drought.

One study shows that eucalyptuses absorb much more nutrition from the soil than do other tree species in order to maintain their fast growth. The shorter the felling interval, the more the nutrition is lost from the forestland. Moreover, the areas and communities that rely on forests as their water resources suffer from severe chemical fertilizer pollution, which causes eutrophication to local lakes and reservoirs. This is because it is necessary to fertilize the planted eucalyptus forests two or three times a year and there is a typical fertilizer loss of more than 40% on mountains. Both fish and birds continue to decrease two years after fertilization begins. In addition, the local water quality becomes noticeably worse, as the water in some wells even turns black, according to one report. Eucalyptuses produce a chemical that inhibits the growth of other plants. That is also one of the reasons for sparse bushes and grass in a eucalyptus forest. Therefore, developing man-made forests consisting only of eucalyptus is unscientific in terms of soil preservation, soil improvement, and biodiversity protection. It is no exaggeration that eucalyptuses may worsen droughts.

Now that we know eucalyptuses have these problems, why do local governments spare no effort to encourage their plantation? That is because their fast growth contributes to government officials' performance evaluation during their short tenures. Hence, government decisions and actions focused on short-term objectives are in opposition to sustainability and scientific development.

2. *Is a Man-Made Rubber Tree Forest Really a "Green Desert"?*

Rubber trees originated from the Amazon Rainforest in Brazil and were introduced into the Xishuangbanna Dai Autonomous Prefecture (西双版纳) in Yunnan Province in 1939. The local plantation area was small for a long period of time until the late 1960s, when rubber tree farming grew rapidly. Nine of the top ten local state-owned farms are rubber tree farms with a total plantation area of 100,000 hectares, nearly all of which used to be part of the local rainforest. The rubber industry saw explosive growth in the 1990s as the price of rubber rose sharply in international markets. At that time, farmers were racing to sell their mountains and land to rubber tree farm owners, who then converted the land into rubber tree forests. Under the slogan of "Rubber Is the Basis of the Country's Economy," Xishuangbanna entered a third round of rubber tree forestation since 2000. Nearly all the areas at altitudes below 1,000 meters have been converted into rubber tree forests, the total area of which has skyrocketed to 410,000 hectares from 77,333 hectares in 1988. Xishuangbanna has lost about 466,667 hectares of virgin rainforest over the past three decades, with only 373,333 hectares of rainforest remaining. The three national nature reserves there have become "isolated islands" surrounded by the planted rubber tree forests.

Such substantial changes in the local ecosystem bring negative effects on the regional climate. The average temperature in Xishuangbanna has risen by 0.4 to 0.5°C in the most recent decade after the large-scale rubber tree plantations began, according to analysis from a geographic information system (GIS). "Such climate change is local rather than regional, as it has not occurred in the neighboring Pu'er (普洱) and Lincang (临沧). The atmospheric circulation is the same. These local changes are related to changes in the land coverage. That is, the effects of solar energy absorption on the local climate," says Ma Youxin (马友鑫), a tropical arboretum expert. In fact, Xishuangbanna has been suffering from water shortages since the large-scale rubber tree plantations began. Seasonal temperature differences have been on the rise along with falling relative humidity (RH) values over the past five decades, according to long-term measurements from the local meteorological bureau. There were 184 foggy days in the city of Jinghong (景洪), the prefecture capital, in 1945, compared with only 22 such days in 2005.[3]

The water loss of a rubber tree forest is three times that of a rainforest with the same area, while its soil loss is 53 times as much, according to one study. Numerous local villages have been suffering from lower groundwater levels since the rainforest was replaced by the rubber tree forest. Pesticides from the rubber tree forests pollute local ponds and reservoirs. Wastewater from rubber production pollutes local rivers, such that some residents living in the rainforest areas now have to buy drinking water.

Harm to the local ecosystem caused by alien species such as rubber trees and eucalyptuses cannot be ignored. More severe ecological disasters will be unavoidable if we do not change our current practices.

III. *Why Do China's Hydraulic Works Fail to Alleviate Droughts or Floods?*

Despite their large number, reservoirs in China fail to alleviate frequent droughts and floods. They were no use in the fight against the severe drought that hit Southwest China in 2010, for example. Furthermore, they caused a lot of problems during the floods across China, even worsening them in some areas.

The role of a reservoir in the fight against floods or droughts depends on how it is built and operated. Theoretically, a reservoir should play a

3 "Why Is Xishuangbanna No Longer Beautiful?" *Beijing Youth Daily*, April 28, 2008.

positive role by storing flood water for use in dry seasons. In other words, it should store more water than is discharged in rainy seasons so as to reduce the flood peak level, while doing the opposite in dry seasons so as to increase the flow rate. Only in this way can it contribute to disaster prevention and reduction. To maximize the financial benefits of power generation and assure dam safety, however, reservoirs tend to make the opposite adjustments by discharging flood water in rainy seasons and storing water in dry seasons. It is due to such an unnatural practice that they never reduce the flood level in rainy seasons, while worsening water shortages in dry ones. One scholar pointed out: "Experts seek trade-offs by carefully designing complex scheduling modes. But the reservoirs' functions are still far from what they expected. Among them, the Sanmenxia Reservoir has given up such functions as storing flood water for use in dry seasons."[4]

In addition, China has long focused more on large-scale hydraulic works than on small ones. Some of the large hydraulic works focus on the financial benefits of power generation, as they worsen water shortages by occupying water that should otherwise belong to the local ecosystems and farmers. In contrast, they fail to play their desired roles in the fight against floods or droughts. Small hydraulic works in rural communities are the ones that contribute to disaster prevention and reduction. Nonetheless, it was during the 2010 severe drought that problems with the small works such as their long-term underinvestment, lack of maintenance, and functional deterioration prevented them from avoiding the disaster.

Droughts and floods are a part of nature, and they are not necessarily disastrous. Human factors tend to be involved in their conversion into natural disasters. The human activities that are partly to blame for turning natural droughts and floods into disastrous ones are exemplified by: building dams on rivers and their branches, such as cascade development; changing watercourses at random; and reclaiming large areas of lakes, wetlands, and flood plains that are complementary to rivers.

The cascade development going on in the Yangtze River basin is characterized by dense hydropower plants built on the Min (岷江), Jialing (嘉陵), Yalong (雅砻), Jinsha (金沙), Dadu (大渡), Lancang (澜沧), and Hongshui (红水) rivers, as well as their branches. Three hundred fifty-six dams have been built or planned on the Dadu River alone and there

[4] Qin Hui, "Why Is China Often Criticized by Other Countries in the Mekong River Basin?" *The Economic Observer*, December 29, 2010.

are about 10,000 reservoirs throughout the Yangtze River basin. There are also about 3,000 reservoirs throughout the Yellow River basin with a total storage capacity of 66 billion cubic meters. This is even bigger than the annual average runoff of the Yellow River (58 billion cubic meters). Additionally, there are around 50,000 water pumping works along the river.

As another example, wetlands have an important function. That is, they provide water to fight droughts and store water to prevent floods. Nonetheless, the total area of wetlands in China has decreased by 16.76%, from 366,000 square kilometers in 1990 to 324,000 square kilometers today. The wetland area in the country was decreasing at a more astonishing rate in the mid-20th century. The total area of lakes and wetlands along the lower and middle reaches of the Yangtze River reduced sharply by 62% in the 1980s compared to the 1950s. As a province that was known for about a thousand lakes, Hubei has only 300 lakes today. As China's largest wetland, the once wild wetland in the Sanjiang Plain (三江平原) in Northeast China has been largely converted into farmland. It is now 80% smaller than what it used to be. On average, twenty lakes disappear across China each year, which implies that more than 1,000 lakes have disappeared over the past five decades.

Various aquatic ecosystems such as rivers, lakes, swamps, and flood plains provide socio-economic and ecological functions. But these ecosystems are suffering from the negative effects of hydraulic works. According to a report by the China Council for International Cooperation on Environment and Development (CCICED), when it comes to their ecological functions, China's aquatic ecosystems have increasingly prominent problems in terms of functional deterioration. This includes river closures, hydraulic features, and habitat changes. Major reasons for such problems undoubtedly include insufficient attention to the ecological functions of aquatic ecosystems and the corresponding defects in public policies and administration.[5]

Qian Zhengying (钱正英) served as China's Water Resources Minister from the mid-1970s through the late 1980s. She led the construction of numerous large-scale hydraulic and hydropower works from the 1950s through the 1980s. She gained a deeper understanding of such works after she became vice chairman of the Chinese People's Political Consultative

[5] CCICED, "On the Policy Framework for Improving Ecological Service Functions of Aquatic Ecosystems," 2010.

Conference. While reflecting on them on different occasions, she reiterated that:

> I made a mistake while I was leading the Ministry of Water Resources. I failed to realize that we should have first ensured there was enough water for aquatic ecosystems... We stressed river control for many years, but our efforts turned out to have dried up the Yellow River. Later on, problems got worse, as inland rivers such as the Tarim (塔里木河) and Hei Rivers (黑河) also dried up. I have gradually understood that there was a problem with the hydraulic works—loose administration and overdevelopment... We would produce 'side effects' if we went too far. Our mistakes should serve as a warning for future hydraulic works.[6]

IV. *Reflecting on Hydraulic Works and Ecosystem Preservation*

Will anyone follow Ms. Qian's suggestions to "first ensure that there is enough water for aquatic ecosystems" and to avoid the mistakes of "loose administration and overdevelopment"? At a time when low-carbon economies are being promoted worldwide, will hydropower developers use the slogan of "energy conservation and emissions reduction" to build more dams, thereby further damaging the environment? Is it possible that they will learn from history and give up the existing hydraulic work building bonanza? Given the increasingly frequent natural disasters and the above-mentioned lessons, should we take the try-my-luck approach? What we urgently need is to reflect on disaster-relevant issues, including:

A. *Hydropower Generation Should Not Be Given More Importance Than the Comprehensive Function of Water Resources*

Hydraulic work construction in China has been focused on hydropower plants over the past couple of decades. Unfortunately, the comprehensive functions of such works are seldom considered. Irrigation and ecological functions were not considered in almost all the cascade hydropower plants in Southwest China in particular. As a result, the consequence is the success of one sector at the cost of all the others.

Natural rivers support healthy ecosystems and provide various functions. Each river has its rhythm and ecological flow. The ecological flow is comprised of the relationship between: (1) seasonal river level changes,

[6] Qian Zhengying, talks during an interview with *Asia Weekly* and some conferences, April 2010.

the water's physical and chemical properties, a river's sand content and flow diversity, and regionally diverse spatiotemporal distribution; and (2) fishes' spawning and migration, bird migration, seed dissemination, and areas where rivers and seas meet. The Nature Conservancy has proposed ecologically sustainable hydropower development through ecological flow research. The main idea is to identify river flow rates compliant with the ecological flow based on the needs in different timeframes. The aim is also to develop scheduling schemes for various reservoirs, thereby achieving optimization between the fight against floods and droughts, and the river's ecological functions. The concept aims to achieve ecological objectives while obtaining proper financial benefits. Sustainability is about considering the balance between financial and ecological benefits, local and global results, and the interests of the current and future generations. Water resource development focused only on particular interests is doomed to failure.

B. *Floods Are Also Resources*

From the drought in Southwest China to the floods across the country, China was in a constant state of water crisis in 2010. In fact, we were still suffering from water shortages even in the periods of flooding. The total water volume of the floods in southern China was about 200 billion tons, or 600 billion RMB worth of water resources. After just finishing the South-to-North Water Diversion Project at a high cost, the dominant viewpoint concerning water was attacked when southern China was hit by a drought shortly after it was swept by floods and part of the available freshwater flew with the floods into the seas. According to Professor Yang Dawen (杨大文) at Tsinghua University, we should shift to flood management from flood control and study flood prevention in combination with other relevant issues. To eradicate water crises, China must become a country with a network of water resources. For the 21st century, it should focus its water resource strategy on the combination of flood and drought issues by using floods as resources.

How do we turn floods into resources? How do we develop effective flood detention systems in river basins? How do we store as much precipitation in soil as possible by combining biological controls with rainwater utilization facilities? It is critical to find the answers to these questions. We should work to restore and rebuild natural forest ecosystems or "green reservoirs." We should also work to rehabilitate aquatic ecosystems such as flood plains, lakes, and swamps, while valuing nature's functions for preventing floods and droughts as well as the storage function of river basins. Only in this way will we no longer be fighting disasters every year. Flood plains are seasonal wetlands and, like lakes, swamps, and forests, are spaces needed for natural water circulation. Nonetheless, reckless urbanization and agricultural development are occupying these spaces, which have now lost their ecological functions. As a result, about 700 million people and 70% of the cities in China are faced with regular floods, droughts, and debris flows.

"Human beings may achieve sustainable water resource management only when they respect the integrity of water spaces," said Wang Yongjie (王永洁), director of Geography at Qiqihar University. In other words, we must stop changing the natural functions of water spaces, including forests, wetlands, lakes, and flood plains. Changing natural water spaces may bring us short-term, huge benefits, but will lead to long-term disasters.[7]

[7] Wang Yongjie, "Water Is One of the Most Active Natural Elements," *China Daily*, November 18, 2010.

SEVERE GEOLOGICAL DISASTERS DURING THE 2010 RAINY SEASON

Fan Xiao

Abstract: The mountainous areas in western China are prone to geological disasters such as collapses, landslides, and debris flows. There has been considerable discussion on the causes of these geological disasters over the past few years. It is believed by some that they are primarily caused by natural factors; and it is believed by others that human factors should never be overlooked. This paper analyzes the causes of geological disasters, with a debris flow that hit Zhouqu County, Gansu Province in 2010 as an example. It also discusses disaster warnings and post-disaster reconstruction.

Keywords: geological disaster, debris flow, disaster warning, post-disaster reconstruction

On August 8, 2010, a debris flow swept the villages of Sanyan (三眼村), Yueyuan (月圆村), and Chunchang (春场村) in Zhouqu County (舟曲县城), Gansu Province. The debris flow originated from two valleys, Sanyanyu (三眼峪) and Luojiayu (罗家峪), north of the county town, part of which was also destroyed by it. It was the deadliest debris flow in China since 1949. And it was only one of a range of "large" and "ultra-large" geological disasters that hit the Chinese mainland in the rainy season of 2010.[1]

About 65 large and ultra-large geological disasters occurred across China in the rainy season (May through September) of 2010, leading to 2,385 deaths and missing people, according to a collection of statistics published by the Ministry of Land and Resources (MLR) and local governments. A total of 30,466 geological disasters occurred across China from January through October 2010, according to MLR figures. As a result, 2,909 persons were killed or unaccounted for, or five times as many as in the same period of 2009. The year 2010 was therefore referred to as a year of the most severe geological disasters since the People's Republic of China was founded in 1949.

---

[1] China's *Regulations on the Prevention and Control of Geological Disasters* specifies: a geological disaster is "ultra-large" if it kills 30 persons or more and causes a direct financial loss of at least 10 million RMB; it is "large" if it kills 10–29 persons and causes a direct financial loss of between 5 and 10 million RMB.

The mountainous areas in western China are prone to geological disasters such as collapses, landslides, and debris flows. In 2010, severe debris flows occurred in all four areas prone to disaster, including southeastern Tibet, southern and western Sichuan, northeastern and western Yunnan, as well as southern Gansu and southern Shaanxi. Examples include the debris flows that hit the Bomi (波密) and Basu (波密) sections of the Sichuan-Tibet Road in southeastern Tibet; Longmenshan (龙门山) in western Sichuan; Gongshan (贡山) in western Yunnan; and the above-mentioned Zhouqu County in southern Gansu. An ultra-large landslide plus debris flow caused heavy casualties in Guanling (关岭), Guizhou Province and showed the huge geological risks in the karst areas of southern China.

I. *Human Factors and Natural Causes for Geological Disasters*

There has been considerable controversy on whether geological disasters are attributable only to natural factors or also to human factors ever since the outbreak of geological disasters represented by the ultra-large debris flow in Zhouqu.

The author believes that the areas prone to geological disasters would still be prone to collapses, landslides, and debris flows even without any effects of human activity or even if the vegetation was in its original state. The disasters result from a natural process of surface erosion and geomorphic cycles. Nonetheless, the incidence and intensity of geological disasters may sharply increase due to the effects of human activity, such as more extreme weather events, destroyed vegetation, and soil erosion. In addition, defects in construction planning, protection work and disaster warnings add to the casualties and property losses.

In Zhouqu County for example, there are frequent geological disasters in the Bailong River (白龙江) basin where the county is located. There are 13 locations prone to landslides and 12 valleys prone to debris flows along the 17-kilometer section of the Bailong River that begins in the county town. Numerous landslides have occurred 5.5 kilometers down from Zhouqu. The largest one had a volume of more than 60 million cubic meters. Another landslide blocked the Bailong River to form a landslide-dammed lake in 1981, flooding part of the county town. There is also another location near the county town where a very large landslide, known as the Suo'ertou (锁儿头) Landslide, occurred. Those locations were all on the 2010 list of main areas for geological disaster prevention in Gansu Province.

Sanyanyu and Luojiayu, the two valleys where the debris flow happened and hit Zhouqu County, are both prone to debris flows. The entire county town was actually built on a fan that resulted from long-term deposition of debris flows. Eleven large debris flows have occurred in Sanyanyu over the past 180 years, causing heavy losses to the county town. The most recent one occurred on August 7, 1994. The largest one occurred on June 4, 1992, when 344 houses were destroyed and 87 people killed or injured, with a direct financial loss of 12.6 million RMB. The others include: a debris flow on May 10, 1989, when 360 houses were destroyed, 51 people injured, and numerous bridges and roads destroyed, with a direct loss of nearly 10 million RMB; a debris flow in the summer of 1961 that destroyed 160 houses, killed or injured 28 people, and suspended local transportation for 40 days; and a debris flow in the summer of 1916, when more than 90 houses were destroyed and 63 people killed or injured.

All these debris flows were tied to rainstorms and the one in 2010, which was the largest, was caused by a combination of multiple factors. Its direct causes included human factors in addition to a heavy rainfall.

First of all, measures were not in place in terms of disaster prevention and a warning system. Some disaster prevention works were built at Sanyanyu in the 1990s, including debris flow retaining and discharge/diversion works. But they were not finished due to a shortage of funds. Second, there was no effective warning mechanism. A rainstorm was the direct cause of the debris flow, which did not occur until the rainstorm had already been going on for some time. Some warnings could have been given during the heavy rainfall, but there were none. Third, urban planning was irrational. There should be sluiceways, including debris flow discharge/diversion works, built in areas that debris flows may run through. Unfortunately, a large number of buildings are located in dangerous areas as a result of continuous urban expansion. Severe damage will therefore occur once a debris flow occurs.

Long-term harms to the natural environment caused by human activity also intensify geological disasters. There were two periods during which human activity had a significant environmental impact across China, especially its western regions. The first one was the deforestation and over-reclamation up to 1998 and the other has been the large-scale hydropower development, mining, urban and industrial-development-area expansion as well as transportation infrastructure construction since 2001.

Forest coverage in the Bailong River basin used to be high, but it has decreased by nearly 133,333 hectares after the large-scale deforestation from the 1950s through the 1990s. As for farming, statistics show that nearly all the slopes with gradients less than 40 degrees in this area have been reclaimed. The county town on the Bailong River is more than 1,300 meters above sea level, while the highest peak near the river is more than 3,800 meters above sea level, leading to a very steep gradient. Nonetheless, farmland stretches nearly halfway to the peak despite the gradient. All these factors have greatly worsened soil erosion.

The local government found another source of revenue shortly after China issued a ban on felling natural forests in 1998. It changed the local wood-based economy into a water- and ore-based one by encouraging hydropower development and mining, leading to higher, rather than lower, environmental impacts.

As for hydropower development, China is conducting cascade development across all the river basins throughout its western mountainous regions. In Zhouqu there are seven cascade hydropower plants on the Bailong River in the county alone. In addition, there are now dense hydropower plants on all its branches. Today, there are 156 hydropower plants already built across the Gannan Tibetan Autonomous Prefecture (甘南藏族自治州) in Gansu Province, and a similar number of hydropower plants that are under construction. Hydropower revenue has become the primary source of revenue for the prefecture, as it represents more than half of the total local revenue. Some experts have written articles over the past few years to discuss environmental protection during hydropower development in Gannan. They made special mention that the local government disobeyed relevant laws and regulations while introducing projects and reviewing project applications. Examples include lowering the entry barrier, allowing projects to be implemented while they are still reviewing applications, delaying planning, failing to conduct environmental impact assessments (EIA), and issuing loose regulations.

Power plants, mines, and transportation infrastructure construction entail excavation on mountains and waste piling. As a result, the mountain structure becomes less stable and a lot of loose waste is piled up in river valleys. These lay a good physical basis for the formation of debris flows and landslides, as geological disasters are likely to occur once there is a heavy rainfall. Such activities as excavation and waste piling were tied to a landslide that killed 23 people at the site of the Jinping Hydropower Plant (金平水电站) in Kangding County (康定县), Sichuan Province on June 14, 2010.

The ultra-large landslide plus debris flow that occurred in Guanling, Guizhou Province on June 28, 2010 was related to the water storage by the Guangzhao Hydropower Plant (光照水电站), a large structure on the Beipan River (北盘江), in addition to a rainstorm as the direct cause. The Guangzhao Hydropower Plant is the largest, leading plant among the cascade hydropower plants on the river. It has a dam that is 200.5 meters high, a total storage capacity of 3.25 billion cubic meters, an installed capacity of 1,040 MW and a total reservoir area of 51.54 square kilometers. It began to store water in late December 2007 before beginning to generate power in 2008. With a higher water level after the reservoir began to store water, the incidence and intensity of earthquakes around the reservoir have increased, according to measurements from the earthquake-monitoring network for the reservoir area as well as research results from experts. As of August 31, 2008, as many as 1,299 earthquakes were detected after water storage began. This includes more than twenty earthquakes with magnitudes of up to 3.2 that were felt by people. Local villagers also reported that they have felt many earthquakes since the Guangzhao Hydropower Plant was built and that a lot of houses have cracked since then. A mountain at the site of the landslide in Guanling cracked in the previous year before the landslide occurred. Local villagers jointly submitted a report to the government but saw no results. All those became potential causes for the landslide a year later.

As for geological actions, most of China's western mountainous regions are eroded regions, where most debris from debris flows, collapses, and landslides are moved down rivers. Dense cascade hydropower plants flood the best part of river valleys in mountainous areas, where humans lose large pieces of land most suitable for farming, living, and urban construction. The resulting resettlement also adds to the environmental burden. In addition, such plants lead to mud deposition, which then causes higher riverbeds and slows down floodwater movement. As a result, they worsen flood threats in mountainous areas and lead to long-term environmental hazards.

A town newly built for the resettled was heavily damaged by an ultra-large landslide in Wangong (万工镇), Sichuan Province on July 27, 2010. This also implies that resettled people have been faced with higher geological risks after they were moved from valleys because of the cascade hydropower plants in the western regions.

Most of the severe geological disasters that occurred in China in 2010 were related to heavy rainfalls, but nearly all of them were affected more or less by human activity. Some Chinese experts have pointed out that geological disasters caused by improper human activities have outnumbered those caused by natural reasons. The geological changes and disasters we are talking about now are therefore more than traditional geological issues, since they are tied to the relationship between human activity and natural actions. We cannot say that human activity has become the leading factor that is affecting and changing the earth (since human beings are unable to change the processes and basic laws of the earth's evolution and geological actions). Nonetheless, frequent and significant changes may have occurred to our environment within a very small geologic time scale because our survival status and prospects have been greatly affected by the feedback mechanism between human activity and the earth's natural actions.

II. *Relocation or Rebuilding?*

Should Zhouqu's county town be relocated or rebuilt? This is a common question asked after severe disasters occur. Like many other towns in the Bailong River basin, the county town was built on a fan resulting from long-term deposition of debris flows. This town was located as a result of long-term natural selection. Are debris flows disasters? Zhouqu would not exist without them. The locations of numerous ancient towns and villages embody our ancestors' wisdom in selecting the locations of their homes based on terrain.

The population of the county town increased to more than 40,000 in 2010 from 21,400 in 1996. The urban area was expanding rapidly and a lot of buildings were constructed on the paths of debris flows, leading to heavy losses. But that is not necessarily reason for relocation. Moreover, it can be said that the county town is sited on a rare ideal place for human residence within the Bailong River basin. Our ancestors might have moved long before if there were any better place to live. It is true that a growing population is a problem, but the number of residents per unit of land area is also multiplying. Countries with very high geological risks and population densities, such as Japan, have created highly developed communities despite the fact that they are threatened by potential disasters such as earthquakes, volcanic eruptions, tsunamis, landslides, and debris flows.

There was much controversy surrounding the relocation of county towns in Wenchuan (汶川) and Qingchuan (青川), Sichuan Province after the earthquake that occurred in 2008. Towns, big or small, came into being because they were at geographical centers, which also served as the hubs of interpersonal interactions and goods distribution for certain areas. In other words, they were located to facilitate transportation, trading, and administration. The most basic function of a city is to facilitate market growth; cities can be relocated or created, but markets sometimes cannot. The proposal to relocate the county towns was actually about relocating government authorities rather than the entire towns. Given their number, it would be impossible for local residents to move elsewhere even if they wanted to. Moreover, Zhuyuanba (竹园坝), or the proposed location of the new Qingchuan county town, is not big enough to house all the residents in the original one. Weizhou Town (威州镇), or Wenchuan's county town, is at a critical location in the upper reaches of the Min River (岷江) in terms of river and road transportation, where it was built about 1,000 years ago. Good conditions for urban construction are available where the Min and Zagunao (杂谷脑河) rivers meet, according to a geological survey. Numerous geological disasters can be avoided as long as both the area of land used for human activity and the city size are within the environmental capacity. If Wenchuan's county town were moved to Yutang Town (玉堂镇), Dujiangyan (都江堰市), or the proposed location of the new county town, then it would be equivalent to setting an "enclave" in the Chengdu Plain far from Wenchuan County. That would be far from the original objective—relocation to avoid disasters.

Zhicheng Town (治城镇) served as Beichuan's county town for about 1,000 years. As the geographical center of Beichuan County, it is in a relatively stable geological environment. In order to get the county town closer to the central city of Mianyang (绵阳) in northwestern Sichuan, the government recklessly moved the county town in the 1950s to Qushan Town (曲山镇) on the central fault of the Longmen Mountains (龙门山). The moved county town suffered multiple geological disasters, including a 6.2-magnitude earthquake in 1958. There have been calls for moving the county town back to Zhicheng since then. Qushan Town was destroyed by the Sichuan Earthquake, which made Beichuan the only county that had to relocate its county town. Nonetheless, Zhicheng was still not considered in the relocation plan. The local mountainous environment on which Beichuan (an ethnic Qiang autonomous county) relies was given up and a large area of good farmland in the Anchang River Plain (安昌江平原) of Yuan'an County (原安县) dominated by the Han Chinese was occupied to build the new county town in an ethnic Qiang style. Moreover, a large industrial development area was built on farmland near the new town. As a result, a large number of farmers living along the Anchang River, who were also among the victims of the Sichuan Earthquake, lost their land and houses before being forced to move. Since the new administration center is outside Beichuan County, some townships and towns dominated by the Han Chinese in Yuan'an County were redefined as part of

Beichuan. Will such relocation be favorable for protecting the traditional Qiang culture and for the Qiang Autonomous County (羌族自治区) to find a culturally compatible approach to sustainability? Careful analysis is needed to answer this question.

The above-mentioned problems reflected by the post-quake relocation are common ones that deserve consideration for future urban reconstruction. City location selection and planning should both fully consider geological environments, the capacity of the land for construction, and the possibility of geological disasters. There are upper limits for city sizes that can be accommodated by certain areas and such limits are referred to as capacity. A city will surely be faced with dangers if it exceeds that capacity by expanding itself into areas unsuitable for constructing buildings. But it does not mean that a city can never be built in an area or, if it is already there, must be relocated as long as it is subject to geological disasters. The point is that its size needs to be controlled along with good planning. Considerations include where the valleys subject to debris flows are, what the scope of their effects is, and what preventive measures are needed.

III. *Which Is More Important, Earthquake Prediction, or Disaster Warning?*

Defects in geological disaster prevention and control have become known due to disasters such as the Sichuan Earthquake and the ultra-large debris flow in Zhouqu. As for earthquake prediction, we are far from making short-term earthquake predictions that meet requirements for three key factors including time, location, and magnitude, apart from mid- and long-term earthquake trend analysis and hazard assessment. Unfortunately, people focus their hopes on earthquake prediction, while overlooking the more important consideration, which is building houses that meet China's *Standards for Earthquake-Resistant Structure Designs* (建筑抗震设计规范). Compliant structures would achieve the objectives of being intact in low-magnitude quakes, being reparable in medium-magnitude ones, and being free from collapse in high-magnitude ones, even if they are on an earthquake fault line. They would also not completely be ruined, even if an earthquake exceeded the upper limit of their resistance; thereby, avoiding heavy casualties. The above-mentioned standards were well implemented in Wenchuan's county town since it was designated as a demonstrative town for earthquake relief. As a result, few buildings were ruined by the Sichuan Earthquake. A lot of local buildings saved lives despite the fact that they were severely damaged and needed to be rebuilt.

Geological disasters such as debris flows are predictable to a high extent. Unfortunately, the public does not pay as much attention to such predictions and warnings as it does to earthquake predictions. Tasks relevant to such predictions and warnings include disaster identification and monitoring, rainfall forecasting, and the issuance of advance notices for evacuation. As for geological disasters such as debris flows, staying away from hazardous areas is critical for disaster prevention and reduction. If buildings are in such areas or timely predictions or warnings are unavailable, then it would be impossible for those buildings to survive severe debris flows, landslides, or collapses regardless of their quality. A lot of geological disasters are directly tied to heavy rainfalls. The accuracy of rainstorm forecasts is still low today, but it is much easier to improve them compared with earthquake predictions. The intensity of geological disasters is proportional to the rainfall. Areas prone to such disasters should be on alert when rainstorms are likely to occur. Major locations of geological disasters should be included in the monitoring system. Monitoring points should be set. Relevant organizations should have their staff members on duty for monitoring during rainy seasons. Otherwise, automatic monitoring and warning equipment should be in place and send warnings before rainstorms occur. Response plans should be developed for threatened downstream areas, where evacuation drills are required. Local residents should receive training on how to save themselves and escape within the shortest possible time. They should know what the routes are and where the shelters are. Drills should be conducted according to response plans. Otherwise, people will become too frightened to know what to do when a disaster occurs. Disaster protection is a complete system.

IV. *How Should Post-Disaster Reconstruction Be Conducted?*

A heavy rainfall occurred in the Longmen Mountains, Sichuan Province in August 2010, causing large-scale debris flows, collapses, and landslides. A series of dammed lakes were formed along the local river and led to an overflow. Houses rebuilt after the Sichuan Earthquake were destroyed or severely damaged. Some towns became "isolated islands" due to destroyed roads and bridges. We must further reflect on problems of post-disaster reconstruction.

Post-earthquake reconstruction is conducted amid worse environmental conditions and much higher geological risks. Unfortunately, there are many more large-scale projects in the Longmen Mountains and other western mountainous areas than before. Projects that were prohibited before the earthquake have been launched in the name of post-disaster reconstruction, including cement, chemical, and transportation projects. Mega projects such as a highway built for tourism in Jiuzhaigou (九寨沟) as well as the Chengdu-Lanzhou and Sichuan-Tibet railways are all being built in areas subject to geological disasters; and they will all run through the Hengduan Mountains (横断山) with an extremely vulnerable ecosystem and high biodiversity. Given the high environmental and social costs that will have to be paid, what will the overall benefits from those projects be? Prudence is obviously needed for this question. Reminders in 2010 included disasters such as a landslide to the Shanghai-Kunming Railway, debris flows to the Sichuan-Tibet Road, Zhouqu, and Yingxiu (映秀), and the broken bridge above the Shiting River (石亭江) as part of the Baoji-Chengdu Railway.

As for the post-earthquake reconstruction, the new settlements are characterized by how they are concentrated. In fact, rural houses in the hilly and mountainous areas across Sichuan Province and even the Chuanxi Plain (川西平原) are usually scattered to adapt to local natural environments and land for building houses. The above-mentioned debris flows in the province in August 2010 have revealed problems such as sluiceways being occupied by newly built towns, underestimation of the continuity and severity of post-quake secondary geological disasters as well as delays in building protective works and their inadequacy. As a result, it is necessary to discuss how to rebuild local villages and towns after the earthquake occurred. Should they be globally concentrated and locally scattered? Or should they be the other way around? They are related not only to disaster prevention and reduction but also to local residents' ways of working and living as well as the protection of living environments, cultures, building styles, and community characteristics in mountainous areas.

Geological disasters will occur for a long period of time after a huge earthquake. We must not overly concentrate rebuilt towns and settlements in the western mountainous areas; nor should we try to achieve high speeds for transportation infrastructure. We should control project sizes and coordinate the construction of geological disaster prevention and control works with the reconstruction of residential areas and transportation infrastructure. We should make sure that our money is spent on the right things. We would save money if we scientifically conduct justification, planning, and decisions in advance rather than doing reconstruction after a disaster occurs.

SOUTHWEST CHINA'S WATER CRISIS: RETHINKING THE URBAN GROWTH MODE[1]

Zhou Lei

Abstract: Based on a case study of droughts in Southwest China, this paper intends to analyze the fundamental reasons behind the disasters, and review the cognitive problems of urbanization patterns, knowledge management systems, water strategies, and the water system. The drastic water crisis now facing China calls for a knowledge convergence that can be preliminarily defined as urban ecology and includes a wide variety of different disciplines, such as geography, meteorology, urbanology, engineering, economics, and political science. In some cases, "soft" knowledge, such as religion, sociology, and anthropology, is needed to solve China's urban ecological problems.

Keywords: droughts in Southwest China, water ecology, cognitive crisis, urbanization, urban hydrology

I. *The Water Crisis behind the Drought in Southwest China*

In the spring of 2010, five provinces in Southwest China suffered severe droughts which affected more than 50 million people and nearly 5 million hectares of crops, including 400,000 hectares with complete crop failure. Furthermore, 20 million people were facing a shortage of drinking water. According to statistics from the Yunnan Provincial Forestry Department, the continued drought affected about 7.5 million *mu* (1 *mu*≈0.067 hectare) of nature reserves in the province, with about 100,000 *mu* suffering severe droughts, seriously hampering Yunnan's biodiversity conservation. The province's forest pest occurrence area reached 3.44 million *mu*, 1.7 million *mu* of which were above a moderate degree.[2]

As was recognized by Chinese scientists and some government officials, the droughts in southwestern China started in September 2009. But

[1] This paper is one of the outcomes of the 2010 special research project of the Friends of Nature Green Paper, which was sponsored by the Delta Environment and Education Foundation.

[2] There are a huge amount of news reports about the drought in the Southwest. The data used in this paper are cited from Netease News, http://news.163.com/special/00014868/drought2010.html.

because of the unique Chinese political conditions and the weak capacity of self-help among the public, the disaster continued to spread for three seasons, until it developed to such a serious level that the government began to pay attention to it.

During the droughts, the negative effects of urbanization were particularly apparent. Due to the improper urbanization mode and resource mobilization mechanisms, the negative effects of the drought have accumulated. Take Yunnan for example, which was surveyed by the author. The southwestern province is suffering from an all-round water ecology dilemma: its megacity development mode—with tremendous energy consumption—misguided the direction of urbanization in this region, and turned Southwest China into a black hole that absorbed energy, water, and other resources. As a result, major local projects emerged to make up a high-risk model of development with a huge volume but inferior structure. The construction and development strategy, featuring Kunming, Lijiang (丽江), Yuxi (玉溪), Dali (大理), Baoshan (保山), and other regional centers, have exhausted the region's ecological capacity and recovering capabilities. The construction mode of megacities resulted in the huge demand for water, soil, air, and energy, and along with the amplifying effect of megacities, continued to cause serious ecological and environmental impacts.

The scholar Zeng Nianchang (曾念长) has pointed out that the drought affecting large areas in five southwestern provinces was soon followed by flooding covering a larger area, which indicates that the natural water regulating system of southern China is suffering a staggering failure. An ecological regulating system, consisting of vegetation, soil, and water, should become part of people's scientific knowledge defended as their ethical bottom line.[3]

In addition, improper urbanization patterns and views on modernization have led to the continuous inter-regional water transfer in southwestern China, such as the water transfer from Zhangjiu River, (掌鸠河) Qingshui Lake (清水海), and Niulan River (牛栏江) to Kunming to address its urban water supply. On the one hand, the dynamic water cycle system was closed, resulting in the ecological degradation of part of the neighboring areas. On the other hand, the process of urbanization did not include urban network reconstruction, efficient water use, water recycling, or the concepts of a water-saving or smart city. It also expanded

[3] Zeng Nianchang, "Severe Drought and Flooding: The Destruction of Water Sources Brings Water Crisis to China," http://www.chinanews.com.cn/ny/2010/07-26/2424911.shtml, July 26, 2010.

the city infinitely, occupying surrounding quasi-agricultural areas in the craze to build the city infrastructure, only to lead to deterioration of the urban environment.

The rigid demand for resources in other major cities in China has in effect plundered water and other natural resources in southwestern China. A major cause of the severe drought in Yunnan Province, for example, is single-crop industrialized planting. Huge amounts of water are needed in the production of rubber, tea tree, eucalyptus, sugar cane, hemp, and bananas in Xishuangbanna (西双版纳), coffee and macadamia nuts in Baoshan, and tobacco, Barbados nuts (Jatropha curcas), and greenhouse vegetables across eastern Yunnan, in order to supply "industrialized" agricultural products to the abnormally developing city. During the drought, a large amount of groundwater was pumped for agricultural production, further exacerbating the severe desertification and groundwater depletion in some areas in central and eastern Yunnan.

When we reflect on the drought in the Southwest, we tend to adopt modern knowledge, while overlooking the oral tradition of social facts and customs, particularly the true wisdom of resource management. Based on the author's observations for several years in the surrounding villages of Dian Lake (滇池), there is a comprehensive, systematic, and complex Dragon religion, which does not only exist in the Han Chinese culture, but also directly relates to the ancient documents of the Yi ethnic group. The pollution of Dian Lake means the collapse of the Dragon belief system and the village water sharing and management mechanism. In a village temple beside Dian Lake, the author found inscriptions from the Qing dynasty, in which the squires mentioned that land reclamation and destruction of reeds in Dian Lake would be subject to heavy penalties. In this sense, the ecological effects of a traditional culture are revealed. There is an ecological awareness, ecological ethics, eco-philosophy, and eco-view of the world behind the importance that ethnic minorities attach to water sources, sacred groves, deity forests, dragon forests, dragon mountain woods, the view and belief in the dragon of the Han people, and the deism and rainmaking customs of the local society. These are all conceptual and spiritual elements that can conserve water.

II. *Revelation from the Drought in Southwest China: Reflections on Urbanization Patterns and Ecological Concepts*

It should be realized that the droughts in the Southwest show a "development syndrome" that takes the form of droughts and poor harvests.

We have to fundamentally deal with the dilemma of China's response to natural disasters, by reflecting on civic education for disaster preparedness, the government's disaster management and scientific decision-making system, and the overall design of the social and natural sciences in China. We should also be aware that the drought is a comprehensive reflection of the disparity between urban and rural China that has plagued the country for many years. Around the drought and other ecological disasters, social mobilization, official resource mobilization, and social science research have been reduced to a state of inefficiency, isolation, repetition, self-interpretation, bureaucracy, and emptiness.

The drought in the Southwest gives us another opportunity to understand the natural and social decision-making system. It also allows us to think more clearly about the functional division of various Chinese geographical regions in all kinds of natural and social science research. We should reconsider a long-term strategy from the perspective of disaster prevention when we divide the country into southwestern, northwestern, northern, eastern, and southern parts.

First, droughts reflect the negative outcomes of a resource-intensive development mode. Technically, China's urban development has been caught up in a vicious circle, taking in a lot of energy while destroying its own circulation system. This leads to a rigid desire for water, yet thirst still remains. From the bionic point of view, to address droughts and other existing ecological problems in China, in addition to increasing the water supply, we should also restore the micro-cycle of water between villages, towns, and cities. This is three-dimensional, circular, intelligent, and moderate, compared to cross-regional water transfer. We need the city to provide its own water supply.

The drought in the Southwest revealed China's water crisis, which can be identified as a cognitive crisis. The current Chinese model of development is based on a huge, detailed, segmented, and hierarchical knowledge system. On the one hand, along the cognitive path of science and inquiry, it constantly generates complex, term-loaded, qualitative "technical knowledge." On the other hand, the ideas that guide the "scientific breakthroughs" and "knowledge adventure" are closely related to sociology, which are often relevant to the history of the oppression of the Chinese people, the utopian collective consciousness, anti-Western new Orientalism, and the demoralization and structural crisis of the Chinese people.

In terms of urban form, in almost any large city of China there is a paradoxical combination: the transportation system has become more sophisticated, yet it takes people more time to commute to work; as the number of buildings rise, people's private and social space becomes increasingly narrowed; with the information system becoming more complex and sophisticated, more and more people lack common sense; the diseased imagination of the individual and the ill constitution of society have caused irrational consumption in the urban population; there is a deteriorating urban natural environment yet a massive pollution treatment system; and there are ever-expanding landmark buildings that shrink urban communities.

Such high-risk urbanization patterns have many dazzling aspects on the surface: high-speed transport and high-speed intercity rail networks that virtually link every part of the country; a multitude of Pan-Asian, Pan-Pearl River Delta, Pan-Northeast Asia, and Pan-Pacific centers (China again embraces "the rise of the Middle Kingdom" at least on a social linguistic level); the boom of the construction of extravagant surreal buildings; the attraction of foreign capital and shopping sprees; and so forth. But upon closer observation, there is a deep cognitive crisis behind the prosperity: like many other countries around the world, the current wave of urbanization and economic improvement in China is built on a growth mode and growth fantasy. In order to ensure an increase and figure delivery in terms of statistics, large numbers of scholars and scientists working for the government begin to create a variety of knowledge, causing losses of the country's limited resources.

For this reason, many issues involving China's urban development and environmental protection are somewhat like a daydream, with many cities intending to become international centers, building infrastructure on a grand scale, and designing architecture comparable to the Great Wall. At the same time, Chinese cities and people move forward stimulated by such a daydream, implementing the idealistic state strategy by consuming its ecological reserves.

III. *Adjustment of the Cognitive Process and Development Mentality to Solve China's Water Crisis*

As for the plight of the environment reflected by the droughts in the Southwest, we need to rethink the development strategy, knowledge integration, city models, and other aspects, and seek a fundamental solution to the crisis. First of all, in terms of a development concept, we should not only build "hard cities" based on extensive growth, but also "soft cities" with improved quality and structure.

In the face of China's urbanization symptoms, we should rethink the contradictions between such terms as development, growth, and acceleration, and consider the possibility of de-growth. As a revolution in the field of political science, economics, and sociology, de-growth bases itself on the environmental stance of anti-capitalism and anti-consumerism. It is not asking the individual to sacrifice the opportunity for happiness or the possibility of social development, but stresses a new "growth logic" featuring moderate consumption, rational design, and appropriate development, while playing down the zeal for hardness, speed, and frequency. It adjusts the investment in architecture and infrastructure, by transferring social resources to such public projects as art, music, family, culture, community, and ethics, enabling people to enjoy their life.

As individuals, we need a new philosophy of life that advocates simplicity, in order to reduce society's consumption of fossil fuels and other natural resources. The zeal for growth and hard construction has brought about the wrong economic model and urban economic system by stimulating excessive consumption to manage the city. The effort to access quality via quantity, however, ultimately ends up with nothing.

The concept of de-growth emphasizes moderation, rationality, and limitation. It advocates a non-consumerist growth mode, which rebuilds the micro-cycle in the city and society, the flow of voluntary behavior, sharing and benefit, and even gift giving.

In fact, Nicholas Georgescu-Roegen constructed a model of de-growth and bio-economy in 1971. He analyzed how mainstream economics and modern physics ignored the limits in physics and the Second Law of Thermodynamics: when entropy increases, the useful energy continues to decrease. For this reason, humans cannot design or pursue an infinite economic growth and social development model. From this perspective, many ambitious plans for zoning and development of functional areas inside Chinese cities are actually a political game rather than a scientific concept of development.

The significance of de-growth is that, it is not only one of the ways to ease urbanization symptoms, but also reconstructs the bond of human feelings, nostalgia, family, ethics, travel, visiting friends, and homesickness between different cities and regions.

In the concepts of future city planning, both southwestern China and economically developed cities should pursue human dignity and overall well-being, rather than sustained growth of economic data. From this point of view, emerging cities like Kunming need to think about their own ecological constraints and natural possibilities in an effort to become international and modern cities, based on which they will design their own eco-future.

The future urban design and water eco-design of Kunming need to build on the ecological base and existing contents of historical, human, social, and religious significance of southwestern China. In the design of the new Kunming, we should not abandon the large number of historical relics, traditional ecological pattern, ecological memory, and ecological imagination for the mere sake of newness. Instead, through the integration of "hard science" and "soft science," we need to think about how to connect water ecology and religion, to preserve the ancient dragon belief and folk life system (i.e. religious hydrology), and create a diversified water culture model of Dian Lake under the guidance of different views of religion and nature.

When we design urban construction, transportation, and community, we should integrate psychology, aesthetics, philosophy, sociology, anthropology, hydrology, engineering, geography, urban design, and architecture, in order to build a "thinking" city, a smart city, and "hydropology." Such multi-disciplinary integration ensures that we combine the advantages of all kinds of knowledge and contain their defects in the construction and water eco-design of a city. With regards to national development strategies, we need to promote knowledge integration, build a disaster prevention system, while seeking a more macro, integral, and sustainable cognitive approach.

On a macro level of national development strategies, the water system in southwestern China is not only a natural resource and geographical concept, but also has important religious, cultural, and historical significance. Rivers and water systems are the basis of the civilization of a country and region, which are material and spiritual ties binding countries. There should be a state-funded, large-scale international research program to systematically study the water system of the region, so as to make a long-term development plan for southwestern China's water security and natural habitats. We should also develop "cultural economics" and livelihood design based on the ecological situation of the region, since the plants, animals, geology, human race, and meteorological resources of Southwest China are enormous in the sense of scientific research. This can be included in China's ecological development plan oriented to the 22nd century, yielding continuous economic benefits. At the same time, based on the design of natural and cultural economics, we will think of the significance of the region in the belief system. This cultural area preserves thousands of religious branches and secular religious thoughts.

Through such studies we can understand the "pan-Southwest" cultural region that extends along the cross-border rivers and ethnic groups in the Southwest.

We should be alerted to the massive projects to channel, seek, and store water in the Southwest after this painful drought. Such storage of rainwater should be encouraged, but it is not an excuse for the excessive development of water sources, sensitive aquatic habitats, wetlands, and vulnerable groundwater.

In general, with the above study, we can formulate more scientific and effective response plans for the drought in southwestern China, as well as other ecological crises in terms of hydrology, meteorology, geology, phenology, botany, and zoology, which establish response systems connecting the general public, government, academics, civil intellectuals, international organizations, and other parties. With the integration of a variety of knowledge, disaster science should be established as a basic national policy of China, because it concerns people's lives and the future of the country.

In fact, after the earthquake on May 12, 2008, a new discipline should have been set up: cross-cultural disaster science in a multi-disciplinary approach. Any disaster that occurs in Southwest China should be documented into a huge database, where existing disciplines should attempt real knowledge integration. Every major natural disaster in the region should become our key to decipher the future, which, in spite of the time pressure, should be compared, analyzed, and integrated with similar knowledge in other regions of China and in other countries, and ultimately lead to a solution with creativity and imagination.

CONOCOPHILLIPS OIL SPILL AND ECO-DISASTERS IN THE BOHAI BAY

Zhao Zhangyuan

Abstract: This article analyzes the total amount of ConocoPhillips oil spill in the Bohai Bay and the extent of pollution. It argues that this oil spill caused more pollution than the BP oil spill and that ConocoPhillips should be held responsible. Taking into account the fact that the Bohai Bay was already the most polluted sea area in China before the oil spill, this article also projects a disturbing prospect of pollution and possible ecological disasters there.

Keywords: total amount of Bohai Bay oil spill, eco-disaster, pollution load, ecological collapse

Two severe oil spills, one on June 4, 2011 and the other on June 17, 2011, occurred on the Bohai Penglai 19-3 Oilfield, China's largest oil and gas field, operated by ConocoPhillips China. The oil spills lasted more than three months and severely polluted a large area of the Bohai Sea. This incident occurred in an important part of the Bohai Sea less than one year after an explosion and a resulting leak that occurred at an oil depot in the Dalian Bay in the Bohai Sea. Disorderly management was shown through news reports, information disclosure, and incident handling after the oil spills occurred. Moreover, this incident is already related to China's long-term energy decision-making. What effects has ConocoPhillips had on the Bohai Sea? That has become one of the public's major concerns.

I. *Severe Damage from ConocoPhillips Oil Spills*

A. *How Much Oil Was Leaked?*

There are multiple versions of the total amount of the spilled oil in the Bohai Bay (one of the three bays in the Bohai Sea). They range from tens of tons to tens of thousands of tons. Different stances led to figures that are different by orders of magnitude. Which figure is relatively objective? The estimation methods and reasons should be disclosed. Here, we recommend a relatively simple and practical estimation method for public reference. This method does not take into account the specific causes for the oil spills, their occurrence, or the controls against them. In addition, the method avoids the historical differences between relevant sectors whenever possible. Instead, it is based on measurements taken from the scene and, for most people, is easy to understand and use.

The amount of the spilled oil was estimated on the basis of factors primarily including:

1. *The Spill Duration*—More than three months beginning from early June.
2. *The Monitoring Result*—The North China Sea Branch of the State Oceanic Administration (SOA) conducted tracking and monitoring activities at the scene using aircraft, ships, and other resources including satellites after the incident occurred. The monitoring result released by this authority served as the primary basis for this estimation.
3. *The Polluted Area*—The polluted area was greater than 5,500 square kilometers and the maximum oil concentration exceeded the standard by 86 times, according to figures published by the North China Sea Branch. The area of seawater worse than Level 4 was greater than 870 square kilometers.[1]
4. *The National Seawater Quality Standard*[2]—0.05mg/L for Levels 1 and 2 seawater, 0.3mg/L for Level 3 seawater, and 0.5mg/L for Level 4 seawater.
5. *The Monitoring Criteria*—Seawater monitoring and sampling must be in accordance with the SOA's *Oceanic Monitoring Criteria* (海洋监测规范), which requires that sampling at the surface (depth<1 meter) and the deep layer (at least 2 meters from the seabed) be made if the seawater is more than 10 meters deep. The concentration value at a sampling point shall be the vertical seawater average at this point.[3]

The released measurements shall each represent the average concentration value of the seawater layers below and above each sampling point, according to the *Oceanic Monitoring Criteria*. The maximum oil concentration

1 Shu Jing, "The SOA Says 840 km² of Seawater in the Bohai Sea Suffers Lower Water Quality," the Finance Section of *Sina*, http://finance.sina.com.cn/g/20110705/174910097722.shtml (accessed July 5, 2011).

2 *National Seawater Quality Standard* (GB3097-1997), China Environmental Science Press, 1998.

3 The State Oceanic Administration, *Oceanic Monitoring Criteria*, The Ocean Press, 1991.

within the 870 square kilometer area of worse than Level 4 seawater exceeded the relevant standard by 86 times. Conservative figures are taken here with the assumption that Level 1 seawater is incompliant; the maximum measured concentration is 4.3mg/L. Given the combined action of tides, winds, and ships travelling in this area, the maximum oil concentration is 2mg/L, which is less than half of the measured value that exceeded the relevant standard by 86. If the average depth of the Bohai Sea is 18 meters, then the total amount of oil is 31,300 tons. Next, we calculated the amount of polluted seawater in the total peripheral area that was polluted. If the concentration is 0.4mg/L, then the estimated total amount of oil is 33,300 tons. The total amount of spilled oil may be about 65,000 tons, according to the preliminary estimation. This estimate represents the objective concentration value in the period ranging from the occurrence of the incident through the measurement conducted by the North China Sea Branch at the scene.

This amount of spilled oil is unusual in China, however conservative the estimate. All the oil will float with currents in this closed sea area for an indefinite period of time, except for the collected oil-seawater blend, which was more than 70 cubic meters in volume, and a small amount of oil on the periphery that will be decomposed by microbes.

B. *Pollution Intensity Much Higher Than in the Gulf of Mexico*

The BP oil spill caused immeasurable environmental damage and financial losses to the United States and even the whole world, as several U.S. states such as Louisiana were forced to declare a state of emergency. Comparatively speaking, the total amount of the spilled oil in the Gulf of Mexico was much larger than the one in the Bohai Sea, but the amount of oil per unit area in the latter was much higher than the one in the former.

About 780 million liters, or about 600,000–700,000 tons, of crude oil leaked during the BP oil spill. That figure was about ten times as large as the amount of oil leaked in the Bohai Sea. Nonetheless, the Gulf of Mexico is second only to the Bay of Bengal in area. It has an area of 1.5 million square kilometers, an average depth of 1,512 meters and a maximum depth of 5,203 meters; the total pollution intensity was estimated to be about 0.42 tons/km^2.

In contrast, the Bohai Sea has an area of only 77,000 square kilometers, an average depth of 18 meters and a maximum depth of 85 meters; given 65,000 tons of oil that leaked into the sea, the total pollution intensity was estimated to be about 0.84 tons/km^2. In other words, the amount of oil per unit area in the Bohai Sea was about twice as large as that in the Gulf of Mexico. If the estimation was made by volume, then the total amount of oil per cubic kilometer of seawater in the Bohai Sea would be about 11 times as large as that in the Gulf of Mexico.

Sufficient attention must be paid to the incident in the Bohai Sea. It seems that this incident was unlike the BP oil spill, which was characterized by a sudden, massive oil leak in a small area, or the explosion at an oil depot in Dalian, Liaoning Province, which was characterized by a gush of oil into a small area of sea. Nonetheless, the oil spill lasted more than three months and the source of the oil leak was at the center of the Bohai Sea, suggesting that the resulting pollution load was large. It can be said that the oil spill caused by ConocoPhillips has led to environmental pressure on China's Bohai Sea too heavy to withstand.

Now that we know the amount of oil pollutants per unit area of the Bohai Sea caused by the oil spills was higher than the one in the Gulf of Mexico, we may review the aftermath of the BP oil spill and the Exxon Valdez oil spill and let it be a lesson to us.

The ecosystem along the coast of the Gulf of Mexico has been suffering an unprecedented disaster since the BP oil spill occurred. The ecosystem characterized by the mangroves and aquatic birds on Cat Island, for example, suffered significant degradation one year ago. The United States has announced the expansion of the local fish sanctuary to 20,000 square kilometers. An expert has pointed out that the pollution may destroy the wetlands and beaches 1,000 miles in length along the coast of the Gulf of Mexico, that the oil clean-up would take nearly ten years, during which time the affected sea area would become unavailable for human use, and that the resulting financial loss would be in the hundreds of billions of U.S. dollars. The disaster will lead to an increase in federal deficit spending measured in "tens, and maybe hundreds, of billions . . . and federal and state revenues tied to Gulf-region businesses will fall." A deterioration of the economic statistics for the U.S. is expected. In addition, the local fishing industry will be injured, vulnerable species will die out, and the food chain will inherit a pollution legacy.

The Exxon Valdez oil spill occurred in Prince William Sound, Alaska, on March 24, 1989, when the Exxon Valdez, an oil tanker bound for Long Beach, California, struck Bligh Reef and spilled about 34,000 tons of crude oil. The spill killed a large number of young herring fish. Epidemics spread among stocks of herring in 1993 since oil pollutants damaged their immune system. The local herring industry, which was once prosperous, collapsed and has not recovered since. The oil spill is expected to affect the ecosystem of the sea area off Alaska for two decades, as some of the marine species have never been seen again since the tragedy occurred.

C. *The Impact of the ConocoPhillips Oil Spill on the Bohai Sea*

Tens of thousands of tons of oil will be floating for an indefinitely long period of time. Tides will expand the polluted area, as the crude oil will spread to the coasts of Tianjin as well as Shandong, Hebei, and Liaoning Provinces over several weeks with decreasing concentrations. Sand and rocks tainted by crude oil are expected to be seen along parts of the coasts. After the pollutants spread to the coasts, the crude oil, the emulsion, and the components solved by the oil dispersants will be gradually released and enter an extremely complex movement and conversion process along the food chain.

First, oil begins a complex series of changes, including spreading, vaporization, dissolution, emulsification, photochemical oxidation, microbial oxidation, sedimentation, the generation of tar balls, movement along the food chain, and other processes, once it enters the sea. Those processes mostly occur in a temporally overlapped manner despite differences in sequence and size. Oil particles reside on beaches, the seabed, and inside the bodies of animals and plants; it is difficult for them to naturally degrade apart from some microbial action after a very long period of time. Fishermen who specialize in offshore fishing in Qikou Village (岐口村), Cangzhou (沧州), Hebei Province found in mid-August that their fishing nets were full of such sticky particles, along with a thick layer of the same things floating in the sea. The fishermen could catch no fish. All the fishing boats used for offshore fishing in this village became idle by September. A large amount of black pollutants appeared on the beach of the No. 3 Sea Bathing Area in Qingdao, Shandong Province on September 4, 2011. More than half of the scallop raised in Laoting County (乐亭县), Qingdao was killed by the pollutants.

Oil is a toxic blend of various hydrocarbons. Aromatic hydrocarbons in gasoline, primarily including benzene, toluene, and xylol, are generally recognized as carcinogens. Some of them may damage cells and narcotize animals.[4] For the nervous system, oil pollutants may cause quasi-schizophrenic onset, hysterical paroxysms, and acute toxic brain diseases; for the urinary system, they may cause chronic poisoning and

[4] Kong Zhiming, *Environmental Toxicology*, the Nanjing University Press, 2004.

nephritis; for the respiratory system, they may cause toxic pneumonia, bronchial inflammations, and lung cancer; for the circulatory system, they may cause cardiovascular injuries. For the blood system, benzene may cause aplastic anemia and leukemia. Oil pollutants are highly destructive to animals and plants, as they may cause dysfunctions. As an example, a layer of oil on the water's surface may suffocate fish and shrimps. Foreign studies show that children living near gas stations or garages are five times more likely to have acute leukemia than other children, while they are eight times more likely to have acute nonlymphocytic leukemia (ANLL) than children who live in the same region but who do not live near gas stations. Residents are even ten times more likely to have cancers in some areas of China.

This oil spill incident is expected to produce the most immediate and greatest effects on the spawning ground in the Laizhou Bay. It will affect fish migration and spawning throughout the Bohai Sea and even part of the Yellow Sea.[5] It will cause at least a sharp decrease of the production among the coastal fish farms in the next several years, making local fishermen suffer heavy financial losses. The seafood quality is more worrisome as some toxic substances difficult to degrade will inevitably enter the food chain and may appear on our tables at any time.[6] As a result, people should be wary of seafood products, especially varieties resistant to pollutants, from the Bohai Sea in the next few years. It makes sense that people fear that the Bohai Sea aquaculture will be destroyed. Customers in the United States, Hong Kong, and other places have been asking about the death of the scallops, and many of them will send investigators to China.

The direct financial losses caused by the oil spill incident to the surrounding areas are actually easy to estimate, and it is not difficult to clean up the oil. Nonetheless, the resulting indirect ecosystem damages will last many years and spread across a very large area. That is why oil spills are known as super environmental killers. In fact, crude oil affects the nearby areas at a very slow rate after it gushes into a bay.

[5] Lin Changshan and Wu Yuming, *Environmental Biology*, Higher Education Press, 1986.

[6] Deng Nansheng and Wu Feng, *Endocrine Disrupters in the Environment*, Chemical Industry Press, 2004.

D. *ConocoPhillips' Responsibility*

What exactly were the causes of the ConocoPhillips oil spill incident? Did it properly handle the incident? Did it act as a responsible company? Given what it did after the incident occurred, ConocoPhillips showed a rather poor attitude.

Water should have been injected layer by layer into Well 23, Platform B at the Penglai 19-3 Oilfield, with attention to the amount and pressure of water injected into each layer to prevent breaks under excessive pressure. Unfortunately, ConocoPhillips injected water with no attention paid to inter-layer differences and happened to run into a fault. The intra-well pressure decreased rather than increase on June 2, 2011, when the amount of injected water increased to 6,000 barrels from 3,700 barrels. But it turned out that ConocoPhillips overlooked the anomaly and continued to inject water for at least seven more days. That was a simple mistake made by an international giant.

The debris backfilling technology was first used on Well C25 to increase the pressure. An environmental impact report said that pressure-reducing sleeves must go down to 350 meters deep. After the drill bit would not go down any further, however, ConocoPhillips had the sleeve stop at a depth of 225 meters and made the Water-injection Well C20 beside it. It turned out that ConocoPhillips broke Well C25, before high-pressure oil came out of Water-injection Well C20. This is known as a blowout. There are reportedly 29 wells at risk throughout this oil field. All the mistakes that occurred as a result of failing to obey relevant standards could have been prevented. The root cause for this incident was obviously ConocoPhillips' non-compliant operations and carelessness.

The above-mentioned incident reminds us of the BP oil spill, as BP paid insufficient attention to various important tests in order to reduce costs, leading to an explosion. In fact, oil extraction in the Bohai Sea area is carried out with conventional methods, and this incident was much less complex than the one in the Gulf of Mexico. But ConocoPhillips allowed it to last more than three months. It was not only a matter of technical capabilities but also, and more importantly, a failure to fulfill its social responsibility.

This was not the first oil pollution incident in the Bohai Sea area, according to an official of the Shandong Provincial Department of Marine and Fishery Industries. On May 13, 2011, the North China Sea Branch received a report from CNOOC Tianjin that an oil spill occurred at the Nanhai No. 1 platform at the 34-1 North Block. A severe oil spill occurred in the same area in 2007, which was reported by a Mr. Sun (孙先生) of a fish farm on Chang Island (长岛). There are photos showing that CNOOC was already discharging oil pollutants into the sea the previous year. The company sent two tugboats to make crude oil floating on the sea surface spread wider, thereby circumventing satellite surveillance. Was this the first such mistake CNOOC has made? CNOOC has never disclosed relevant information to the public. Moreover, it did not tell the public of the total amount of the spilled oil in this incident, and the rough amount it disclosed was smaller by orders of magnitude than some experts' estimates.

II. *Gloomy Prospects for the Heavily Polluted Bohai Sea*

A. *Pollution Sources around the Bohai Sea*

The Bohai Sea is among the top 12 closed seas in the world, as it connects with the Yellow Sea only through a narrow strait. Seawater exchange between the Bohai Sea and the Pacific Ocean is impossible because of the Korean Peninsula and the four main islands of Japan, leading to very poor water exchangeability and very limited pollution carrying capacity. It is very difficult for pollutants to spread to the Yellow Sea. It would take several decades for sea currents to exchange all the seawater.

Industrial pollution in the Bohai Economic Rim is one of the major pollution sources of the sea. Over the past few years, a huge amount of wastewater has flowed into the Bohai Sea with the fast economic growth of the Bohai Economic Rim. There are more than 100 sewage outlets around the Bohai Sea, into which about 2.8 billion tons of sewage flow, or nearly half of the total amount of pollutants into China's seas. A huge amount of treated and untreated industrial wastewater flows into the seas from China's rivers or land surface.

Oil spills from ships and oil exploitation activities constitute another major pollution source in the Bohai Sea. With large oil-bearing structures, the Bohai Sea area has abundant oil reserves. Located around the sea are the Shengli Oil Field (胜利油田), Dagang Oil Field (胜利油田), and the Liaohe Oil Field (辽河油田), in addition to the offshore oil fields. Like a natural oil tanker, the Bohai Sea has become a second Daqing Oil Field in China. As a result, this area suffers frequent oil spill incidents, leading to much more severe oil pollution than elsewhere in China, where the most severe oil pollution, more than 63 times higher than China's Level 1 seawater standard, is in the part of the Bohai Sea off Tianjin.

B. *Frequent Oil Spill Incidents in the Bohai Sea*

Large oil-bearing structures, numerous oil fields, and oil transportation are among the reasons for frequent oil spill incidents in the Bohai Sea area. Some examples include: the Panamanian-registered tanker Mandalay that sank in the water area west of the Changshan Channel (长山水道) in 1992, spilling more than fifty tons of oil; the tanker Zhepuyuyou No. 31 (浙普渔油) that sank in the Laotieshan Channel (浙普渔油) after it collided with another ship in 1996, spilling 476 tons of lubricant; and the Maltese-registered tanker Tasman Sea carrying 80,000 tons of crude oil that collided with the Chinese ship Shunkai No. 1 (顺凯) in the Bohai Bay on November 23, 2002, spilling a large amount of crude oil and leading to severe pollution (a 170 million RMB claim was filed to the insurer for the foreign shipping company). Over the past eight years, 81 oil spills have occurred in the sea area off Dalian, Liaoning Province, including seven oil spills caused by large ships with more than 600 tons of spilled oil. Every year, more than thirty incidents occur in which ships sink, run aground, collide with others and/or spill oil in the sea areas off the cities of Yantai (烟台), Weihai (威海), Dongying (东营), and Weifang (潍坊).

A severe oil product leak occurred at the China Marine Bunker (Dalian) Company (中燃大连) around the end of winter 2010. Tank No. 3 at the Dalian Crude Oil Terminal exploded while oil products were being injected into it from a tanker. Workers at the Tianjin Huishengda Petrochemical Technology Company (辉盛达公司) and Shanghai Xiangcheng Company (祥诚公司) injected desulfurizer that contains strong oxidants into the oil pipeline after the Liberian-registered crude-oil tanker Cosmic Jewel suspended the oil offloading operations, causing a chemical explosion inside the pipeline and a fire. As a result, a large amount of base oil flew through ditches and sewers into the Dalian Bay. There was crude oil dozens of centimeters thick offshore and the polluted area increased rapidly to hundreds of square kilometers.

C. *The Bohai Bay Is the Most Polluted Sea Area in China*

The Bohai Sea has been receiving large amounts of pollutants for numerous years, leading to continuously worsening environmental quality. Early monitoring results show that the excessive major pollutants at most monitoring stations have led to seawater worse than Level 3. All the monitoring stations in the cities of Dandong (丹东), Yingkou (营口), Panjin (盘锦), Jinzhou (锦州), Dongying (东营), and Weifang (潍坊) have detected pollutants that exceed national standards. Nutrient salts constitute the leading pollutants, among which inorganic nitrogen is increasing at an annual rate of 18% in terms of concentration. The percentage of monitoring stations that have detected pollutants that exceed national standards is increasing at an annual rate of 26%. The concentrations of oil pollutants are increasing at an annual rate of 7%. All this suggests that the sea areas off most coastal cities around the Bohai Sea have been heavily polluted. The contents of heavy metals detected in the local seabed sediment are up to 2,000 times as high as the national standards. Sometimes there is no dissolved oxygen in the water that flows into the sea from some rivers. According to China's *Report on Marine Environmental Quality* (海洋环境质量公报), in recent years the overall pollution of the Bohai Sea has been continuously worsening. Originally dominated by Level 1 seawater, the Bohai Sea saw the areas of Levels 2 and 4 seawater increase in 2010 alone to 1.92 and 3 times as large as in 2006 respectively. With unprecedented pollution, the Bohai Sea has actually become a sewage reservoir.

Eutrophication occurred in the Bohai Sea many years before, in addition to nearly twenty red tides, which covered thousands of square kilometers in the area in the last seven years. The Bohai, Laizhou, and Liaodong Bays are among the areas frequented by red tides, which cause massive deaths of fish, shrimp, and shellfish and the resulting heavy financial losses to local people. Some of the local people have been forced to leave their hometowns around the Bohai Sea because of ecological problems. Fishing resources that are being exhausted have led to a worsening trophic structure. Even the most adaptive species, the Yellow-eyed Crab, has now died out. Nearly two-thirds of the marine species in Laizhou Bay have died out, and the stocks of the remaining species are also shrinking exponentially. The vulnerable Bohai Sea ecosystem is already on the verge of collapse.

Today, local governments around the Bohai Sea have very poor capacities of environmental management and coordination. Recent explosions and the Typhoon Muifa-caused jetty destruction at the Dalian Fujia Dahua Petrochemical Corporation in the past few years have begun drawing public attention to the huge environmental threats behind the GDP growth of the petrochemical industry.

D. *The Prediction of the Future Bohai Sea Ecosystem*

Given the current pollution state of the Bohai Sea and the local governments' management capacity, it is predicted that the Bohai Sea will become a dead sea very soon and that the people living around it will be faced with a severe ecological disaster. Given lessons from the BP oil spill and other similar incidents, the upcoming ecological disaster in the Bohai Sea area will likely have the following characteristics:

1. Animals and plants such as birds, fish, and shrimp will gradually die in large numbers; most offshore fish farms will see reduced production and poor sales of sea-food products; some of them will shut down and add to the ecological refugees.
2. Species, especially already endangered ones, will die out at faster rates; local vegetation will degrade; seafood varieties will be on the decline.
3. The food chain will inherit this "legacy of pollution," as various pollutants that have entered the food chain will make a direct contribution to higher incidences of diseases in Shandong, Hebei, and Liaoning Provinces, the city of Tianjin, and even inland regions.
4. The Bohai Sea will have noticeably weaker functions or even lose them, including in offshore aquaculture, marine wildlife reserves, tourism, bathing beaches, sea food processing, salt-making, desalinization, and in spawning and feeding grounds.[7]
5. The degrading ecosystem will ultimately become out of balance; diseases caused by virus mutation will occur from time to time; the incidences of diseases among the public will obviously increase, posing serious threats to human well-being.

III. *Reflection on the Bohai Sea Management Crisis*

A. *Serious Management Inaction*

The Bohai Sea oil spill incident was among the few major sudden pollution incidents in China. The public should have been informed of this incident at an earlier time. The Chinese state leaders reiterated that there should be no excuse for non-disclosure or delay, but it was not until nearly a month after the oil spill occurred that the SOA held a briefing under pressure from the public.

In contrast, the BP oil spill was reported by the media to the whole world several hours after it occurred. U.S. President Barack Obama went to visit the scene shortly after the incident occurred and expressed his

[7] Zhao Zhangyuan, *China's Strategy of Managing the Offshore Marine Environment by Region and Level*, China Environmental Science Press, 2000.

anger over it on many occasions. Moreover, he proposed that the U.S. Congress review a bill to sharply increase the ceiling of the resulting compensation or even set no ceiling for it, which could financially ruin BP. The U.S. public urged Obama to shift attention to poor regulation over deep-sea drilling, which was the root cause of the problem, promoting the development of more effective laws and regulations. Obama as the decision maker correlated the U.S. reliance on oil with climate challenge as a long-term challenge, enabling the passage of a climate change act and pushing comprehensive new energy policies.

In contrast, the investigation into the Bohai Sea oil spill in China has been under way for five months and the total amount of the spilled oil has yet to be released. Nobody has announced the total amount of compensation, not to mention the possibility of a senior government official going to visit the scene of the incident. It now seems as if this incident did not occur at all. There are conflicting versions of explanations for this incident, many of which are illogical. The regulator's breach of duty has infringed the Chinese public's environmental rights. Again, the incident has revealed long-standing problems with China's marine environment management regime.

B. *An Unsuitable Marine Environment Management Regime*

For many years, China's offshore marine resources have been managed by administrative bureaus. Nonetheless, such a method usually leads to lack of management passing between regulators since China's seas have not been administratively demarcated. As environmental problems worsen in the Bohai Sea, conflicts will become increasingly intensified. Nine ministries and commissions are involved in sea area management in China. Agricultural non-point source pollution, for example, is within the jurisdiction of the Ministry of Agriculture (MOA); pollution flowing into the seas from rivers and shoal rehabilitation are within the jurisdiction of the Ministry of Water Resources (MWR); near-shore forests are within the jurisdiction of the Ministry of the State Forestry Administration (SFA); and municipal sewage discharge and treatment plants are within the jurisdiction of the Ministry of Housing and Urban-Rural Development (MHURD). The ministries often fight for power and, once pollution-relevant disputes occur, attribute to others the responsibility for their own actions. The environmental authority, in particular, tends to keep silent on such disputes.

The former State Environmental Protection Administration (SEPA) developed the "Blue Sea Initiative" (碧海行动计划) ten years ago. Unfortunately, this initiative ended up disappearing after spending tens of billions of *yuan*. Nine ministries and commissions including the National Development and Reform Commission (NDRC) as the leader, the SOA, SEPA, MWR, MOA and the Ministry of Finance (MOF), along with the local governments of Shandong, Hebei, and Liaoning Provinces as well as Tianjin, jointly developed the *Overall Environmental Protection Plan for the Bohai Sea* (渤海环境保护总体规划). Ironically, they produced a result similar to the one of the "Blue Sea Initiative" and remained unable to solve any problems. It generally takes eight to ten years and huge amounts of money for China to develop laws and regulations such as the marine environmental law, the national seawater quality standard and the functional division plan for the offshore marine environment, often with no result. Moreover, some important national environmental standards are quietly issued without serious discussion.

We should learn lessons from one after another oil spill in order to improve our management capacity. China is in urgent need of creating interregional marine environment protection and management regimes that suit economic development in its coastal regions and marine characteristics, and the accountability regime in particular. China has not created complete mechanisms for responding to and compensating for oil pollution. It needs to make improvements in responsiveness, the ability to clean up oil, and the post-incident compensation mechanism.

C. *The Public's Call for Compensation*

A 2004 document issued by the MOA specifies that pollution incidents in the fishing industry caused by oil spills, pollutant discharges, and waste dumping shall be investigated and assessed by the fishery administrations of the competent county-level or higher local governments or the fishery supervisors under which they work and shall be handled under the *Marine Environment Protection Law* (海洋环境保护法). Pollution incidents where animal and plant resources are damaged shall be handled by the above-mentioned fishery administrations, which may also raise compensation requirements on behalf of the Chinese government to those who are to be responsible for such incidents.

The Bohai Sea oil spill has drawn much attention from the Chinese public. Signs of oil pollution have been seen around the Bohai Sea since July 2011. The offshore fish farms in Shandong and Hebei Provinces, in particular, suffered massive deaths of fish, shrimp, and shellfish, bringing about heavy economic losses. Scallops farmed in 23,333 hectares of sea areas in Laoting County, Tangshan, Hebei Province alone had a death rate of 50–60%. The resulting financial loss amounts to about 300 million RMB and more than 160 fish farmers are involved, according to preliminary estimates. The affected local fish farmers called on the authorities to protect their rights. The Xinhua News Agency, mainstream newspapers and magazines, as well as numerous NGOs in China went to visit the Bohai Sea oilfield and the fish farms around the sea while demanding ConocoPhillips, ConocoPhillips China, CNOOC, the SOA and its South China Sea Branch, and even the Ministry of Environmental Protection (MEP) take responsibility. Among the NGOs, Green Beagle Environment Institute (Green Beagle), Friends of Nature (FON), Green Earth Volunteers (GEV), and other environmental NGOs went to the scene and held workshops on the incident. They called on government agencies to take action and provided constructive advice. The public's call for compensation fully embodies their strong awareness of social responsibility and environmental protection.

It was under pressure from the public that the SOA held a news conference and required that the Penglai 19-3 Oilfield: suspend its drilling, injection, and production operations; continue to identify and eliminate the sources of the spilled oil; clean up mineral oil-based drilling mud in a timely manner; rewrite a report on its marine environmental impact; develop an oil-deposit pressure reduction scheme with CNOOC to assure oilfield safety and protect the oil deposits; and report information in a timely manner.

IV. *Be Resolved to Address the Marine Ecosystem Crisis*

What will the Bohai Sea be like, a blue sea or a dead one? Neither the "Blue Sea Initiative" nor the *Overall Environmental Protection Plan for the Bohai Sea* has produced satisfactory results in terms of the Bohai Sea's rehabilitation. The primary problem is actually with the authorities rather than legislation or technology, as their inefficiency and inaction have greatly reduced the public trust in the Chinese government. When it comes to pollution, the Bohai Sea has entered a critical phase and we must be resolved to take powerful measures in order to address the following major environmental problems:

1. Control the economic development pace. We should no longer focus on GDP growth and we must develop a circular economy. As far as the Bohai Sea is concerned, we should reduce and even stop exploration at a

time when we have no access to proven technologies, thereby preserving the oil resources for our offspring. It is never too late to explore oil when we have access to proven technologies.

There have long been excessive heavy chemical industry projects around the Bohai Sea, as China's top three giants—China National Petroleum Corporation (CNPC), China Petrochemical Corporation (SINOPEC) and CNOOC—along with a dozen international giants such as Shell, Rosneft, the SK Group, and Mitsubishi Chemical, have all set up facilities there. There are already four ten-million-ton-level oil refineries and one-million-ton-level ethylene projects in Tianjin alone. The total investment in the petrochemical industry in the Bohai Economic Rim has exceeded 1 trillion RMB and the total annual production capacity of local refineries has exceeded 100 million tons, according to incomplete estimates. Given this trend, the Bohai Economic Rim will definitely become China's first region to see the collapse of an inner-sea ecosystem.

2. Conduct serious supervision and examination, hold companies responsible for illegal pollution, and punish them more severely. All relevant organizations should implement measures to prevent oil spills. Any organization that causes ecological damages too heavy to remedy should be deprived of all its assets for compensation. The ecological damages caused by the Bohai oil spill, for example, are already too heavy to remedy, so we must hold the companies responsible for them and should file claims against the major one so as to financially ruin it.

3. Replace the oil tanks and pipelines at about 100,000 gas stations across China as soon as possible; regulate all the gas stations and oil depots; create automatic underground monitoring systems for all the major oil depots. Conduct risk assessment on oil depots in environmentally-sensitive areas to eliminate hazards. Make sure that oil depots are located far from such areas.

4. Enhance public participation, which is an effective means of overcoming bureaucracy and corruption and preventing management inaction. The public should be encouraged to monitor environmental protection and those who provide information on illegal actions that cause ecological damages should be rewarded, thereby creating a good atmosphere in which all people care about and protect the marine environment. There are good examples to follow when it comes to addressing marine pollution. As an example, both the Seto Inland Sea and the North Sea were heavily polluted to the extent that fish and shrimp almost disappeared, but the local ecosystems have been restored after rehabilitation efforts were made.

QUESTIONING THE CHROMIUM SLAG POLLUTION INCIDENT IN QUJING

Zeng Xiangbin

Abstract: The chromium slag pollution incident in Qujing, Yunnan Province in August 2011 calls our attention to the seriousness of such pollution and the insufficiency of pollutant neutralization in China. Although the local government adopted remedies after the incident occurred, questions remain. Meanwhile, the gaming among the government, businesses, and public environmental rights continues. A public interest litigation filed by the Friends of Nature (FON) and other environmental non-governmental organizations (NGOs) has been heard by the intermediate court in Qujing, making it an important and much anticipated environmental public interest case.

Keywords: chromium slag pollution, integrated control, environmental public interest litigation

I. *Chromium Slag Pollution Incident in Qujing, Yunnan Province*

On August 12, 2011, the *Yunnan Information Daily* (云南信息报) published an article called "Sheep and Pigs Die and Water Has Become Undrinkable after 5,000 Tons of Highly Toxic Chromium Slag Leaks." The article said:

> The Chachong Reservoir (叉冲水库), on which some of the residents in Yuezhou Town (越州镇), Qujing (曲靖市), Yunnan Province rely, has been severely damaged in the past few months as a result of illegal slag dumping by the Yunnan Luliang Chemical Industry Company (陆良化工) (Luliang Chemical). Washed in by rainwater, more than 5,000 tons of highly toxic chromium slag has brought hexavalent chromium into this 300,000 square meter reservoir, where its concentration was once 200 times higher than the national standard.

The electronic version of this newspaper was deleted shortly after being published, but the information it contained was then disclosed on some microblogs, drawing immediate public attention to the chromium slag pollution.

Chromium slag is a highly toxic industrial waste that comes from producing chromium and its salts. Chromium changes into ionic chromium, which is soluble in water and rapidly becomes extremely toxic hexavalent chromium if it is not turned into insoluble forms. Hexavalent chromium primarily causes chronic illness in human beings, as it accumulates in the liver, kidneys, and glands after it enters the human body through the alimentary canal, respiratory tract, skin, and mucous membrane. Hexavalent chromium is absorbed by cells of tissues and organs after humans and animals drink water that contains it. It is very difficult to remove chromium slag once it enters water or soil, posing potential threats to future generations. It takes 100 years to degrade toxins in hexavalent chromium-contaminated water in a natural environment.

Hexavalent chromium is listed by the international community as one of the eight most hazardous chemicals for human beings. It is generally recognized as a carcinogen. Chromium-containing wastes are on China's *National Catalogue of Hazardous Wastes* (国家危险废物名录). The *Environmental and Technical Specifications on Chromium Slag Pollution Prevention and Control* (铬渣污染治理环境保护技术规范) issued by the former State Environmental Protection Administration (SEPA) has stringent requirements on chromium slag piling, excavation, transportation, and storage, including: chromium slag storage sites shall be staffed with full-time managers; any and all operations within a chromium storage site shall receive approval from its managers; measures shall be taken to prevent rainwater from flowing into a chromium storage site; chromium slag shall be transported along routes that are far from environmentally-sensitive areas such as residential and water source areas wherever possible; packaging shall remain intact throughout the transportation; and the *Multi-Copy Receipt-Based Management Measures for Hazardous Wastes Movement* (危险废物转移联单管理办法) and other relevant requirements shall be implemented.

Pollution incidents caused by illegal chromium slag dumping have occurred around the world, and severe chromium pollution happened in Japan and the United States in the 1970s. Xinglong (兴隆村), a cancer-hit village near Luliang Chemical, became publicly known after the chromium slag pollution incident in Qujing was reported on the Internet. It remains uncertain whether there is any tie between the high cancer rate in this village and the chromium slag in question. Nonetheless, it is beyond doubt that both this chemical plant and the chromium slag have become sources of fear and nightmares for the villagers, whose cattle and sheep died or fell ill after drinking the contaminated river water and whose paddy fields were heavily polluted. Local people must now buy bottled water.

II. *Seeking the Truth about the Chromium Slag Pollution*

Although it remains to be seen whether there is a cause and effect relationship between the local cancer cases and the chromium slag, the truth about the illegal chromium slag dumping has been squeezed out bit by bit thanks to Chinese Internet users' attention to this incident. According to the official explanation:

> Luliang Chemical began moving the chromium slag without approval on April 28, 2011. Wu Xinghuai (吴兴怀) and Liu Xingshui (刘兴水), as the carriers, failed to move the chromium slag to the Xingyi (兴义市), Guizhou Province-based Sanli Fuel Company (三力燃料有限公司). Instead, they dumped 141 truckloads, or 5,222.38 tons, of chromium slag in Qujing's Sanbao Town (三宝镇), Ciying Township (茨营乡), and Yuezhou Town. As a result, 75 farm animals were killed; 40,000 cubic meters of water in the Chachong Reservoir, and more than 3,000 cubic meters of water in a nearby valley were heavily polluted. There were no casualties caused by the poisoning.[1]

Nonetheless, some local people said that the already identified chromium slag is not all that has been illegally dumped, since it is very difficult to identify dumping locations scattered throughout the mountainous area, for which the author has found some evidence in relevant documents. According to written records from the Environmental Surveillance Team of the Yunnan Provincial Department of Environmental Protection:

> This company signed a chromium slag supply contract with the Xingyi, Guizhou Province-based Sanli Fuel on May 28, 2011. Sanli Fuel was to pay 273,088 RMB to Luliang Chemical for 2,730 tons of chromium slag. The license plate numbers of the trucks carrying chromium slag were Yun D42655 and Yun D21375. A total of 5,272.16 tons, or 174 truckloads, of chromium slag were transported in a 28-day period from April 28, 2011 to June 12, 2011.[2]

There is a clear difference between 174 and 141 truckloads.

China began paying attention to pollution problems caused by chromium slag in 2005, when the National Development and Reform Commission (NDRC) and the former SEPA jointly issued the *Integrated Regulation Scheme for Chromium Slag Pollution* (铬渣污染综合整治方案). Under

[1] Qujing Municipal Environmental Protection Bureau, "Report on the Illegal Chromium Slag Movement and Dumping by Luliang Chemical Industry Company," August 29, 2011.

[2] The Environmental Surveillance Team of the Yunnan Provincial Department of Environmental Protection's written record of a site investigation, August 13, 2011.

this program, the treatment of chromium slag stored by Luliang Chemical (140,000 tons was planned for Phase 1) was one of the ten projects whose implementation began in 2005. Luliang Chemical therefore submitted two applications for the treatment of more than 280,000 tons of chromium slag. The 140,000-ton Stored Chromium Slag Neutralization Project (Phase 1) received 20.2 million RMB from the NDRC under the Central Government's Investment Budget Plan for Resources Conservation and Environmental Protection Projects (资源节约和环境保护项目中央预算内投资计划) in October 2006.[3] With a designed annual chromium slag neutralization capacity of 50,000 tons, the Phase 1 facility was put into operation at the end of 2008 with approval of the Yunnan Provincial Department of Environmental Protection. But the facility's actual treatment capacity is only 20,000 tons per year. In February 2010, Luliang Chemical received 28.6 million RMB from the Chinese government for the 148,400-ton Stored Chromium Slag Neutralization Project (Phase 2).[4] The Phase 2 project has a designed annual chromium slag treatment capacity of 60,000 tons, but it was still under construction at the end of 2011. In other words, Luliang Chemical received subsidies from the Chinese government in the name of chromium slag neutralization, but it has made slow progress since then.

As for the above-mentioned integration control program, the Chinese government has paid 48.8 million RMB to Luliang Chemical's chromium slag neutralization project, but the company has treated only a small amount of chromium slag, with the vast majority still being stored near the Nanpan River (南盘江). "The use of money for irrelevant purposes and technical bottlenecks were among the major reasons for the failure of the last round of control efforts," said Bao Xiaodong (鲍小东), a reporter from the *Southern Weekend* (南方周末), in an article titled "The Legacy of Chromium Slag."[5]

In addition, the exact amount of chromium slag produced by Luliang Chemical is an important question. Luliang Chemical stored 140,000 tons of chromium slag, according to Appendix 2 of the Summary of Chromium Slag Stored across China in the *Integrated Regulation Scheme for Chromium Slag Pollution*. After the occurrence of the chromium slag

[3] Document FGBHZ (2006) No. 2272.

[4] Document FGBHZ (2010) No. 331 (a written record about an onsite investigation made by Yunnan Province's Environmental Surveillance Team on August 13, 2011).

[5] Bao Xiaodong, Shen Nianzu, and Zhang Lu, "The Legacy of Chromium Slag," the *Southern Weekend*, October 20, 2011.

dumping incident, municipal and provincial authorities announced that Luliang Chemical had stored 148,400 tons of chromium slag. But is this the truth?

On December 9, 2009, the Qujing Municipal Environmental Protection Bureau issued an administrative approval decision letter in which it decided not to grant an official Sewage Discharge Permit to Luliang Chemical. It noted that it was because Luliang Chemical had disposed of only 30,000 to 40,000 tons of stored chromium slag by the end of 2009 through the 140,000-ton Stored Chromium Slag (Phase 1) Neutralization and Resources Utilization Project, with about 280,000 tons of stored chromium slag yet to be disposed of and the second phase of the project yet to be kicked off.[6] It therefore required that Luliang Chemical speed up the disposal of the 280,000 tons of stored chromium slag and reclaim the entire 140,000 tons of chromium slag allocated to Phase 1 for cement production. It can thus be seen that Luliang Chemical has at least 250,000 tons of chromium slag to be disposed of since it stored more than 280,000 tons of chromium slag and had neutralized only 30,000 to 40,000 tons of chromium slag by the end of 2009. Moreover, this number does not include chromium slag produced later.

This estimate matches the amount confirmed by a competent environmental authority after multiple examinations made at the site. According to a written record dated July 25, 2011 about the onsite examinations made by the environmental surveillance team of Luliang County Environmental Protection Bureau, a Luliang Chemical official in charge of the site confirmed: about 4,500 tons of chromium slag was still stored and covered by fabric at the company's old factory area; about 29,000 tons of chromium slag were stored in the new factory area; nearly 5,000 tons of chromium slag (containing large amounts of soil and rocks) moved back from Qilin District (麒麟区) were stored in the open air in the factory area; chromium slag stored at the site opposite to the No. 2 slag repository was covered by fabric; and about 270,000 tons of chromium slag were stored near the Nanpan River. A record of the onsite examination made at Luliang Chemical by the county's environmental protection bureau on the afternoon of June 1, 2010 showed that 284,000 tons of chromium slag was stored by the Nanpan River and that more than 4,000 tons of chromium

[6] Qujing Municipal Environmental Protection Bureau, "Decision Letter on Granting Administrative Approval (QHK (2009) No. 7)."

slag was stored in the old factory area without the three required protection mechanisms.[7]

It is easy for us to conclude that the remaining amount of chromium slag stored by Luliang Chemical is at least 250,000 tons without including the newly produced chromium slag. Luliang had stored at least 300,000 tons of chromium slag by the time the Chinese government published the *Integrated Regulation Scheme for Chromium Slag Pollution* in 2005 to collect statistics on chromium slag stored across China. But official documents published by governments at all levels fail to provide any exact amount. How can they develop a regulation plan if they do not even know the amount of stored chromium slag?

III. *A Look at the Government's Regulation from the Perspective of the Chromium Slag Pollution Incident*

Regulators had different attitudes toward the chromium slag incident before and after the incident became public. Before August 12, 2011, they handled this incident by fundamentally following the traditional process of environmental law enforcement. The local environmental authority made investigations at the illegal dumping locations and took measures only for such locations without considering the compensation to the victims, regulatory problems with the chromium slag storage sites, or the large amount of chromium slag stored by the Nanpan River. It was not until the incident was revealed by the media that the local authorities began paying much attention and handling it at a fast rate.

In his book titled *Three Decades of Environmental Law Enforcement: Have We Been Successful?* Professor Wang Jin (汪劲) summarized the administrative factors that affect environmental law enforcement with the phrase "law enforcement in a quandary," saying: "The environmental authorities generally avoid handling minor problems rather than major ones at times when the public's environmental rights conflict with the interests of governments and companies. In other words, they deal with the public in a perfunctory manner where possible in order to maintain the governments' authority and companies' interests."[8]

---

[7] Luliang County Environmental Protection Bureau's a written record of an onsite environmental investigation in Yunnan Province, July 25, 2011.

[8] Wang Jin, *The Blue Book on Environmental Law Enforcement in China (1979–2000)*, 227–228, Peking University Press, August 2011.

The chromium slag storage areas of Luliang Chemical are among the major hazardous wastes repositories in Luliang County. They were already registered by the Chinese government in 2005 as part of its efforts to reduce chromium slag pollution. Luliang Chemical received monetary support from the Chinese government in 2006 for the purpose of neutralizing chromium slag. Unfortunately, the results are disappointing. Luliang Chemical must have illegally moved chromium slag more than once. Does the local environmental authority not have any idea how the multi-copy receipt-based transportation management system is being implemented? How could it have had nothing to do with the pollution incident?

The local environmental authority had long known that Luliang Chemical was illegally moving chromium slag elsewhere but, for unknown reasons, it took no action against the company. A written record of an onsite investigation by the country's environmental surveillance team on March 4, 2011 shows: "The multi-copy receipts for chromium slag movement are unavailable, and the intra-county movement documents were not written in accordance with relevant requirements (some documents do not have all the required signatures, for example)."[9] The team then required that "Luliang Chemical manage chromium slag in accordance with environmental requirements." The environmental surveillance team had obviously found out that Luliang Chemical was not effectively implementing the multi-copy receipt-based management system in the process of moving chromium slag. Nonetheless, it did not recommend any practicable measurements to be taken against such illegal activity. It was therefore natural that Luliang began moving chromium slag to the neighboring Guizhou Province from April 28, 2011 onward. We have no idea whether the Luliang County Environmental Protection Bureau would have issued a penalty notice if the drivers hired by the company had really moved the chromium slag to Guizhou Province in accordance with the contract signed with Sanli Fuel rather than dumping it in the mountainous Qilin District.

It is certain that Luliang Chemical had begun illegal chromium slag movement before 2011. The Luliang County Environmental Protection Bureau examined the Luliang Chemical's slag repositories by the Nanpan River on July 29, 2010. It wrote in an environmental safety surveillance record that "There are signs that the chromium slag stored at these

---

9 Luliang County Environmental Protection Bureau's a written record of an onsite environmental investigation in Yunnan Province, March 4, 2011.

repositories has been greatly reduced." But the bureau made no further investigation despite the existence of such signs. The environmental authorities in Qujing and Luliang County conducted a total of six examinations in the period ranging from June 13, 2011, when the illegal chromium slag movement and dumping took place, through July 25, 2011. But again, the examinations were only done in a perfunctory manner, as they did not even mention how the required three protection mechanisms should be implemented at those repositories by the Nanpan River, let alone take any other substantial moves.

Things have changed since this incident was revealed by the media on August 12, 2012. The Qujing Municipal Committee of the Communist Party of China (CPC) and the Qujing Municipal Government held a meeting to discuss measures the very evening the *Yunnan Information Daily* published their report. Yuezhou and Sanbao Towns in Qilin District each held a meeting on August 14 and 15, 2011, informing the public of how they would handle the pollution caused by the illegal chromium slag dumping and listening to their opinions and recommendations. At the meetings, the local officials also answered questions about compensation, pollutant removal, and water quality monitoring. They compensated the affected farmers for the resulting direct financial losses and paid relevant organizations for the resulting expenses, with a total amount of 786,000 RMB. According to a report, only 95,000 RMB was paid to the affected farmers for the loss of 77 heads of livestock while the remaining amount was for transportation costs, electricity expenses caused by water pumping, and other costs that occurred as a result of handling this incident. In the report signed by Qujing officials, it was required that "All the direct and indirect losses caused by the pollution be accurately assessed and be compensated for by individuals and organizations that have caused this incident; those who have caused this incident will be held responsible." Nonetheless, the local governments have so far made no material response to Xinglong Village near Luliang Chemical, a village that is said to have more than thirty cancer patients.

Yunnan Province's Environmental Surveillance Team went to Luliang Chemical on August 13, 2011 for an onsite examination. The Yunnan Provincial Department of Environmental Protection issued a letter of opinion on the incident's handling on August 16, 2011, saying that "According to an investigation, this company intentionally caused this pollution incident. The company must finish neutralizing all the chromium slag stored in its old repository and the new one in its current factory in the period of production suspension. It must not resume production without approval from the Qujing Municipal Environmental Protection Bureau. There remain serious environmental safety hazards at the old chromium slag repository." However, the local government allowed Luliang Chemical to resume production in November 2011.

Moreover, changes have occurred to the chromium slag repositories by the Nanpan River. The slag there is now covered by asbestos sheets; the retaining walls have become higher than before the incident; and drilling and concreting were carried out alongside the road by the Nanpan River to prevent pollutant infiltration. All these have been reported by China Central Television's (CCTV) News Probe and Topics in Focus. The chromium slag repositories were even manned with security guards to prevent lawyers who sought the truth for the purpose of public interest litigation from taking photos.

At a news conference held in Qujing on September 4, 2011, Chen Jun (陈军), a member of the Standing Committee of the Qujing Municipal CPC Committee and deputy mayor of Qujing, summarized the lesson from this incident and admitted that the illegal chromium slag dumping, had exposed problems such as some companies' failure to fulfill their environmental obligations and regulators' failure to fulfill their duties. He said that "The environmental authorities failed to supervise companies obeying requirements specified in the *Multi-Copy Receipt-based Management Measures for Hazardous Wastes Movement*, thereby giving the carrier a chance to illegally dump the chromium slag and cause this pollution incident. There were aspects that the environmental authorities did fail to regulate, bringing about grave repercussions and serious lessons."

IV. *Remedies for the Chromium Slag Pollution Incident*

On September 27, 2011, the Qujing Municipal Government issued the *Implementation Scheme for Chromium Slag Pollution Reduction and Control* (铬渣污染治理实施方案), the most complete chromium slag reduction/control plan developed by the local government ever. This scheme specifies objectives and tasks including: finish the cleanup and the improvement and meet the acceptance requirements by January 2012; finish neutralizing the 25,000 tons of chromium slag newly stored by Luliang Chemical, the chromium slag moved back from the dumping locations, and the polluted soil by the end of December 2011; and finish neutralizing the 148,400 tons of chromium slag originally stored by Luliang Chemical by the end of December 2012. In other words, the new and old slag is to be neutralized by the end of 2011 and 2012 respectively. But the question is: since the local government has never carefully checked the amount of the chromium slag, how can it make sure that these planned objectives will be achieved?

When it comes to speeding up the construction of neutralization facilities, the above-mentioned scheme specifies that: The aim is to neutralize at least 10,000 tons of the new chromium slag by the end of 2011; accelerate the implementation of the Phase 2 chromium slag neutralization project to make sure that the facilities be put into trial operation within the year; and neutralize at least 64,000 tons of chromium slag by the end of 2012. Additionally, the scheme specifies: 1) Use advanced, applicable technologies available from iron and steel companies in the city, including the high-temperature reduction, sintering, and smelting process, to neutralize chromium slag; make sure that 15,000 tons of newly stored chromium slag will be disposed of by the end of 2011; and 2) Neutralize at least 100,000 tons of the originally stored chromium slag by the end of 2012. Why had they not started using the high-temperature reduction, sintering, and smelting process at an earlier time if it is really able to dispose of such a large amount of chromium slag per year?

Behind this plan is a major precondition—money. The plan is beautiful, but money is more important. After Luliang Chemical spent the 48.8 million RMB in subsidies for chromium slag neutralization, the Qujing Municipal Environmental Protection Bureau submitted a request to the Yunnan Provincial Department of Environmental Protection on October 12, 2001, asking it to grant Luliang Chemical additional subsidies for chromium slag neutralization, again with reasons that sound decent:

> It is difficult and costly to dispose of the 148,400 tons of chromium slag stored by the Nanpan River for historical reasons. Nearly 250 million RMB would be needed based on an estimated cost of 1,660 RMB per ton. As the company that is responsible for disposing of the chromium slag from Luliang Chemical, the Yuegang Group (越钢集团) would need 18.1 million RMB to build new, closed chromium slag repositories, modify its equipment and facilities, purchase monitoring equipment, provide personal training, and prepare protection for chromium slag-related operations.

So it asked the Yunnan Provincial Department of Environmental Protection "to grant 200 million RMB in subsidies to Luliang Chemical for chromium slag neutralization."[10] The bureau submitted a request to the Qujing Municipal Government on October 27, 2011, saying that:

---

[10] Qujing Municipal Environmental Protection Bureau, "Request for Granting Subsidies to Luliang Chemical Industry Company for the Purpose of Chromium Slag Neutralization (QHQ (2011) No. 264)."

> Luliang Chemical will dispose of 9,900 tons of the 25,000 tons of newly stored chromium slag at its own cost. The Yuegang Group will be responsible for disposing of 15,100 tons. Luliang Chemical will sign a disposal agreement with the Yuegang Group. Given a disposal cost of 1,500 RMB/ton, the total expense will be 226.5 million RMB and will all be borne by Luliang Chemical ... There is 148,400 tons of chromium slag stored by the Nanpan River, among which, Luliang will be responsible for disposing of 63,400 tons at its own cost. The Yuegang Group will be responsible for disposing of 85,000 tons. Given a disposal cost of 1,500 RMB/ton, the total expense will be 127.5 million RMB. Luliang Chemical will bear the cost of 36 million RMB; the Luliang County Government will bear the cost of 20 million RMB; the municipal environmental protection bureau will seek 20 million RMB from the provincial and central governments; as for the remaining 51.5 million RMB, the Yuegang Group will share 20 million RMB, and we ask the municipal government to help with the other 31.5 million RMB.[11]

There are reasons to worry that the above-mentioned approach is the typical Chinese way of handling pollution incidents—a company illegally discharges pollutants and makes profits before being assisted by the local government in pretending that they are innocent. The company is therefore able to keep its profits while the governments at upper levels have to pay for pollution handling and the resulting compensation. Moreover, the local environmental authority asks the governments at upper levels for money on behalf of the company, which in turn will start disposing of the pollutants only after it receives money and even may not be active in disposing of it after it does receive money. China's *Environmental Protection Law* (环境保护法) specifies that those who cause pollution shall be held responsible for minimizing its consequences, but the current fact is that those who cause pollution make money. The subsidies for chromium slag disposal are being spent, but it seems that nobody is responsible for or held responsible for achieving the desired objectives.

When it comes to chromium slag pollution across China, the availability of technologies and money is only a superficial reason for the problem. Compared with Japan and the United States, China lacks laws in this aspect. In Japan, the *Waste Management and Public Cleansing Law* specifies that "businesses shall appropriately manage the waste left as a result of their business activities."[12] The U.S. *Comprehensive Environmental*

---

[11] Qujing Municipal Environmental Protection Bureau, the *Request for Instructions on the Money and the Scheme Needed to Neutralize the Chromium Slag Stored by Luliang Chemical Industry Company by the Nanpan River* (QHQ (2011) No. 288).

[12] The Japan Federation of Bar Associations (JFBA), *Typical Environmental Cases in Japan and Comments*, 135, China University of Political Science and Law Press, 2011.

Response, Compensation, and Liability Act (CERCLA) is more stringent as it specifies:

> Both past and current owners and operators of contaminated facilities as well as persons who transported or arranged the transportation of wastes to such facilities and who received wastes shall all be held responsible for the resulting contamination; these persons shall be held responsible for cleanup costs, natural resource damages, and the costs of federal public health studies; the polluters shall ultimately be responsible for the costs of cleaning up such facilities, even if the federal government has paid them.[13]

Chromium slag pollution is an environmental challenge for the entire world. The U.S.-based United Chemical Corporation shut down in the mid-1980s because it was unable to solve its environmental problems in a timely manner. Today, all the developed countries have reduced their respective chromium salt production capacities and turned to the developing ones for such products. In 2005, the NDRC and the former SEPA sought to achieve tight control over China's chromium salt production, as they required in the *Integrated Regulation Scheme for Chromium Slag Pollution* that:

> ... the chromium salt industry policies be researched and developed on the basis of satisfying domestic market demand; determine a reasonable chromium salt production capacity to avoid excessive development; prevent the gaining of economic benefits at the sacrifice of the environment. Given the excessive production capacity, massive import of chromium minerals, and the fundamental balance between supply and demand in China, chromium salt production should aim at satisfying the domestic market. Chromium salt export should not be encouraged. New or expanded chromium salt production projects should be under tight control.

Unfortunately, this policy is being ignored by local governments that focus their attention on economic growth. Luliang Chemical, for example, has a steadily increasing chromium salt production capacity, as it sees no obstacle caused by slow chromium slag disposal. In November 2007, the company was allowed to begin pilot production for a project involving annual production capacities of 1,000 tons of vitamin K3 and 30,000 tons of chromium powder. By November 19, 2008, it had increased the vitamin K3 production to 2,600 tons while increasing the chromium powder production to 70,000 tons.

---

[13] Lü Zhongmei and Wang Lide, *Environmental Public Interest Litigation—China vs. America*, 158, The Law Press, 2009.

In February 2011, China's State Council gave an official reply to the *Twelfth Five-Year Plan for Comprehensive Heavy Metal Pollution Prevention and Control* (重金属污染综合防治十二五规划), which was China's first ever plan specific to five heavy metals including mercury, chromium, cadmium, lead, and arsenic. For companies specializing in producing such metals, the Ministry of Environmental Protection (MEP) stressed that:

> Companies that have not received environmental impact assessment and whose pollution treatment facilities have not been accepted in accordance with requirements must all suspend production and implement improvements; companies near drinking-water sources must all be closed; companies whose pollution treatment facilities are not normally functioning and that have long been discharging/emitting pollutants must all suspend production and implement improvements; companies where serious safety hazards have been found must all suspend production and implement improvements, and those that fail to meet relevant improvement requirements must all be closed; environmental authorities must not issue letters of consent to any company's request for getting listed or refinancing within two years after the company causes serious environmental damages.[14]

Given our understanding of the chromium slag pollution incident in Qujing, we have reasons to worry about how these "must all" requirements can be met.

V. *Environmental Public Interest Litigation against the Chromium Slag Pollution Incident*

Friends of Nature (FON), an environmental NGO, set up a team of public interest lawyers shortly after the illegal chromium slag dumping was reported by the media. On August 26, 2011, members of the team went to the scene for evidence. They planned to focus their efforts on Luliang Chemical, the stored chromium slag, and the cancer-hit village, including reaching all the stakeholders of this pollution incident, such as Luliang Chemical employees, occupational disease patients, Xinglong Village officials and farmers, cancer patients, officials of the local environmental protection bureau and county government, and the president of Luliang Chemical. The villagers, the occupational disease patients, and the employees were willing to have interviews with the lawyers and helped

[14] Wang Xi, "The State Council Gives an Official Reply to the *12th Five-year Plan for Comprehensive Heavy Metal Pollution Prevention and Control*," the *Southern Weekend*, February 22, 2011.

them explore the entire scene, enabling the lawyers to secure important evidence.

On October 19, 2011, the intermediate court in Qujing officially heard a public interest case about the chromium slag pollution. This was China's first ever public interest litigation filed by environmental NGOs (FON and the Green Volunteer League of Chongqing in this case). Article 108 of China's current *Civil Procedure Law* (民事诉讼法) requires that the plaintiff be directly related to the case in question, which obviously no longer applies to the current situation. The local court's hearing the case to which the environmental NGOs are not directly related is a significant advancement. On October 24, 2011, the Xinhua News Agency reported that the National People's Congress had released the *Draft Revision of the Civil Procedure Law* (民事诉讼法修订草案), which specifies that "relevant government agencies and NGOs may file lawsuits with the courts against activities that hurt public interests, such as environmental pollution and infringement on consumer rights." This draft is now under discussion.

Later, the Qujing Municipal Environmental Protection Bureau joined the environmental NGOs as a co-plaintiff against Luliang Chemical, for causing the chromium slag pollution, and Luliang Heping Technologies Company (陆良和平科技有限公司), as its shadow company. The plaintiffs demanded: the defendant immediately stop illegal chromium slag storage that causes environmental damages; the defendant immediately eliminate hazards by taking effective measures to eradicate the environmental pollution caused by its dumping and storing chromium slag; such measures should be assessed by a third-party organization and supervised by the plaintiffs and the court, and information should be disclosed to the public; the defendant compensate the environmental losses caused by the chromium slag pollution (the compensation amount is set at 10 million RMB for the time being and is subject to the judicial assessment report); this compensation be transferred to a public interest account opened by the plaintiffs for restoring the ecosystem polluted by the chromium slag and, under the joint supervision by the plaintiffs and the court, be used to restore the ecosystem damaged by the defendant; and the defendant bear any and all the reasonable expenses resulting from this litigation and execution. The court session originally scheduled for December 12, 2011 has been postponed as a result of the large number of evidence submitted by both sides. How this litigation will come to an end in 2012 has become a noteworthy issue.

Yunnan Province is one of China's few pilot provinces for the purpose of environmental public interest litigation, suggesting that this province has advanced ideas and effective actions in the process of becoming a resource-saving and environment-friendly region. The fact that the chromium slag pollution incident occurred in Qujing does not mean that Yunnan Province is the most polluted province in China. There is several million tons of chromium slag stored across China. A report says that: "In August 2011, Greenpeace conducted an investigation in Tianjin and found that the hexavalent chromium concentrations in the water near the exterior walls of Tongsheng Chemical Plant (同生化工厂) and in the nearby manholes exceeded the Level 5 water standard by 7,529 and 28 times respectively." Back in 2007, Cao Shuliang (曹树梁), director of the New Material Institute of the Shandong Academy of Sciences (山东省科学院新材料研究所), found that "Around one hundred billion cubic meters of water would be heavily polluted even if only 15% of the hexavalent chromium contained by the millions of tons of chromium slag stored across China entered local water systems."[15] The mound of chromium slag in Tianjin would have received no attention if the chromium slag pollution incident had not occurred in Yunnan Province. Being faced with such a huge amount of stored chromium slag, China has started a new round of chromium slag disposal.

Holding polluters responsible for environmental pollution through public interest litigation is a positive approach. How have Luliang Chemical's illegal chromium slag storage and dumping affected and damaged the local environment and ecosystem? Given the lack of relevant standards and technical constraints in China, how to accurately determine the consequences of chromium slag pollution has become a big challenge. According to legal theories concerning environmental litigation, it remains difficult to determine the cause and effect relationship in addition to the need for identifying damage-relevant facts through assessment. Fortunately, the *Tort Law* (侵权责任法) follows the principle of "reversal of the burden of proof," which exempts plaintiffs from the responsibility of providing evidence. As shown by this chromium slag pollution incident, China has a long way to go in terms of chromium slag pollution regulation, while the fight among stakeholders may have just begun.

---

[15] Bao Xiaodong, "The Legacy of Chromium Slag," the *Southern Weekend*, October 20, 2011.

THE DALIAN PX INCIDENT: ON INSTITUTIONAL SAFEGUARDS FOR INTEGRATED DECISION-MAKING IN ENVIROMENTAL PROTECTION

Wang Shekun

Abstract: The Dalian paraxylene (PX) incident, together with a series of other similar ones, demonstrates the indispensability of institutional safeguards to achieve an integrated decision-making which takes both economic development and environmental protection into consideration. Firstly, the economic entities and decision makers should be held legally accountable for environmental protection. Secondly, detailed legal procedures, such as environmental impact assessment (EIA), should be completed and implemented. Thirdly, political reform should be deepened and the performance assessment system should be improved so as to provide officials incentives to include the environmental factor in their economic decision-making. Fourthly, information transparency should be ensured and the public should be encouraged to participate in economic decision-making, serving as the external supervisor for environmental considerations.

Keywords: economic decision-making, environmental impact assessment, institutional safeguards, Dalian, PX

I. *Why Are People So Angry about a PX Plant?*

In August 2011, Typhoon Muifa not only breached the protective dyke of a chemical plant at the Jinzhou Industrial Complex of Dalian (大连金州开发区), it also unveiled the mystery of one of the manufacturers located there: the PX project of the Dalian Fujia Dahua Petrochemical Company (大连福佳大化石油化工有限公司). What enraged local residents was that the PX project, driven out of Xiamen by their residents' peace walk in 2007, had been quietly launched in Dalian, out of public sight.

On August 14, 2011, the residents of Dalian gathered in the People's Square in the city center and demanded that the Fujia PX plant be moved out and people involved in the launching of the project be held responsible. It was a relief to them that the municipal government responded quickly. On the very day of the protest, authorities in Dalian ordered the Fujia PX project to be halted and relocated. It was a result that fairly satisfied the Dalian residents and temporarily gave the Communist Party of China (CPC) Dalian Municipal Committee and the city government some rest. Nevertheless, with a view of the PX incident in Haicang District of

Xiamen (厦门海沧) in 2007, we cannot help but ask: "Why are people so angry about a PX plant?"

PX, an abbreviation for Paraxylene, is an aromatic hydrocarbon. This crystalline and colorless liquid features good chemical stability and is insoluble in water, and it is the raw material for synthetic polyester fiber, resin, pesticides, coating, and coloring materials. The Material Safety Data Sheet (MSDS) shows that PX has a low toxicity, which is higher than salt but lower than alcohol. The WHO's International Agency for Research on Cancer (IARC) defines PX as a "Group 3 Carcinogen," which means that the evidence available now cannot prove its carcinogenic effect. In other words, PX is not highly toxic.

Then why has this low-toxic substance led to such fierce public opposition? The root cause may lie in the way the Fujia PX project located itself in Dalian. The Fujia PX plant was approved by the National Development and Reform Committee (NDRC) in December 2005. It began construction in October 2007 and was fully completed in November 2008. It had a trial production in May 2009 and went into full production in June 2009. It covers an area of 80 hectares and is one of the largest single aromatic hydrocarbon projects in China, with a working capital of 9.5 billion RMB, an annual output of 26 billion RMB, and tax revenues of about 2 billion RMB.

However, as early as 2006, after an inspection and evaluation of the environmental risks of the key petrochemical enterprises in China, the former State Administration of Environmental Protection reported that the Dagushan Peninsula (大孤山半岛), where the Fujia PX plant is located, was dotted with too many petrochemical plants and had a high environmental risk in the area. What is more noteworthy, residents of Dalian were kept in the dark during the whole period from its construction in October of 2007 to its start of production in June of 2009. It was only when the typhoon destroyed its nearby protective dam that this mysterious project was brought to light.

In fact, the Fujia PX incident is not the only problem in Dalian. There have been a series of environmental disasters along with the fast and triumphant advance of "Greater Dalian": a subway collapse, explosion of oil tanks, chemical plant leak, demolition of traditional buildings, the loss of the "Garden City" Award, and wanton land reclamation. Along with these incidents, there have been some brilliant economic indexes: in 2010, Dalian's GDP, fixed asset investment, and its expected municipal revenues were respectively 2.4 times, 4.6 times, and 3.3 times of those in 2005. It is not difficult to see that the development of "Greater Dalian" has been a lopsided one, a development oriented only to GDP. What has aroused the anger and opposition of the residents of Dalian is not simply a PX project, but what this PX project represents—an economic policy that excludes environmental factors, and an economic policy-making procedure that lacks transparency.

II. *Substantive Law on an Integrated Decision-Making including both Economic Development and Environmental Protection*

The periodic chaos in Dalian in recent years, with the PX incident being a typical example, reflected the absence of an integrated decision-making which takes both economic development and environmental protection into consideration. There has been a negligence of environmental factors both in the economic decision-making and in its implementation.

To achieve this integrated decision-making would require substantive law to define the legal obligations of decision makers to include environmental factors in their economic decision-making. As a matter of fact, the current Chinese law is fairly complete in this respect. The Constitution clearly says that "the state protects and improves people's living environment and the ecology, prevents pollution, and other public disasters." This is the constitutional foundation for the state, the biggest and ultimate decision maker, to take the responsibility to protect the environment.

The *Environmental Impact Assessment Law* (环境影响评价法) defines the same legal obligation of economic decision makers on the micro level as well as to take environmental factors into consideration before they approve a project. For instance, it stipulates:

> Where the document for evaluation of the environmental effects of a construction project is not examined, or is examined but not approved, by the examination and approval department specified by law, the department for examination and approval of the said project shall not be allowed to approve construction of the project and the construction unit shall not be allowed to start construction.

As a matter of fact, the Fujia PX project, in itself, could have complied with the environmental guidelines. That was the reason why the Dalian Environmental Protection Bureau declared in 2009 that: "The Fujia Dahua PX plant is in line with the state's industrial policy and its effects on Dalian's economy and ecological environment are in conformity with the state's general guidelines."

Nevertheless, if one saw beyond this single project and analyzed it within the whole frame of the city's development, one would discover the general environmental risks, even when every single project within the city complies with the environmental requirements. After all, a project is only the end product of a series of decision-making. A single project that meets the requirements can only mitigate the environmental damage, but cannot protect the environment or avoid its macro effects, indirect effects, secondary effects, or cumulative effects on the environment. In comparison with single projects, the environmental impact of the government's policy and planning is much more extensive and enduring, and it is much more difficult to tackle its adverse impact. If there were no prevention of environmental problems from the start of economic planning, more political and economic costs would have to be paid.

Therefore, it is important to define the legal responsibility to protect the environment of the economic decision makers on the macro level, who, for example, are responsible for the approval of planning. The *Urban and Rural Planning Law* (城乡规划法) stipulates:

> In making and implementing urban and rural plans, attention shall be paid to following the principles of overall planning for urban and rural areas, rational layout, conservation of land, intensive development and planning before construction, to improving the ecological environment, promoting conservation and comprehensive utilization of resources and energy, to preserving cultivated land and other natural resources and historical and cultural heritage, to maintaining the local and ethnic features and traditional cityscape, to preventing pollution and other public hazards, and to meeting the needs of regional population development, national defense construction, disaster prevention and alleviation, and public health and safety.

Meanwhile, the *Environmental Impact Assessment Law* also states:

> When making arrangements for formulating plans for the utilization of land for construction in or development and utilization of certain areas, river basins and sea areas, the relevant departments under the State Council, local people's governments at or above the level of the city divided into districts and the relevant departments under them shall see to it that environmental effects are evaluated in the process of formulation and that in the plan a chapter is devoted to such effects or an explanation on the effects is given.... The examination and approval authority shall examine and approve no draft plan without a chapter devoted to or an explanation made on environmental effects.

Obviously, the Chinese substantive law has already defined the legal obligations of economic decision makers on different levels to protect the environment. Yet, these are only guidelines. To ensure their implementation requires the safeguard of procedural law.

III. *Procedural Safeguard for Integrated Decision-Making*

At present, an Environmental Impact Assessment (EIA) is only a procedure to achieve integrated decision-making considering both economic development and environmental protection. EIAs analyze and assess, before the decision-making related to a plan or a construction project, the possible environmental effects and remedies once this decision is implemented, and the final decision on the original policy is made according to the assessment results. The *Environmental Impact Assessment Law* defines procedures for plans as well as projects.

There are three key points in the EIA of projects. First, the construction unit shall entrust a competent EIA institution to prepare a document, analyzing the possible effects that the construction project may exert on the environment and putting forward measures to prevent or mitigate adverse effects. Second, the construction unit shall submit the EIA document of the project to the environmental protection department for approval. Lastly, if a construction project fails to make an EIA document, or if its EIA document is not examined, or is examined but not approved by the department specified by law, the department for examination and approval of the said project shall not be allowed to approve construction of the project, nor shall the construction unit start construction.

There are three key points in the EIA of plans as well. First, the authority that draws up a plan shall organize an EIA document to be prepared, analyzing the possible effects produced after implementation of the plan and putting forth measures to prevent or mitigate adverse effects. Second, the EIA document shall be submitted together with the draft plan to the authority for examination and approval of plans. A special group shall be organized for the examination and they shall submit a written report. Lastly, the EIA results and the written report of the examination shall be an important basis for the decision-making of the authority for examination and approval of plans when considering whether the draft plan is to be approved or not.

If the Fujia PX project had strictly followed the procedures of the above-mentioned EIA of projects, and if the developmental plan of the Dagushan Petrochemical Industrial Complex (大孤山石化产业园区) in Dalian, where the PX plant is located, or the industrial plan, of which the PX project is a part, had strictly followed the procedures, the subsequent public uproar could have been averted.

However, this was not the case. There has never been any EIA of the whole area since the Development Zone of Dalian (大连开发区), of which Dagushan Petrochemical Industrial Complex is a part, was approved for construction in 1984. Before it obtained the approval from the environmental protection department in March 2008, the Fujia PX project had already started its construction in October 2007. Before it was examined and approved for a trial production by the Department of Environmental Protection of Liaoning Province in April 2010 and passed the environmental protection check conducted by the Ministry of Environmental Protection for completed construction projects in November of the same year, it had already started its full production in June 2009.

One cannot help asking how the Fujia PX plant could easily have broken both the substantive law and the procedural law set up to ensure integrated decision-making. It can be argued that this is all rooted in the government, the most important economic decision maker. The government's blind pursuit of economic benefits has overwhelmed the public need for environmental benefits. And, the current performance assessment system and the political system have given those officials a convenient tunnel to bypass the environmental rules.

IV. *Institutional Safeguard for Environmental Impact Assessment in Economic Decision-Making*

There have been about 30 laws on environmental protection since the *Environmental Protection Law* (*trial*) (环境保护法(试行)) was established in 1979. There are also countless regulations and rules on different administrative levels for the same purpose. Nevertheless, the mass production of environmental protection laws has not achieved as good an effect as expected. The PX incident in Dalian demonstrates again what is lacking is not laws or regulations to ensure an integrated economic decision-making, but a necessary respect for the law and a willing identification with the environmental protection laws.

Since the reform and opening up beginning in 1978, "Development Is Everything" has been regarded as a universal guideline. Consequently, "Everything Is for Development" is simply understood as "Everything Is for GDP," and GDP has become a political aim for most leaders both in the party committee and in the government. This economic priority has led to a performance assessment system of the officials only oriented to economic indexes. Environmental protection includes great capital investment, which could hardly be rewarded in tangible forms within a short time, nor could it be reflected in the growth of GDP. To maximize personal benefit, the officials would eventually do their very best to push forward the GDP growth when they are in power. Environmental protection policy, which seems to run counter to GDP growth, is thus rarely adopted in their administrative agenda.

This point can be easily seen in the Dalian PX incident. As early as 2006, Liaoning Province had drawn up an industrial developmental plan; that is, "to depend on projects to advance the development of the petrochemical industry." Dalian, which has a natural advantage in this industry, became the site to carry out this plan. A whole petrochemical industrial chain has been established on the Dagushan Peninsula of Dalian, with Western Pacific Petrochemical (西太平洋石化) in the lead and such projects as PX, PTA, and PET as extensions. The total value of its industrial output has exceeded 100 billion RMB. This line of decision-making has obviously focused only on economic factors. The subsequent environmental risks have been ignored. With the great support of the CPC municipal committee and the city government in Dalian, both the petrochemical industry complex zone and the PX project have been constructed and put to use. Their EIA was merely a formality.

In 2006, a new regulation entitled *Integrated Assessment of Local Party Committee and Government Authorities and Officials with a View of Scientific Development* (*Trial*) (体现科学发展观要求的地方党政领导班子和领导干部综合考核评价试行办法) was put into effect, which marked an important and conceptual change in the assessment of officials. The importance of sustainability is highlighted. But this regulation only provides some guidelines for the official performance assessment system. No specific and hard rules are given about the proportion of environmental protection in this new assessment system. To implement this regulation, it still requires local party committees and governments to make specific policies; otherwise, it would only be lip service. Nevertheless, the "GDP Priority" inertia will remain for quite a period of time. The local party committees and governments may still lack initiative to protect the environment.

V. *The Public Participation and Supervision in Integrated Decision-Making*

The chance landing of Typhoon Muifa offered the residents of Dalian an opportunity to showcase their public power. It also showed the enormous role played by the public in the information age to ensure a consideration of environmental factors in economic decision-making.

When their basic living needs are met, the public starts to demand more and more for their environmental benefits. As the government is still indulged in the GDP inertia, most people believe it a basic right to "breathe in fresh air." This spontaneous and conscious act of environmental protection from the public stands to safeguard the environmental benefits neglected by the officials. Yet, the public paid a high cost to achieve their aim.

In fact, in the process of city planning and economic decision-making, the government should attend to public opinion to achieve democratic and scientific decision-making. On the one hand, in a socialist country where the power belongs to the people, the residents of the city have the right to choose their urban developmental direction and to air their opinions on matters where their vital interest is involved. On the other hand, the government has the legal obligation to gather public opinions in their city planning and economic decision-making. For instance, the *Urban and Rural Planning Law* stipulates:

> Before submitting an urban or rural plan for examination and approval, the authority in charge of its formulation shall, in accordance with law, publish the draft of the plan and solicit opinions from experts and the general public by holding appraisal conferences or hearings, or by other means. Publication of the draft shall remain for at least 30 days. The authority in charge of formulation of the plan shall fully consider the opinions of the experts and the general public and, when submitting the materials for examination and approval, attach an explanation on its adoption of the relevant opinions and the reasons.

The *Environmental Impact Assessment Law* has a similar article:

> As for a special plan, which may cause adverse effects on the environment and may have a direct bearing on the rights and interests of the public in respect to the environment, the authority that draws up the plan shall, before submitting the draft of the plan for examination and approval, hold demonstration meetings or hearings, or solicit in other forms the comments and suggestions from the relevant units, specialists, and the public on the draft report on environmental effects, except where secrets need to be guarded as required by state regulations.

However, although the law protects the public right to information, the public participation in most cases is still in its elementary stage, often in the forms of "participation after the fact," or "passive participation." The reason is that the government officials in China have long considered themselves "shepherds" of the people, whose "political consciousness" is not strong enough in their eyes, so that they want or need to do everything themselves to take care of the people. The subsequent mode is to make a decision without any public transparency, and the public only has the right to know the results (sometimes even this right is deprived) with little right to oppose the results.

Had the government published the information with honesty at the stage when the Fujia PX project was still being submitted for examination and approval, had they listened to the public opinions and made decisions accordingly, the subsequent public protest could well have been avoided, and the enormous financial loss and waste of resources would not have occurred.

It is foreseeable that the public has already become a main engine to push forward the environmental protection cause. If decision makers keep ignoring the environmental factors in their economic decision-making or the increased needs of the public for environmental benefits, people will remain worried and ask themselves: "After Xiamen and Dalian, where will similar incidents occur next time?"

PART TWO

ENVIRONMENTAL POLLUTION AND ECOLOGICAL PROTECTION

THE ENVIRONMENTAL DEGRADATION OF THE YELLOW RIVER

Wang Jian

Abstract: Ecological changes in the Yellow River source region, including glacier retreat, drying-up of streams, lake/wetland shrinkage, disappearance of swamps, desertification, and pasture degradation, are primarily natural phenomena. The population of rats along the Yellow River has increased dramatically in recent years due to their growing resistance to rodenticides and a decline in their natural predators as a result of hunting and pasture damage. The rapid rate of grassland shrinkage and degradation has adversely affected water runoff formation in the Yellow River source area. This is primarily caused by fencing and other human activities. Wetland destruction in the lower and middle reaches of the Yellow River has directly led to biodiversity loss and eco-landscape disappearance. The Yellow River has long been exploited, while its basic functions as an ecosystem have largely been overlooked, leading to serious consequences for the natural environment.

Keywords: the Yellow River source area, ecosystem degradation, wetland destruction, biodiversity, ecosystem stability

There is now an unprecedented conflict between human demands and the conservation of the long-overburdened Yellow River. The river is faced with increasingly severe water problems and potential disasters. Overexploitation of wetlands has resulted in immeasurable damage; the stability of the river's ecosystem is threatened due to wetland disappearance. This in turn has adverse effects on human survival. Apart from the benefits brought about by hydraulic work along the Yellow River—such as flood prevention, power generation, and irrigation—there are a number of negative effects on the river basin's natural water circulation, where industrialization is accelerating ecosystem degradation.

I. *Options for the Yellow River Source Area*

The Yellow River begins in the Qinghai-Tibet Plateau and then flows through the North China Plain before emptying into the Bohai Sea. The 1.6-million-year-old river runs through nine Chinese provinces and autonomous regions, where more than 40 major tributaries and more than 1,000 smaller streams feed into it. With a basin area of 750,000 square kilometers and a length of 5,464 kilometers, the Yellow River has always held political, economic, and cultural importance for China. Considered the cradle of the Chinese nation, the Yellow River has played a vital role in the development of one of the world's most ancient civilizations.

The land in the upper reaches of the Yellow River is critical for maintaining the stability of the river's ecosystem, but it is now faced with a serious ecosystem crisis. Environmental changes such as glacier retreat, drying up of streams, lake/wetland shrinkage, the disappearance of swamps such as Xingxiuhai (星宿海), increasing desertification, and pasture degradation are all accelerating.[1]

The Yellow River source area covers large swathes of wetlands and snow- and ice-covered mountains. Most of the vegetation which grows at high altitudes (>4,600 meters above sea level) and low temperatures is of the alpine variety and has been relatively unaffected by human activity. Water levels remain stable due to slow evaporation and steady precipitation. The water content of the soil, measured at one meter below the surface, is on average 30%.[2] When the water content of the wetlands exceeds this level the extra water flows into streams, forming the source of the Yellow River.

These streams zigzag across the wide floodplain in the valley where they are supplemented by water from the western glaciers. The wetlands in the source area contribute to one-third of the Yellow River's water. Maduo County (玛多县), located in the source area, is known as the "County of Lakes." The numerous lakes in this area are the region's main feature.

The Eling (鄂陵湖) and Zhaling lakes (扎陵湖), spanning an area of 608 square kilometers and 542 square kilometers respectively, are the largest of the lakes in the area. There are more than 80 lakes here, including more than 40 larger than 1 square kilometer and 15 larger than 10 square kilometers.

Many smaller lakes and ponds used to be located in the Xingxiuhai swamp area. Surrounded by alpine meadows, these smaller bodies of water were called Xingxiuhai (Chinese for "a sea of stars") because the glassy surface of the lakes sparkled in the sunshine and reflected the night sky. In Tibetan, they are known as the Mayong, meaning "a shoal of

1 Wu Yuhu, "Causes for Ecosystem Degradation," based on a recording from an interview in 2010.

2 "This Is Where the Yellow River Came from: Liu Changming, Academician of the Chinese Academy of Sciences, Talks about Shouqu Wetland," *Man and the Biosphere*, February, 2010.

peacocks," because the ponds resemble the spots on a male peacock's tail feathers. Also, this section of the Yellow River is known as Maqu, which means "Peacock River."

The Yellow River source area on the Qinghai-Tibet Plateau is sparsely populated. On average there are only 1.5 people to every three square kilometers, which means the impact of human activity on the environment is very small. However, as an area that contributes a great amount of water to the Yellow River, any changes in the water generation rate can impact the whole of the Yellow River.

The freeze-thaw process of the tundra and glaciers has changed in recent years due to changes in temperature, lower precipitation, and faster evaporation. The water content of the soil horizon has significantly decreased due to glacier retreat, seasonal soil freezing, the reduction of frozen soil thickness, later freezing, an expanding thawed layer, and earlier ice-out. The drying-up periods in streams running from Zhaling Lake to Huangheyan (黄河沿) are getting longer.

The most dramatic change is the disappearance of the Xingxiuhai swamp area. The permafrost has melted due to climate change, leading to rapid downward infiltration of surface water. In other words, the geological conditions necessary to support swamps and lakes has been lost.

The aquatic environment of the Yellow River source area actually began to deteriorate rapidly in the 1970s, and the lakes in the area decreased in number. It is becoming more common on the Qinghai-Tibet Plateau for lakes to shrink in size over time and the saline content in freshwater lakes to increase. Glacier retreat means there is a decreasing supply of fresh water, which will have a severe impact on the upper reaches of the Yellow River and the runoff in the river basins in northwest China.

The Yellow River source area is becoming more arid. This aridity is characterized by the seasonal drying up of springs and streams, lake/wetland shrinkage, and sward removal due to repeated freeze-thaw cycles. As a result, water coverage has fallen from around 90% to 30% or even less. With sparse vegetation and exposed sand, microorganisms in the soil are on the decline in terms of numbers and activity; soil nutrient conversion is hindered; secondary salinization occurs; and the brownish black humus horizon becomes exposed, which is the cause of black sandstorms.

There has been an explosion in the rodent population across the grasslands. The increase in rat holes per square meter has led to the erosion of soil and vegetation. In addition to competing with the livestock for pasture, rats dig up plant roots and soil, leading to sward damage, subsidence, and soil erosion. The eroded ground becomes covered by weeds, which in turn leads to an overall deterioration of the grassland. In extreme cases, the surface soil horizon is so eroded it becomes barren. The proliferation of rat holes also increases water evaporation, accelerating grassland shrinkage and preventing runoff formation.

The increase in the rat population is widely believed to be due to the extensive use of rodenticides in the 1980s. Natural predators of the rat, such as weasels and foxes, were killed as a result of eating poisoned rats. The reduced number of other natural predators such as hawks and foxes is seen as another major cause for the rapid increase in the size of the rat population.

Structural damage and degradation of an ecosystem are complex processes. There is a delay between the cause of environmental damage and its symptoms, as there is with the damaging consequences of human activity and its impact. The natural environment is a delicate balance and an interdependency of all living things within it. It can take a certain amount of time for the effects of the damage to become apparent. An ecosystem has a natural resistance to external interference; this contributes to the delay in the impact of damaging human activity. Additionally, limited scientific knowledge often means that recognition of the effects on the ecosystem comes only after irrevocable damage has been done. It seems unavoidable that humans will pay the price of their damaging activities as they will ultimately be the victims of a degraded environment.[3] The changes happening across the Yellow River source area show the extent of the damage. The issue of water shortages, due to dwindling fresh water supplies, has become a major challenge for environmental protection.

Prevention of overgrazing in the Yellow River source area is one example of environmental conservation measures taken. In 2005, China invested 7.5 billion RMB to launch an ecosystem protection initiative across the Sanjiangyuan (Chinese for "the source of three rivers") National Nature Reserve (三江源国家自然保护区), where 106,667 hectares of wetlands were given highest priority. The primary aim was to protect the alpine wetlands and their biodiversity by allowing them to naturally rehabilitate.

Unfortunately, such conservation measures have not solved all the problems. The steel fences erected to protect the wetlands from human activity have divided the natural rolling pastureland into pieces. The livestock can graze on pasture within their owners' fenced areas but have no

[3] Duan Changqun and Yang Xueqing, *Ecology—Something both Destructive and Constructive*, Huaxia Publishing House, 2005.

access to those of neighboring herdsmen. This eliminates potential conflicts between neighbors but is disastrous for the wildlife of the alpine wetlands. There are only a few hundred Tibetan antelopes in the source area, and they are prey to wolves. The unintended consequence of the fencing is that it has made it more difficult for the antelope, as well as cattle and sheep, to escape predators.

Erecting fences across the wetlands has involved the spending of a huge amount of money on measures that were meant to protect biodiversity but in reality endanger wildlife. Professor Wu Yuhu (吴玉虎) of the Northwest Institute of Plateau Biology at the Chinese Academy of Sciences (CAS) has been quoted as saying that he wishes he could remove all the fences across the Yellow River source area immediately.

The settlement program now in progress is forcing herdsmen to settle down. These large-scale fencing projects have sharply reduced the scope for grazing their livestock. The areas around these new settlements and drinking-water sources have suffered the fastest rate of pasture degradation and desertification due to overgrazing by livestock.

Such developments have resulted in rapid changes to the alpine ecosystem, which is also under threat from industrialization. The local population has lived a nomadic lifestyle in this area for thousands of years. It is a way of living in harmony with nature through a traditional and ecologically friendly culture that satisfies the principle of sustainability. Unfortunately, this way of life is being regarded as something outdated and is being abandoned across the Yellow River source area. As with the wetlands and pasture land in this area, the local nomadic population is at risk of disappearing in the face of aggressive commercialism and industrialization.

II. *The Fate of Wetlands in the Lower and Middle Reaches of the Yellow River*

The stability of the Yellow River ecosystem relies on the landscape-diversity and biodiversity of its wetlands. Natural wetlands across the Yellow River basin comprise rivers, swamps, paleo-channels and inter-river areas. Their biodiversity is attributable to various aquatic animals/plants and rare birds. These wetlands effectively serve as gene banks.

Wetlands in the lower and middle reaches of the Yellow River are the result of flood silt deposition. These include Henan Province's Sanmenxia Reservoir Area Wetland (三门峡库区湿地), Mengjin Wetland

(孟津湿地), Zhengzhou Yellow River Wetland (郑州黄河湿地), and Liuyuankou Wetland (开封柳园口湿地), as well as the Dongping Lake Wetland (山东东平湖湿地) and the Yellow River Estuary Wetland (黄河口湿地) in Shandong Province.

The wetlands play a crucial role in water volume adjustment in the Yellow River. Most of these swamps are peat swamps. With very large pores, peat acts like a sponge and is highly water absorptive. Meadow peat generally has a water content of 300–600% and sphagnum peat has an even higher water content of up to 2,000%.

Today, the Yellow River is faced with wetland shrinkage, dry-up, and even disappearance. As China keeps tight control over land use, flood plains along the Yellow River are now receiving much needed attention in some regions to try to reverse the effects of damaging activities.

A large number of brick kilns have been built across the flood plains, along with a growing number of sand quarries along the river. Scores of these quarries have been built in an area of more than 53 hectares of wetlands between Bapanxia (八盘峡) and Wujinxia (八盘峡), two gorges on the Yellow River that are 150 km away from each other. The Zhengzhou Yellow River Wetland in Huayuankou (花园口) has been illegally reclaimed. Farmers signed contracts with the authorities to use the land to cultivate 2,250 kg of wheat per hectare. They then removed several hundred hectares of vegetation to open sand quarries or fish farms.

A lot of the wetland in the 45-kilometer-long section of the Yellow River that runs through Lanzhou, Gansu Province has been used for planting cash crops and dumping garbage, which then results in dirty water and stench. Numerous structures have been illegally built in the wetlands, causing the original vegetation to disappear and the ecosystem functions to deteriorate.

Water in lakes, ponds, and pools in the middle reaches of the Yellow River is muddy, bluish black, and stinky. These water bodies have become severely eutrophicated and swampy. Some wetlands have been drained to make way for housing developments, tourism projects, forestation projects, or even rubbish and sewage facilities. These developments are being built on land that would otherwise exist for the natural and essential purpose of draining flood water. This has led to increasing conflicts between developers and wetland protection.

Illegal dumping on the wetlands and even into the river itself is occurring in sections of the Yellow River that run through Baode County (保德县), Xinzhou (忻州), Shanxi Province, in the name of reinforcing the existing levee. This poses dangers on the other side of the river in Fugu County

(府谷县), Shaanxi Province, during the rainy season. In recent years, there have been frequent water-related disputes between the two counties. Unfortunately, portions of the wetlands along the Yellow River have been claimed and used by both counties for development and short-term gain.

Many poorly planned water diversion projects in the middle and upper reaches of the Yellow River have contributed to changes to the local ecosystems in the river basin, leading to lake and wetland shrinkage or even drying up further downriver. Most of the affected areas have become sources of sand and salt storms.

The freshwater Hongjiannao Lake (红碱淖) spans an area of 54 square kilometers and lies between the Kubuqi (库布其沙漠) and Maowusu deserts (毛乌素沙漠) of the Ordos Plateau (鄂尔多斯高原), through which the Yellow River runs. The local government of Yijinhuoluo Qi (伊金霍洛旗), in the Inner Mongolia Autonomous Region, has invested about 100 million yuan to build a dam for the Zhasake Reservoir (扎萨克水库) on the Yingpan River (营盘河), four kilometers upriver from Hongjiannao. It has also built a reservoir on the Manggaitu River (蟒盖兔河). The building of these dams has led to drying up further downstream. Hongjiannao is drying up due to the arid climate, leaving more than 10,000 Relict Gulls (listed as endangered and Class 1 state-protected birds) without a habitat.

The 153,000-hectare Yellow River Delta Nature Reserve (三角洲湿地自然保护区) is a wetland of international importance and designated for the protection of aquatic and coastal ecosystems, in accordance with the *China Biodiversity Protection Action Plan* (中国生物多样性保护行动计划). The delta is home to more than 800 kinds of aquatic creatures, including state-protected fish and whales such as lancelets, finless porpoises, and weever fish, as well as 296 kinds of birds, including 108 listed on the *China-Japan Migratory Bird Agreement* (中日候鸟保护协议). Among these are Class 1 state-protected birds including Red-crowned Cranes, Hooded Cranes, Snow Cranes, Great Bustards, White Storks, Oriental Storks, Black Storks, as well as Golden- and White-tailed Eagles. There are also 49 kinds of Class 2 state-protected birds such as the Whooper and Whistling Swans, Common Cranes, and Oriental Honey Buzzards. Numerous aquatic birds such as herons and ducks also reside at the reserve. Such a variety of species and the natural conditions for their growth are primarily attributable to the continuous flow of Yellow River water and sand into the sea.

The Yellow River Delta is an important habitat for birds due to the combination of its tidal flats and diversity of aquatic plants. It is an essential stopover for migratory birds on their flight north in the Spring and south in the Autumn. About six million birds pass through this area each year. Power companies such as Shandong Luneng (鲁能), Shenhua Guohua (国华能源), Datang International (大唐国际), Huaneng Renewables (华能新能源), and Beijing Chenyuan (北京辰源) are now building large-scale wind turbine projects. Datang International has already put its wind generators into operation. The wind farms pose an immense threat to the survival of these birds.

Many large-scale chemical and industrial projects related to the oil industry have also been started in the Yellow River Delta. The nature reserve is shrinking to make way for such projects. Around 70 chemical companies are releasing effluent and harmful gases into the reserve and threatening the survival of the local wildlife such as fish, invertebrates including shrimps and crabs, intertidal zone creatures, and shellfish in the estuary area.

Overexploitation of water resources throughout the Yellow River wetlands has damaged the ecosystems supported by natural water circulation in the river basin. It has caused a series of environmental problems such as dry-up and lake/swamp shrinkage. The pollution caused by chemical companies is likely to destroy the wetland ecosystem, animal/plant populations, and genetic diversity of the Yellow River Delta. Once this has occurred, no current technology will be able to undo the damage.

Damage to the Yellow River wetlands has already resulted in biodiversity loss and eco-landscape disappearance. From the perspective of biodiversity, wetland disappearance results in the disappearance of animal/plant populations. From the perspective of the eco-landscape, damage to the wetlands leads to the disappearance of vital functions such as flood mitigation, runoff reduction, flood water storage for use during droughts, pollutant degradation, and climate adjustment. The stability of the Yellow River basin ecosystem is being dangerously undermined by the disappearing wetlands. Poorly planned resource exploitation has led to a sharp decrease in these areas. The losses caused by wetland degradation are immeasurable.

III. *Aquatic Ecosystem Crisis Caused by Hydraulic Works*

Making the Yellow River run clear has been the Chinese people's dream for centuries. China built the Sanmenxia Dam with the aim of achieving this. Unfortunately, the Yellow River did not run clear. Instead, the dam silted up and caused water to overflow from the Jing and Wei rivers (泾渭河), flooding nearby areas.

Today, we are again relying on dams to address the problem of flooding in the Yellow River basin. A number of huge dams and reservoirs have been or are being built in the middle and upper reaches of the Yellow River, including those at Longyangxia (龙羊峡), Lijiaxia (李家峡), Liujiaxia (刘家峡), Yanguoxia (盐锅峡), Bapanxia (八盘峡), Daxia (大峡), Qingtongxia (青铜峡), Gongboxia (公伯峡), Xiaoxia (小峡), Chaishixia (柴石峡), Laxiwa (拉希瓦), Suzhi (苏只), Wujinxia (乌金峡), Zhiganglaka (直岗拉卡), Tianqiao (天桥), Wanjiazhai (万家寨), Longkou (龙口), and Qikou (碛口). And eleven other dams will be built at Tage'er (塔格尔), Guancang (官仓), Saina (塞纳), Mentang (门堂), Tajike (塔吉柯), Jiabusi (加布斯), Ma'erdang (玛尔挡), Cihaxia (茨哈峡), Banduo (班多), and Yangqu (羊曲), all of which are up river from Longyangxia.

The cascade development in the main channel of the Yellow River source area should be planned with extreme prudence. Hydropower development is not the primary objective of the plan for the source area. No more hydropower plants should be built at Maqu. Building the world's largest hydropower plant would serve no meaningful purpose and would be ecologically unsound. The disadvantages of hydropower plants in the middle and upper reaches of the Yellow River outweigh the advantages because the evaporation rate is greater than the rate of precipitation in this area.

The building of more than 3,000 reservoirs on the Yellow River means that the lower and upper reaches no longer constitute a single system.[4] Most of the river water is stored behind the dams in the middle and upper reaches of the river. The flow in the lower reaches has sharply decreased due to around 50,000 smaller water diversion projects in the middle and upper reaches, not to mention the large-scale water diversion, supply, and irrigation projects. These include those for diverting Yellow River water into the cities of Taiyuan, Datong, and Shuozhou in Shanxi Province, for diverting Datong River water into Qinwangchuan (秦王川), and for the Yanhuanding Water Lifting Project (盐环定扬水工程).[5] The lower reaches will consequently suffer from huge fluctuations in flow between

---

4 Lin Xiuzhi and Su Yunqi, *Yellow River Water Resources Exploitation and Its Effects on River Dry-Up*, China Environmental Science Press, 1997.

5 Chen Huijun, "Disasters from Yellow River Dry-Up," the SEPA Seminar on the Environmental Impact of Yellow River Dry-Up and Solutions, 1997.

rainy and dry seasons. The Yellow River has already experienced drying up in its lower reaches, and the section in Shandong Province is facing a long-term dry up. No water will flow in the lower reaches except for muddy flood water in years when there is above average rainfall. This will pose a severe threat to the lower reaches and the delta in terms of economic development and ecosystem management.

Provinces and autonomous regions along the Yellow River have been expanding farmland areas without a unified plan, increasing the total area of river water irrigated farmland to 7.33 million hectares from 800,000 hectares. Various large-scale, water-consuming projects in the energy, power, chemical, smelting, and coking industries have been launched one after another, while the amount of water consumed for industrial, agricultural, and domestic purposes has increased to 30 billion cubic meters from 12.2 billion cubic meters. Those provinces and regions race to divert the river water at full throttle in dry years, leading to severe water wastage and a sharp increase in the amount of diverted river water. Most regions in the river basin are failing to meet their demand for water despite the fact that local water resources are already overexploited. As a result, oases are forming in the upper reaches while the lower reaches are left without water.

The lower reaches of the Yellow River are suffering severe water shortages due to an excessive number of dams and poorly planned water diversion projects in the middle and upper reaches. The land area around the estuary increases on average by between 2,000 and 2,667 hectares per year. But now, seawater has encroached by 10,000 meters, causing a land area reduction of one million hectares. The SINOPEC Shengli Oilfield has been forced to close 200 wells, leading to an annual loss of 1.5 billion RMB.[6]

By building dams, we gain certain benefits in terms of flood protection, irrigation, power generation, and shipping, but we are also creating more fundamental problems such as farmland loss, resettlement, land salinization, water quality deterioration, siltation, resulting geological disasters, and biodiversity loss.

The total reservoir capacity throughout the Yellow River basin has reached 66 billion cubic meters, which already exceeds the annual average runoff of the river, or 58 billion cubic meters. The hydraulic works are having an adverse effect on the natural water circulation in the river

[6] Fan Xiao, *Ecological and Social Crises amid Large-Scale Hydropower Development*, China Environmental Science Press, 2005.

basin. There is an arid climate in the middle and upper reaches of the Yellow River, with an average evaporation of 984.7 millimeters.[7] Several billion cubic meters of water would evaporate each year, even if the annual average evaporation loss was only 1–2% from the more than 3,000 reservoirs. The huge loss of water through evaporation is equivalent to losing a number of large reservoirs every year. Moreover, the continuity of the water flow, the riverbed, and the local aquatic ecosystems have all suffered considerable damage.

Water circulation in the flood plains is closely connected with the rivers. The fertile land in the Yellow River flood plains was originally created by floods. When it comes to the living environment, human beings would be well advised to stay away from the plains. As a result of the misguided belief that 'man can conquer nature' there is now a population of around 1.8 million living in the flood plains. Major concerns extend to levee building, the resulting costs that continue to increase, and the risk of levee collapses due to a Yellow River overhead (where the river bed in the lower reaches is elevated higher than the land alongside the river). Jia Rang (贾让), a Chinese scholar during the Han Dynasty, proposed that the best measure for river harnessing was to resettle the affected population so as to set aside sufficient space for flood drainage. He even proposed a compensation mechanism, utilizing the fund for levee building as resettlement compensation.[8] In contrast, we are now building dams on the Yellow River, causing large swathes of land to lose flood water supply and water circulation and isolating flood plains from the rivers. This has resulted in severe groundwater level reduction and rapid lake/wetland/vegetation shrinkage. These changes in water circulation have brought about severe damage to the integrity, biodiversity, and ecosystem stability in the lower and middle reaches of the Yellow River.

Soil erosion in the Loess Plateau, through which the Yellow River runs, has increased due to the effects of economic activity. The siltation rates at numerous reservoirs on the Yellow River are higher than expected, leading to shorter dam life cycles. Reservoirs built several decades ago will be scrapped in the near future, and both the dams and the reservoirs will then be left as huge ruins. During reservoir building justifications, we often hear how solid and safe a dam will be. Nobody has ever

---

7 Liu Shaoyu, "Preliminary Research on Yellow River Water Resources Exploitation and its Effects on River Dry-Up," China Environmental Science Press, 1997.

8 Ma Jun, *Water Crisis in China*, China Environmental Science Press, 2005.

explained what happens with the massive concrete ruins of a dam after it is scrapped and how the Yellow River will be affected by these defunct concrete structures.[9]

The Yellow River's primary problem is the silt, which has long been embanked to prevent floods. We may be able to make assurances that the river will not overflow its banks in the next decade or two, but what do we do when the riverbed rises by five meters in five decades time and ten meters in ten decades?[10] The Yellow River will then run high above the Huanghuaihai Plain (黄淮海平原). Current efforts to harness water power in the Yellow River do not address these potential problems. As the riverbed rises, the hazards will become ever greater.[11]

The Yellow River is a river with no future. So what will the future look like? What will the Yellow River look like? Nobody knows.

---

[9] Jin Hui, *Scientific Development and River Harnessing*, Huaxia Publishing House, 2005.

[10] Liu Shukun, "Hydraulic Work Modernization and the Emergence of New Hydraulic Works in China," *Water Resources Protection*, 2003.

[11] Xiao Liangzhong, *The Relationship and Coordination between Dam Building on Large Rivers and Potential Resettlement & Cultural Heritage*, Huaxia Publishing House, 2005.

AN URGENT CALL FOR SOLUTIONS TO CHINA'S HEAVY METAL HEALTH RISK

Yang Chuanmin

Abstract: The impact of heavy metal pollution in 2010 was much more serious and extensive than the 2009 social unrest in Fengxiang County, Shaanxi Province and Wugang County, Hunan Province, which was caused by excessive blood levels found in children. As in today's China, Japan enjoyed years of economic prosperity between the 1960s and 1970s during which the country suffered four episodes of Big Pollution Diseases. Three of the incidents were related to heavy metal contamination, which prompted Japan's legislature to pass relevant laws. Today, China's pollution problems are no less serious than what Japan faced forty some years ago. However, China lags behind in its response to pollution in terms of both legislation and damage control. This article aims at exploring solutions to China's heavy metal pollution problems by comparing the health risk exposure situations in typical mining areas with such problems in China and Japan.

Keywords: heavy metal pollution, cadmium pollution, the Itai-itai disease, Environmental Pollution Prevention Act of Japan, soil pollution prevention and control

I. *Extensive Heavy Metal Pollution in 2010*

The contamination accident at the Zijin Mining Group's (紫金矿业) site in Shanghang County (上杭县), Fujian Province and the excessive levels of heavy metal found in tobacco, as reported in the media, were typical episodes of point source and area source pollution. The incident at Zijin Mining exposed the environmental risks at large-scale mining companies. As a public company, Zijin Mining Group failed to promptly disclose information about the contamination caused by a sewage tank leak at its Zijinshan Copper Mine. An investigation of nearby villages showed that mining polluted the Ting River (汀江) and groundwater. In Bitian Village (碧田村), for example, statistics from the village's secretary indicated that the village's cancer mortality rate was around 2.7%, more than twice that of the cancer mortality rate of 1.29% for rural China.[1]

[1] Chen Zhu, "Investigation Report on the Third National Survey on Causes of Death," Peking Union Medical College Press, 2008.

In adjacent Wuping County (武平县), there are at least three mining companies, causing high incidences of malignant tumors. Statistics from the villagers showed that in the past five years the region's average cancer mortality rate was three to four times that of China's rural average.[2] Investigations showed that the villagers' groundwater was severely contaminated. Rust appeared in wells overnight. Family members living together sometimes contracted tumors of the same type. However, no study has been conducted to research the relationship between high incidences of cancer and mining in the area.

Another incidence of heavy metal pollution that caused public concern was typical of area source pollution. In October 2011, studies released by the International Tobacco Control (ITC) Project showed that 13 brands of Chinese cigarettes tested in 2005 and 2006 contained three times more the level of arsenic, cadmium, lead, and chromium than Canadian cigarettes. The researchers estimated that the heavy metal might have come from the soil.[3]

A number of available regional research reports suggest the seriousness of heavy metal pollution in China. For instance, a typical regional heavy metal investigation in 2005, with the cities or provinces of Guangdong, Jiangsu, Zhejiang, Hefei, and Dalian participating, showed that nearly 40% of the farmland on the Pearl River Delta contained excessive levels of heavy metal. Ten percent of the land showed severely excessive amounts.[4]

China is rich in mineral resources. The country has witnessed wider and deeper heavy metal pollution in recent years. The mining process has become a major source of heavy metal pollution in rivers, affecting the lives of nearby villagers. As victims of the polluted water, villagers are at a disadvantage compared to the mining companies. Rural communities are usually the major production bases of grain. Therefore, if the problem of heavy metal pollution in rural China remains neglected, it will affect the life of every Chinese should various heavy metals enter the food chain through grain.

---

[2] Yang Chuanmin, "The Zijin Nightmare," Southern Metropolis Daily, September 1, 2010.

[3] Zhang Weihua, "Chinese Cigarettes Accused of Containing Excessive Tar and Deceiving Smokers," *Jinghua Daily*, November 1, 2011.

[4] Yang Chuanmin, "40% of the Vegetable Fields in the Pearl River Delta Show Excessive Amounts of Heavy Metal," *Southern Metropolis Daily*, March 24, 2005.

II. *A Comparative Study of the Relationship between Heavy Metal Risks and Health in Two Mining Areas*

Differences in rural and urban economic patterns and types of heavy metal pollution in soil require different solutions. Prevention and control of heavy metal pollution should start from control of point source and area source pollution. Prevention and control of mine pollution usually affects rural communities.

This article will conduct a comparative analysis of the prevention and control measures taken at Japan's Kamioka Mine where the Itai-itai disease broke out and China's Dabaoshan Mine (大宝山矿区). Heavy metal pollution took place at both Dabaoshan Mine in China's Guangdong Province and Kamioka Mine in Japan's Toyama Prefecture. The former specializes in such metal sulfides as iron and copper, the latter specializes in lead. The decision to choose Dabaoshan Mine was made not only because it was exposed to heavy metal but also because it has been studied by many Chinese scholars. The author visited Dabaoshan Mine in 2005 and 2009. In 2010, the author visited the Kamioka Mine in Japan. The two mines presented different epidemics. Kamioka Mine had a high incidence of Itai-itai disease while Shangba Village (上坝村), where Dabaoshan Mine is located, had a high incidence of esophageal cancer and liver cancer. The correlation between these illnesses and environmental pollution has been proven to different degrees.

A. *Correlation between Mine Pollution and Health*

Pollutants in both mines included cadmium, which accumulates in human kidneys and causes loss of calcium in bones. In addition, Dabaoshan Mine was also found to be polluted by lead and arsenic. The two mines showed the following similarities:

1. *Both Mines Are Situated Upstream Near Important Rivers Populated by People Engaged in Agricultural Production.* Dabaoshan Mine is located near the Hengshi River (横石河), a secondary tributary of the North River. There are four adjacent villages along the river: Shangba Village, Yanghe Village (阳河村), Tangxin Village (塘心村), and Liangqiao Village (凉桥村). Kamioka Mine is situated upstream on the Jinzu River whose lower reach flows through several agricultural zones, including Nakacho. The agricultural zones' water source in both countries—including rivers, wells and groundwater—was polluted.

2. *The Villages' Agriculture and Other Economic Production Were Affected.* An investigation by Professor Lin Chuxia (林初夏) from South China Agricultural University showed that cadmium pollution at the Dabaoshan Mine area was most prominent. Among the 15 samples tested by the professor, the mass fraction of cadmium in young stems of sugar canes, bananas, lettuces, bitter gourds, eggplants, chilies, water spinaches, sweet potato leaves, and rice were 149, 187, 7.7 to 29, 6.6 to 10.5, 15 to 24, 7, 15 to 59, 33, and 2 to 5.7 times the standard value.[5, 6] Moreover, a study on the diversity of large invertebrates in the Hengshi River showed that not a single benthic species was found in the 25 kilometer range between the tailing dam and the polluted lower stream. In contrast, at least 36 species of benthic fauna were found in the unaffected upstream.[7]

3. *The Initial Compensation Claims Made by Local Residents Were Related to Their Loss in Agriculture and Fishing.* The color of the section of the Hengshi River that runs through Dabaoshan Mine turned brick red. Villagers said that their feet itched and became cankerous after washing them in the river. There were no fish in the river. Local agriculture and fishing were affected. Fish in the Jinzu River was also affected at the Kamioka Mine. Fishing constitutes an important component of local economic income. Therefore, the initial compensation was centered on losses in the fishing industry.

4. *Health Risks Did Not Present Themselves as Immediately as Economic Losses.* These mines were exploited for decades or even a century, during which the morbidity of villagers downstream increased. According to statistics from the village committee of Shangba Village of Guangdong Province's Qujiang County (曲江县), in this village of 3,461 people, more than 250 died from cancer over the past twenty years. Each year as many as ten people are added to the village's cancer population. The village's cancer mortality rate was more than twice the figure released by the Ministry of Health in 2008. Peasants living along the Jinzu River were diagnosed with Itai-itai disease after decades of exploitation at the Kamioka Mine. Japan's

---

[5] Lin Chuxia, Lu Wenzhou, Wu Yonggui, et al., "The Environmental Impact of Dabaoshan's Water Disposal II: Agricultural Ecosystem" [J]. *Ecology and Environment*, 2005, 14 (2).

[6] Yang Chuanmin, "Shangba Village's Redemption and Hope," *Southern Metropolis Daily*, November 18, 2005.

[7] Wu Gonggui, Lin Chuxia, et al., "The Environmental Impact of Dabaoshan's Water Disposal I: Downstream Aquatic Ecosystem, Ecology and Environment," 2005, 14 (2).

Ministry of Health and Welfare deduced that the first Itai-itai disease patient appeared in 1911 in the Jinzu River area. Since the 1970s when the Ministry of Health and Welfare started its patient identification process, a total of 195 patients were diagnosed with Itai-itai disease and another 404 suspected cases were subjected to medical observation.

5. *Heavy Metal Pollution is Related to Large Scale Exploitation of Natural Resources.* Kamioka Mine is one of Japan's most important mining areas for lead and zinc. The Mitsui Business Group which owns Mitsui Mining is one of Japan's big business consortiums. Wengyuan County of Shaoguan (韶关翁源县), where Dabaoshan Mine is located, is an important heavy metal industrial base in South China's Guangdong Province. Dabaoshan Mining Company—a provincial, stated-owned, limited liability company—operates the Dabaoshan Mine. Although it is located in Qujiang County, it is owned and managed by the Guangsheng Asset Management Company (广晟资产经营有限公司) of Guangdong Province, rather than Wengyuan County where Shangba Village is located. In contrast to the influential and powerful mining companies, villagers in both counties were disadvantaged. The influence of people who were victimized by the pollution was minimal.

6. *Relevant Scientific Research Took Place in Both Mines Where Pollution Occurred.* However, it is highly difficult to affirm the relationship between pollution and the victims. Although Shangba Village witnessed a high incidence of malignant tumors, it was still very difficult to define the relationship between diseases and environmental pollution. The Itai-itai disease found among women in the countryside of Nakacho was considered an endemic in the 1950s. In the early 1950s, rheumatoid arthritis experts who came all the way from Tokyo diagnosed Itai-itai symptoms as vitamin D deficiency.

Summary: The similarities in heavy metal pollution at both mines suggested that heavy metal pollution is often associated with large-scale exploitation. The mines are usually situated upstream or at the water origins. Agricultural and fishing communities often emerge in the transitional areas on the plains or in the mountains. If the mines are close to the village, pollution brought by mining will probably enter smaller water systems through mountain springs, affecting rivers and soil in the region and ultimately affecting nearby people's health via the food chain. The author later had similar findings in Bitian Village near the Zijin Mine in Shanghang County of Fujian Province.

B. *Differences in Solutions to Heavy Metal Exposure Risks and Their Impact*

Although the two places shared many similarities, solutions varied hugely and the problems were solved to different extents. There has been no remarkable increase in the number of Itai-itai disease victims in the past forty years in Japan's Jinzu River area. A total of twelve people were diagnosed with the disease between 1999 and 2009. Of those patients, currently only five are alive. On the other hand, the tumor situation at Dabaoshan Mine continues to worsen and has not been controlled effectively. In terms of heavy metal pollution exposure and its solutions, the two places differ in the following aspects:

1. *Scientific Evidence Links Heavy Metal Pollution to Health Damage.* At the Jinzu River area, a local doctor named Noboru Ogino invited Professor Kobayashi Jun to test the water in the Jinzu River with chemical methods. After testing victims' remains and plants in the area, Yoshioka Kinichi, a doctor of agriculture, identified cadmium as the pollutant. All the research results were published in academic papers establishing a fairly tight chain of evidence for the correlation between cadmium pollution and health damage. The findings were a prerequisite for solutions to the Itai-itai disease.

To the contrary, heavy metal pollution at Dabaoshan was more complex. Test results showed that in addition to cadmium pollution, traces of lead, zinc, and arsenic were also found in the water. The acidity of the water increased the concentration of the heavy metals.[8] Compared with the cadmium pollution in the Jinzu River, no solid evidence links were established between mine pollution and health despite the fact that Dabaoshan led other regions in China in such research.

2. *Evaluation Programs to Determine Environmental Pollution and Health Damage.* The identification of Itai-itai victims confirmed the correlation between pollution and health. Four criteria were set up in the process: (1) The patient had been living for a long time in the cadmium-contaminated area; (2) The victim was not born with the disease, but rather acquired the illness in adulthood; (3) There had been damages to the kidney; (4) The patient was suffering from osteomalacia, which can serve as an important diagnostic basis because it is a very rare disease.

---

[8] Yang Chuanmin, "The Zijin Nightmare," *Southern Metropolis Daily*, September 1, 2010.

Although there has been much media coverage on the high incidence of malignant tumors in the Shangba Village, no evaluation program has been created to confirm that the cancer victims were also victims of environmental pollution. A woman from Shangba Village lost three husbands to cancer. Two of her husbands were from the village while another one was from a different village.

3. *Participation of Lawyers Helped the Patients.* A total of 345 lawyers or 3% of Japan's total practicing lawyers helped Itai-itai patients to seek retribution. Thanks to the media and public awareness of the lawsuits, Mitsui Mining had to agree to cover all medical expenses for the victims. In addition, it agreed to accept onsite inspection by resident investigation groups. Kunihiro Takagi head of the Itai-itai Disease Residents Association confirmed that the investigation group has been carrying out their duties up to now.[9]

4. *At Dabaoshan Mine, Compensation for Health Damages Was Very Limited.* One public interest lawyer tried to provide pro bono service to the villagers at Dabaoshan. The attempts were not successful. Statistics showed that the villages in the area paid less than 100 RMB to each victim as their medical compensation. One of the large remediation measures that Shangba Village took was that the government built a reservoir at a higher place to save water and allow the villagers to have uncontaminated water.

5. *The Formation of a Systematic Solution.* Compensation and medical treatment plans could only solve short-term problems. The area must be remediated in order to relieve its environmental health risks for the long-term. Back in the 1970s, the Jinzu River area started to cover its land with soil from other places to improve soil quality. The recovery plan in Shangba village was only confined to the areas that were allocated for experiments by the Guangdong Institute of Soil Science.

Summary: With the concerted efforts of lawyers, citizen groups, medical workers, and scientists, Japan found a solution to protect victims of cadmium pollution. Meanwhile, onsite inspection by citizen groups helped to regulate the mining operations. In contrast, Dabaoshan Mine in Guangdong Province, which also witnessed a high incidence of cancer, has not come up with a solution that is beneficial to the local residents, nor has it created a plan to solve its soil pollution problems.

---

[9] Yang Chuanmin, "The 'Pain' of the Century: Shadow of Itai-itai Disease and the Price to Repair Cadmium Pollution," *Southern Metropolis Daily*, October 20, 2010.

III. *The Future of China's Control of Heavy Metal Health Risks*

It is easy to pollute, yet hard to remediate. This is especially true for heavy metal pollution due to its high stability. Therefore, prevention and control are important. The economic boom experienced by Japan in the 1960s and 1970s were similar to what China has been through for the past decades. Serious environmental pollution took place all over Japan and gave rise to the Four Big Pollution Diseases, namely Itai-itai disease, Minamata disease, Niigata Minamata disease, and Yokkaichi Asthma. All except for Yokkaichi Asthma had something to do with heavy metal pollution.

Propelled by these public nuisance incidents, Japan passed the *Environmental Pollution Prevention Act* in 1968, which established atmospheric pollution, water pollution, noise, tremors, earthquakes, and stench as public nuisances. In 1970, Japan's Anti-pollution Committee added soil pollution to the list. In the same year, the *Agricultural Land Pollution Control Act* was enacted in Japan, regulating soil pollution control, agricultural land, and agricultural products.

It has been indicated that one of the key methods to solving heavy metal pollution in Japan was to prompt relevant legislation after incidents of pollution took place. In the past few years, there has been social unrest in China initiated by heavy metal pollution such as lead found in children's blood and blood cadmium. Meanwhile, a revision to China's *Soil Pollution Control Act* (土壤污染防治法) is in progress.[10] Given the seriousness of known pollution, the government decided to focus on controlling the issue while preventing further pollution at the same time.

2010 saw many big events which were related to heavy metal pollution control. In 2011, environmental protection agencies from around China completed the country's soil and heavy metal pollution investigation, which lasted for five years. Information released by scholars who participated in the project showed a clear understanding of the geographical distribution of China's heavy metal pollution.[11] However, the project report was not published immediately. In 2011, led by the Ministry of Environmental Protection, *A Comprehensive Plan for the Prevention and Control of Heavy Metal Pollution (2010–2015)* (重金属污染综合防治规划(2010–2015年)) was close to completion. The Minister of Environmental

[10] Yang Chuanmin, "The 'Pain' of the Century: Shadow of Itai-itai Disease and the Price to Repair Cadmium Pollution," *Southern Metropolis Daily*, October 20, 2010.

[11] Yang Chuanmin, "A Clear Picture of China's Soil Pollution Emerges," *Southern Metropolis Daily*, October 20, 2010.

Protection, Zhou Shengxian (周生贤), announced that the plan would be submitted for State Council approval before implementation.

China needs to learn from other countries' and regions' practices in dealing with soil contamination. Apart from Japan, Germany and France have ways to handle soil pollution prevention and control. Taiwan enacted its *Soil and Groundwater Pollution Remediation Act* (地下水和土壤污染整治法) which gives clear guidance for the main body of remediation, technology adaptation, and result evaluation.

It is worth noticing that regardless of prevention or control, information on soil pollution should be made public. For instance, a residential project in Wuhan, in China's Hubei Province was reported to be built on an old chemical factory. A belated environmental assessment report concluded that the project land showed antimony pollution and partial organic pollution, which presented health risks.[12] Had the information been public, the project would not have been executed.

The draft of China's Twelfth Five-Year Plan further stresses the goal to improve people's living condition and deemphasizes the pursuit of high GDP growth. Health is at the center of concern. Heavy metal pollution is so severe that it has already threatened the health of the Chinese people. There is an urgent call for systematic solutions to heavy metal pollution which best suit China's conditions.

---

[12] Yang Wanguo, "Developer in Wuhan Built Economically Affordable Housing on Heavily Polluted Land," *The Beijing News*, November 30, 2010.

XIAONANHAI: ELEGY FOR A NATURE RESERVE IN THE NEW ROUND OF RIVER DEVELOPMENT

Friends of Nature

Abstract: Due to hydropower development on the Yangtze River, the Upper Yangtze Rare and Endemic Fish National Nature Reserve was forced to change boundaries and was thus faced with losing its ecological functions. Represented by non-governmental organizations, the public is actively participating in preventing over-exploitation of hydropower and protecting the rare and distinctive fish species in the Yangtze River.

Keywords: Xiaonanhai Hydropower Station, management of reserves, Upper Yangtze Rare and Endemic Fish National Nature Reserve, hydropower development on the Jinsha River, public participation, sectoral interest

The Yangtze River has the richest aquatic organisms in China. However, its freshwater ecosystem and fish resources are threatened by human activities such as flood storage, land reclamation, water diversion, pollution discharge, over-fishing, and navigation. Within the water system of the Yangtze River, the rare fish species in the upper reaches are particularly abundant, with a markedly higher proportion of China-endemic fish than the middle and the lower reaches.

Currently, hydropower development has moved from the Three Gorges (三峡) up to the Jinsha River (金沙江) section and the upper tributaries, with towering dam heights and a high density of cascades rarely seen elsewhere in the world. During the public review process of the environmental impact assessment for hydropower projects, many experts warned: the superimposed effect of over-exploitation of hydropower would destroy the riparian habitat of aquatic organisms in the upper Yangtze River and cause an irreversible adverse impact on the rare and distinctive fish species.

In 2009 and 2010, due to the possible construction of the Xiaonanhai Hydropower Station (小南海水电工程), the area and function zone of the Upper Yangtze Rare and Endemic Fish National Nature Reserve (长江上游珍稀特有鱼类国家级自然保护区) was to be altered. The local government spared no effort in promoting the approval and construction of the hydropower station, while environmental protection organizations, experts, and the public insisted on protecting the reserve from being damaged. "Xiaonanhai" became a popular word in the media. The back and forth between environmental protection and development reflects the difficult conditions faced by nature reserves, which are responsible for protecting China's ecological safety, given the radical development of hydropower.

I. *Development of Cascade Hydropower Stations on the Upper Yangtze River*

According to the cascade development plan recommended in the *Integrated Planning Report of the Main Course of the Jinsha River* (金沙江干流综合规划报告), which was compiled by the Yangtze River Water Conservancy Commission (长江水利委员会) in October 2003, hydropower development on the Jinsha River is to be composed of nineteen cascades, among which are six projects—Jin'anqiao (金安桥), Guanyinyan (观音岩), Wudongde (乌东德), Baihetan (白鹤滩), Xiluodu (溪洛渡), and Xiangjiaba (向家坝)—that have been launched recently on the middle and lower reaches.

According to the *Briefing on Integrated Utilization Planning of the Yangtze River Basin* (长江流域综合利用规划简要报告), the number of cascade hydropower stations on the middle and lower reaches of the Jinsha River has been increased to twelve (see Figure 1), and the total installed capacity is 58.58 GW. According to the plan, there will be four cascades on the lower reaches from the mouth of the Yalong River (雅砻江) to the city of Yibin (宜宾). These are Xiangjiaba, Xiluodu, Baihetan, and Wudongde. Meanwhile the middle reaches from Benzilan (奔子栏) to the mouth of the Yalong River will have one reservoir and eight cascades, namely, the Upper Hutiao Gorge Reservoir (上虎跳峡水库), Liangjiaren (两家人), Liyuan (梨园), Ahai (阿海), Longkaikou (龙开口), Jin'anqiao (金安桥), Ludila (鲁地拉), and Guanyinyan.

Following the development of the Jinsha River, a large-scale cascade development of major tributaries of the upper Yangtze River has also been launched. Among them, there are already ten cascade hydropower stations planned on the upper reaches of the Min River (岷江) upstream from the Dujiang Weir (都江堰), with another seven planned for the middle and lower reaches. There were originally twenty-two cascades on Dadu River (大渡河), but later Shawan (沙湾) and Angu (安谷) were added, bringing it to twenty-four. The Wu River (乌江) has twelve, and the Jialing River (嘉陵江) has seventeen cascades.

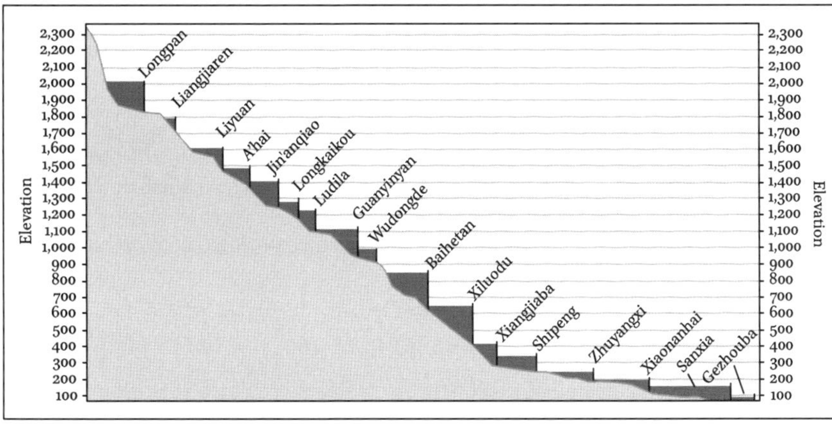

Figure 9.1: Cross-section of Hydropower Stations (including those being constructed or planned) from Longpan Hydropower Station on the Upper Yangtze River to the Gezhouba Hydropower Station.[1]

While "channelized" development has been planned on the main course of these tributaries, their own tributaries at all levels are also experiencing a dense cascade development. Just look at the major tributaries for example: there are fifty-six cascades in the Jinsha River Basin, including rivers such as the Ganqu (岗曲河), Pudu (普渡河), Niulan (牛栏江), Heng (横江), and Baishui (白水江); Furong (芙蓉江) has ten cascades while Maotiao (猫跳河) has six, and both rivers are within the Wu River Basin; the main course of the Fu (涪江), a tributary of the Jialing River, has thirty-one cascades, while Huoxi (火溪河) and Huya (虎牙河), both of which are tributaries of upper Fu, have four and three cascades, respectively; Ba (巴河), a tributary of the upper Qu (渠江), has five; in the Min River Basin, there are nine cascades on Mabian (马边河), eighteen on Qingyi (青衣江), one reservoir and eight cascades on Zagu'nao (杂谷脑河), and two reservoirs and five cascades on Heishui (黑水河); as for the Dadu River Basin, Wasigou (瓦斯沟) has one reservoir and seven cascades, Suomo (梭磨河) eight, Xiaojinchuan (小金川) seventeen, Tianwan (田湾河) two reservoirs and four cascades, Nanya (南桠河) seven, and Guanliao (官料河) seven; and in the Yalong River Basin, Jiulong (九龙河) has six cascades, and Muli (木里河) has one reservoir and six cascades.

[1] Source of image: The Nature Conservancy (China), 2010.

Cascade development is good for maximizing the benefit of power generation. However, with such dense development, the formation and operation of cascade reservoirs will produce a superimposed effect far larger than the impact of building a single hydropower station. If all these cascades were constructed, there would be a huge, irreparable impact on aquatic organisms, particularly on the rare and distinctive fish species in the upper Yangtze River.

II. *Hydropower Development in Recent Years Has Been Constricting the Living Space of Distinctive Fish Species in the Yangtze River*

There are 338 types of pure freshwater fish species in the Yangtze River Basin, accounting for one-third of the nation's total fish species. Among them 162 types are found only in this region, while 265 are seen nowhere else on the world, accounting for 60.36% of the total China-endemic species. The construction of hydropower stations has already imposed a very negative impact on the living conditions of these precious fish species, with their species groups and populations dropping significantly.

A. *Gezhouba Project: The Direct Cause of the Extinction of Chinese Sturgeon*

A large number of distinctive species are living in the upper reaches of Yangtze River. They are seen only in China, and have long been adapted to the ecological conditions there. Back in the 1970s the amount of rare and distinctive fish species in the upper reaches of the Yangtze River was still quite large, and many of them were major commercial river fishes of Southwest China's Sichuan Province, taking up a big share of the total catch. Since the 1980s, the number of species groups has shrunk drastically, the value of fisheries has been lost, and genetic resources have decreased rapidly. In 1989, the completion of the Gezhouba Hydropower Station (葛洲坝水电站) blocked the migration route of the Chinese sturgeon (Acipenser sinensis), leading to the species' extinction in the upper reaches of the Yangtze River.

B. *Three Gorges Project: A Nature Reserve for Rare Fish Species Established in a Compromise*

In 1994, the construction of the Three Gorges Dam started. Experts said about 40 types of fish species, 40% of the distinctive fish species in the upper reaches, would feel the adverse impact of the dam. Moreover, since the area of habitat would shrink by one-fourth, there would inevitably be a corresponding decrease in the number of species groups.

To protect the rare and distinctive fish resources in the upper Yangtze River and reduce the negative impact of the Three Gorges Dam upon these fish species, a nature reserve was to be established—as part of the environmental protection plan of the Three Gorges Project—on the upper reaches of the river from Luzhou (泸州) to Yibin's new town. In 1996, with approval from the People's Government of Luzhou and the administrative office of the Yibin District, a nature reserve was established both in Luzhou and Yibin. In 1997, with approval from the People's Government of Sichuan Province, these two reserves were merged into the Provincial Nature Reserve for Rare and Distinctive Fish Species on the Hejiang-Leibo Section of the Yangtze River (雷波段珍稀鱼类省级自然保护区). In April 2000, with approval from the State Council, this protected area was upgraded to a national reserve.

C. *Cascade Hydropower Development on the Jinsha River: Decapitating the Nature Reserve*

However, in the very year when the national reserve was established, hydropower development started to expand, as scheduled, to the upper section of the Jinsha River, with two planned dams, Xiangjiaba and Xiluodu, being inside the reserve. Article 32 of the *Regulations on Nature Reserves* (自然保护区条例) explicitly forbids the construction of any production facilities in the core and buffer zones of a nature reserve. Yet the 161 meter Xiangjiaba Dam is right inside the core zone of the national nature reserve, while the 278 meter Xiluodu Dam is less than 2 kilometers away from the boundary of the core zone. These two dams are so close they turn the core zone of the fish protective reserve into a reservoir area, destroying riparian habitat, flooding 90% of the spawning field of Chinese paddlefish (Psephurus gladius) and 50% of Dabry's sturgeon (Acipenser dabryanus), and threatening the life of the Largemouth bronze gudgeon (Coreius guichenoti).

To give way to hydropower development on the Jinsha River, the then-State Environmental Protection Administration (SEPA), in response to a report by the Hydropower and Water Resources Planning and Design General Institute (国家水电规划设计总院), issued its reply on issues of the Jinsha River Phase I Project and the national nature reserve (Environmental Protection Office Letter No. 2000249) requesting the institute organize research on the impact of the project upon the nature reserve and on the alternative solutions for the project and the reserve. In February 2003, the People's Government of Sichuan Province submitted to the State Council its request for approval on adjusting the area and function zone and changing the name of the reserve (Sichuan Government No. 20033). In July 2004, the relevant governmental agency submitted to SEPA a solution for making the adjustment. In April 2005, the State Council gave its approval, changing the name to the Upper Yangtze Rare and Endemic Fish National Nature Reserve.

III. *Xiaonanhai: Jeopardizing the Last Fish Sanctuary in the Yangtze River*

While the hydropower development is in full swing, many fish experts consider the reserve as the primary, most important, and last sanctuary for the rare and distinctive fish species of the upper Yangtze River.

However, it was only three years after the adjusted national reserve was zoned that hydropower development started to come once again. It was proposed that, according to the 1990 *Integrated Utilization Planning of the Yangtze River Basin* (长江流域综合利用规划), there would be Xiaonanhai, Zhuyangxi (朱杨溪), Shipeng (石硼), and several other large hydropower stations on the main course of the Yangtze River upstream from the Three Gorges and downstream from Xiangjiaba. Among them, the demonstration stage for the Xiaonanhai project has begun. As the key power supply for the local grid, the dam will have a crest elevation of 206.5 meters, 1.5 kilometers downstream from Luohuang County (珞璜镇) in Chongqing, directly inside the experimental zone of the national fish reserve.

Because the dam would affect the area of the nature reserve, requested in a document (Chongqing Government No. 2008210) by the People's Government of Chongqing, the Ministry of Agriculture organized a seminar for a group of experts in February 2009 to study Chongqing's report on the impact of the Xiaonanhai project on the fish reserve and the measures to mitigate the adverse impact. At the seminar, the expert team pointed out the reserve was a remedial measure to mitigate the negative impact of hydropower projects on fish species, and thus had played an important and irreplaceable role in protecting the fish species and fishery resources of the Yangtze River. They also noted that the river section at Xiaonanhai was an important habitat and ecological passage for the rare and distinctive fish species in the reserve, with about 15 billion fish roe passing through this water area, and that the construction of the dam would cause some degree of impact upon the aquatic resources and the functioning of the reserve.

Before this seminar, experts from the Nanjing Institute of Environmental Sciences, a research body under the administration of the Ministry of Environmental Protection (MEP), had given a graver prediction after a study on the impact of the hydropower project: the superimposed effect of Xiaonanhai and other cascade hydropower stations would significantly change the water environment of the river ecosystem, and cause devastating impact upon the rare and distinctive fish species living there.[2]

Experts from the MEP, based on a study, believed the Xiaonanhai Dam would be a huge physical barrier cutting off the continuity of the river sections upstream and downstream from the dam and blocking the migratory passage of fish. When completed, the project would inundate both the buffering and experimental zones of the fish reserve, destroying the spawning field and habitat of many rare and distinctive fish species, stagnating the riparian habitat of the flooded areas, significantly changing the hydrological conditions and water regime downstream from the dam, and causing evident impact upon the habitat of this region. Once the three cascades of Xiaonanhai, Zhuyangxi, and Shipeng were constructed, both the core and experimental zones of the reserve would be gone, leaving only a 41.9 kilometer buffer area. The reservoir area would inundate 30 spawning fields and concentration spots of these fish, while the structure and functions of the reserve would be almost lost and destroyed.

However, according to public reporting, the expert team at the seminar has only vaguely suggested solutions be proposed based on the *Regulations on Nature Reserves* and other related laws and regulations. In other words, the experts from the Ministry of Agriculture have merely "vaguely" discussed the impact upon the fish, leaving all the problems to the MEP.

Apparently, this suggests another adjustment of the area of the reserve. However, to adjust or not to adjust, that is already a problem. Without adjustment, the functions of the reserve would hardly be sustained given the rounds of encroachments. But even if it were to be adjusted, how should it be? And where? The immense Yangtze River, including the main

[2] Qin Weihua, Liu Lujun, Xu Wanggu, Wang Zhi, Jiang Mingkang, "Predictions on the Ecological Impacts of the Xiaonanhai Hydropower Project upon the Upper Yangtze Rare and Endemic Fish National Nature Reserve," *Journal of Ecology and Rural Environment*, 2008, 24.

course and the numerous tributaries, has already been planned with a density of cascade dams.

IV. *Dispute on Adjusting the Nature Reserve of Wild Fish*

A. *Experts Warn about the Development of Xiaonanhai Project in Open Letters*

On May 6, 2009, eight professionals who have long been concerned with wildlife in the Yangtze River and the ecological protection of Southwest China, including Lü Zhi (吕植), a professor of conservation biology in Peking University, and Xie Yan (解焱), director of China Species Information Service (中国物种信息服务中心), jointly signed an open letter saying the superimposed effect of hydropower over-exploitation would destroy the riparian habitat for the concentration of aquatic organisms in the upper Yangtze River, and cause significant impact on the rare and distinctive fish species in the river.[3]

In this open letter, the experts expressed their worry about the hydropower development on the Yangtze River, particularly about the threat the construction of Xiaonanhai would pose upon the fish reserve. They warned: "To avoid a disaster to the rare and distinctive fish species, we call for urgent action to balance the relationship between development and protection, curb the over-exploitation of hydropower, and protect rare and distinctive fish species." The experts also acknowledged the municipality's demand for energy, but believed the demand could be fulfilled by coordinating the large hydropower projects upstream and downstream from the city.

B. *Environmental Protection Organizations Write to the MEP for Transparency of Expert Review Meetings*

On October 28, 2009, Green Earth Volunteers (绿家园志愿) and five other environmental protection organizations wrote an open letter to the review committee of national nature reserves "appealing to all experts to

---

[3] Lü Zhi, et al., "Experts: Over-Exploitation of Hydropower Must be Stopped to Save Distinctive Fish Species in Yangtze River," *China Economic Times*, May 6, 2009.

be prudent with academic conscience and cast a vote that will withstand the test of history."[4]

On November 9, on behalf of six environmental protection organizations, Friends of Nature (FON) sent an application to the MEP and a request to the ministry's Department of Nature and Ecology Conservation (自然生态保护司) that each of these organizations be allowed to send a representative to attend the annual meeting, to be held that month, of the review committee of national nature reserves.

The reason these organizations paid such attention to the meeting is that this was to be the last effort to stop the Xiaonanhai project from getting the nod. Rumor had it that, thanks to the intense lobbying by the Chongqing municipal government, the Ministry of Agriculture, in charge of the fishery sector in China, was ready to give the green light. Once the MEP nods its head to make the adjustment, there will no longer be a bottleneck for the hydropower project to get the approval. What worries the environmental organizations is that the adjustment will inevitably destroy the riparian habitat of the river section, leaving the nature reserve in name only.

C. *A Response from the MEP: Temporarily Shelve the Controversial Issue*

On November 10, FON received a phone call from the MEP's Division of Nature Reserves and Species Management (自然保护区与物种管理处) in the Department of Nature and Ecology Conservation, confirming the receipt of the application and thanking the organization for its concern. The division said there would not be anything related to the Xiaonanhai project at the meeting, and thus suggested the environmental organizations not attend; if needed, the division would pass on the opinion of the organizations on the adjustment of the reserve at the meeting; while at the same time, the ministry was willing to make suggestions on whether public participation should be introduced into similar meetings in the future.

The information was a great relief for the environmental organizations and experts who had been concerned about the ecology of the Yangtze River Basin. The reserve was safe for the time being, but they could hardly be optimistic about the future. Ma Jun (马军), head of the Institute of

[4] These six environmental protection organizations are Green Earth Volunteers, Friends of Nature, Green Watershed, Green Han River, Institute of Public and Environmental Affairs, and Chongqing Green Volunteers Union.

Public and Environmental Affairs (公众环境研究中心), said, "OK, they are not going to talk about Xiaonanhai and the reserve this year, but what about next year? Hydropower projects have a long construction cycle, and one year is by no means a problem. Besides, there is always the possibility that a meeting will be held at any time to discuss this issue. It is difficult to predict the future."

D. *Adjustment of the Reserve Is Almost Certain*

After more than a half year of silence, the concern of many experts and environmental organizations finally became reality at the end of 2010. Although many organizations had filed applications to the MEP for information disclosure on the adjustment of the reserve and the review process of the Xiaonanhai project, they were never considered stakeholders, nor had they been informed promptly. On November 24, 2010, FON received a call from the ministry's Department of Nature and Ecology Conservation saying the adjustment of the fish reserve had been approved at the annual meeting for reviewing national nature reserves, held from November 19–21, and that the result would be submitted to the State Council for approval. Information from the call also indicated the construction of Xiaonanhai project had not been mentioned at the meeting as the reason for adjusting the reserve, which is different from the case in 2009, but the adjustment of the area and site had coincided with what had been proposed for the hydropower station. As a result, on December 1, 2010, FON, the Institute of Public and Environmental Affairs, and five other environmental organizations jointly sent a letter to the MEP asking for a public hearing on the adjustment of the fish reserve because the project was significantly related to public environmental interests and a hearing would fulfill the public's right to know and facilitate public participation in the discussion.[5]

The passing of the adjustment proposal at the expert review meeting means the reserve has lost the last line of defense. Before the meeting, environmental organizations had contacted the Chongqing municipal government and the China Three Gorges Corporation (三峡总公司) (the builder of the Xiaonanhai project), both of which had held a positive attitude about the hydropower project.

---

[5] These seven organizations are Friends of Nature, Institute of Public and Environmental Affairs, Shan Shui Conservation Center, Darwin Natural Knowledge Union, Green Watershed, Green Earth Volunteers, and Peking University Center for Nature and Society.

It was with their strong promotion that the Ministry of Agriculture gave a green light to the project in 2009, on the precondition that the review committee of national nature reserves must pass the adjustment proposal of the fish reserve.

In fact, the lobbying capability for hydropower development is far beyond the capacities of the environmental organizations. In December 2010, knowing that these organizations were asking for a hearing for the adjustment of the fish reserve, the deputy director of the Chongqing Municipal Commission of Development and Reform went to Beijing in person to meet these organizations "one by one asking them to understand the demands of development in the municipality."[6]

Another reason to be worried is that, once the fish reserve is adjusted, Zhuyangxi, Shipeng, and other hydropower projects, in addition to Xiaonanhai, will be put on the agenda. The five dams of Gezhouba, Three Gorges, Xiaonanhai, Zhuyangxi, and Shipeng will be connected in a line not far from the two other huge dams, Xiajiaba and Xiluodu, which are now under construction. As a person in charge from the Shan Shui Conservation Center (山水自然保护中心) puts it, "the last living space for distinctive fish species in the upper Yangtze River will be lost."[7]

E. *Environmental Protection Organizations Are Making Concerted Efforts While the Public Is Making Its Voice Heard for Protection*

One and a half months after the expert review meeting, the MEP posted on January 14, 2011 its No.1 announcement on its website. The announcement, dated January 4, gave a list of seven national nature reserves which had applied for an adjustment, and among them was the fish reserve. The document, open to public review until January 21, indicated the efforts made by the environmental organizations and experts might come to nothing.

Even so, the environmental organizations took action again. On January 18, 2011, FON sent a letter to the MEP asking for information disclosure on three aspects concerning the adjustment of the reserve: the application for adjustment, the comprehensive investigation report on the adjusted

[6] Meng Dengke, "First Snow of 2011 Falls on Xiaonanhai," *Southern Weekly*, January 20, 2011.

[7] Du Yueying, "Xiaonanhai Hydropower Project in Yangtze River Basin Almost Certain," *China Economic Times*, December 7, 2010.

area, and the conclusion and minutes of the annual review meeting for national nature reserves.

The application was accepted by the ministry. Environmental organizations, including FON, The Nature Conservancy, and Green Earth Volunteers, then joined hands to create a feedback collection plan for the public to know about the adjustment as well as the ecological protection of the Yangtze River, and express their views. Within a few days, more than 100 people sent their viewpoints nationwide via email, fax, and letters.

On January 20, FON once again contacted the ministry asking for an extension of time for public review on the adjustment. On the same day, a person in charge from the ministry responded that the request was reasonable and that the time for public review would be extended for three days to January 26.

On January 21, FON posted an open letter to the representatives of the National People's Congress and the members of the Chinese People's Political Consultative Conference. In this letter entitled "*Many Rare and Distinctive Fish Species in Yangtze River Are Losing Their Homes*" FON pleaded for their attention to the decision-making process of the adjustment in two aspects: 1) to provide suggestions to the State Council that the approval process on the adjustment of the fish reserve be stopped and that alternative solutions be sought; and 2) to enhance the seriousness and transparency of the review meeting on national nature reserves. "Public participation is the important complementation to the full play of the key role of the review committee. However, given the current situation, there is still much room for improvement," said the letter.[8]

That same day, The Nature Conservancy held a press conference in Beijing to announce their opinion on the adjustment and the Xiaonanhai project, while at the same time proposing their alternative solution for protecting the ecology and addressing the power demand of Chongqing: a cooperative and sharing relation could be established between the municipality and the China Three Gorges Corporation on the four dams of the lower Jinsha River (Wudongde, Baihetan, Xiluodu, and Xiangjiaba) under construction or contemplation. Chongqing could invest in these four dams instead of the Xiaonanhai project to get the corresponding share of the ownership and benefits. By adjusting the flood storage capacity of these four reservoirs and making use of the flood storage and detention areas to take part of the flood risk, the corporation might be able to further

[8] Han Lewu, "Rare Fish Species in Upper Yangtze River Losing Their Home? Xiaonanhai Hydropower Station vs. National Reserve," *Legal Daily*, January 25, 2010.

increase the total installed capacity of these four hydropower stations. A feasibility study on the possibility was under way. The municipality might be able to get more installed capacity with the same scale of investment, although more negotiation and coordination would be needed between the municipality and the hydropower corporation.

This solution has evident ecological benefits: dams and reservoirs are avoided in the fish reserve, so is the negative impact upon the rare and distinctive fish species in the Yangtze River. Meanwhile, the solution protects all the investment from the Chinese government and China Three Gorges Corporation in the reserve as the compensation measures for the Three Gorges and other cascade projects on the lower reaches of the Jinsha River, and safeguards China's declaration of "sustainable hydropower" development in the Yangtze River Basin.

On January 24, many experts and scholars who had long been devoted to nature reserves and biological research, including Wang Song (汪松), president of the Chinese National Committee of the International Union of Biological Sciences, Xie Yan, secretary-general of the International Society of Zoological Sciences, and Weng Lida (翁立达), former director of the Yangtze River Valley Water Resources Protection Bureau (长江流域水资源保护局), jointly signed a letter to the MEP, pleading with the ministry to protect, strictly according to the law, the only national fish reserve on the main course of the Yangtze River, and save the precious resources of aquatic biodiversity for future generations.

However, in late 2011, to fulfill the need of constructing the Xiaonanhai Hydropower Station, the boundary of the national nature reserve was eventually modified, shortening the downstream part of the reserve by 22.5 kilometers where dams were planned. In early 2012, a ceremony to commemorate the beginning of construction of the station was held on Zhongba Island, Chongqing.

V. *Future Observation: Xiaonanhai Is Not Alone*

There is still no final outcome of the positioning between the hydropower development of Xiaonanhai and the national fish reserve. Public participation in environmental protection, which has gone beyond the fight for natural resources between the development side and the protection side, had brought the end of 2010 and the ensuing 2011 to a new small climax.

The fish reserve in question has already been adjusted within the 11 years since its inception, and now a second adjustment is under way. How can the integrity of national nature reserves, as well as their unique ecological values, be safeguarded? Also doubtful are the seriousness and authority of the demonstration and establishment of reserves in China. A bad example has been set for the adjustment of a national nature reserves in the future.

If the fish reserve is adjusted for the construction of Xiaonanhai project, then the seriousness and authority of Chinese laws and regulations on nature reserves will be jeopardized. There is the possibility that, in the future, the provinces of Sichuan, Guizhou, and Yunnan might make a similar request, leading to a domino effect and a complete collapse of this fish reserve.

"Adjustment of national nature reserves should be strictly controlled to prevent unreasonable adjustment from damaging the functions of the reserves," said Li Ganjie (李干杰), chairman of the Review Committee of National Nature Reserves (自然保护区评审委员会) and the vice minister of environmental protection, when chairing the review meeting in 2010. It is a conceptual retrogression to ask for adjusting an existing reserve by arguing that the reserve causes constraints upon the development of the local area or a single project while at the same time ignoring the enormous risk that might be incurred by such adjustment.

More important, such a dilemma will not be alone in the decision-making process of Xiaonanhai project. Similar problems can also be found in the dense development of the four cascade dams on the lower reaches of the Jinsha River, the eight cascades on the middle reaches, as well as on the main course and tributaries of the Min, Dadu, Yalong, Wu, and Jialing Rivers. In the past two decades, to give way to hydropower development, the habitat of the rare and distinctive fish species in the upper Yangtze River has been pushed from Gezhouba to the Three Gorges, to Xiluodu and then to Xiaonanhai. And if the Xiaonanhai project is approved, it might be once again pushed to the Min River. Recently, rumor has it that, one or two cascades might be constructed in the experimental zones of a reserve in the lower reaches of the Min River.

Fish habitats are being torn apart by the planning and development of hydropower stations. Experts have racked their minds to propose remedial measures in the environmental impact assessment of every dam. But will these seemingly reasonable remedies save the ecology of the Yangtze River that is being pushed into despair?

PART THREE

URBAN ENVIRONMENTAL ISSUES AND LIVABILITY

LIVABILITY: A NEW WAY OF THINKING ABOUT URBANIZATION IN CHINA

Li Bo

Abstract: As China rapidly urbanizes, the important task of maintaining a sustainable economy and social harmony will fall on the shoulders of cities. From megacities to townships, all face serious challenges from "urban disease." Chinese cities are the main consumers of resources, the key link for a low-carbon economy, and the main battlefield for pursuing justice and fairness in making environmental decisions. Livability should become an important benchmark for urban planning experts, real estate agents, and government officials. The task of measuring cities' livability urgently demands citizens' involvement, and a consideration of their perspectives. This paper attempts to discuss the relationship between livable cities and the current environmental challenges facing China, analyze livability indicators, and suggest that non-governmental environmental protection organizations should become involved in the discussion on building livable cities and pay greater attention to China's urban ecological footprint.

Keywords: livable city, livability indicator, inclusive development

I. *Livability Is the Bottom Line for City Development*

The concept of a livable city was put forward by the United Nations in 1976 during an event concerning the construction and study of habitable environments. However, this concept has been a topic of discussion among China's academia—in particular the urban planning circle—for less than ten years. This can be explained by a global tendency for the topic to be taken seriously by societies only when faced with major environmental disruption. Over the course of the three most recent Five-Year Plans, China's level of urbanization has grown to nearly 50% from less than 18%. Today, the urban population grows by approximately 18 million people annually and is expected to hit 910 million by 2030, while the floating population is expected to grow to over 300 million. Both national and international experts expect that China will achieve a 79% urbanization rate over the course of the next 30 to 40 years. These numbers hold great potential for employment and business opportunities; however, the inherent social and environmental costs cannot be ignored. Urbanization at such a rapid pace and on such a vast scale is nearly unparalleled in human history. Joseph Stiglitz, winner of the Nobel Prize in Economics, predicted in a 1999 speech that China's urbanization following the reform and opening up period would be one of two main events to have a profound impact on human society.[1]

History shows that rapid urbanization typically forces people to compromise on both living space and quality. There are six main symptoms of this: 1) ever-growing challenges to transportation—with ever-crowded streets, poor traffic efficiency, and investments in public transportation lagging far behind the development of the private auto industry; 2) economic growth and trash production volume are directly correlated, but trash processing and recycling policies have trouble keeping up. This results in severe threats to public health from the burning of refuse and the use of landfills; 3) public green space shrinks and, sadly, the public gets used to the lack of nature and becomes indifferent to it. This results in a gradual reduction in the attraction of natural resource protection; 4) the burden of investing in and maintaining public services transfers to common users, which results in unequal allocation of economic gain, especially in terms of housing, education, clean water, and medical resources; 5) erosion of a city's culture and traditions results in loss of completeness, ownership, and pride; and 6) the public finds itself unable to become informed of or involved in city planning and are not given opportunities to monitor its implementation.

All these factors leave the basic social units—individuals and families—with no choice but to passively adapt to this environment. This causes increasing frustration, reduced quality of life, and a dwindling sense of happiness. Today these symptoms are quite evident at different levels in medium to large cities and conditions continue to deteriorate.

Professor Lian Yuming (连玉明), a noted urban expert, once pointed out that livability was the bottom line for city development, and was the "essence" of a city's brand. A city cannot maintain a sustainable economy if it is not built on a bottom line of livability and within a red line of environmental protection, no matter how good the economy at present.[2] When he interpreted the *CPC Central Committee's Proposal for Formulating the 12th Five-Year Program for China's Economic and Social Development (2011–2015)* (中共中央关于制定 "十二五" 规划的建议), Professor

---

[1] Li Jinkui, "*China: Challenges of an Accelerated Urbanization,*" China Architecture & Building Press, 2007.

[2] Tai Guihua, "*Livability and Environmental Protection Are the Bottom line and Redline,*" China Environment Paper, November 29, 2007.

Lian had some observations on a city's value judgments and future trends in development. He said:

> First, urbanization will become a basic driver for economic transformation. Second, quality of life is the core of a city's comprehensive development and inclusive growth. Third, cultural soft power will become an important symbol for a city's overall increase in competitiveness. Public culture is a real need for city dwellers and is their fundamental privilege as well. Fourth, the 'megacity disease' brings risks and challenges to China's urban development. Cities in China are facing a key moment in the stages of accelerated development, city value enhancement, and urban and rural transformation. This key moment also means that cities are experiencing an outbreak of 'urban diseases.' During the 12th Five-Year Program period, there will be more complicated elements and uncertainties that will bring both predictable and unpredictable risks and challenges. Fifth, a modern social governance model based on a core of fairness and justice is the key for urban social service management innovation.[3]

In short, the scale and speed of China's urbanization mean that this is an unprecedented social and environmental project whose social costs cannot be taken lightly. The 12th Five-Year Program aims to stimulate domestic demand and maintain sustainable economic growth. During this period, the question of low-carbon livability will require an answer from China's urban development trends. This is not only a matter of dealing with global warming, but also as a key project if China wants to develop a more competitive green industry and an economic model for recycling.

II. *The Definition and Indicator Index of Livability*

Despite a lack of consensus on the precise definition of a "livable city," general discussion and observation elicit a basic context. First, a livable city is a place where people feel happy, secure, and free to live. This should be based around the experiences of *individual's* own judgments, rather than that of organizations or commercially oriented "experts." Gu Wenxuan (顾文选), the secretary-general of the Chinese Society for Urban Studies (CSUS) (中国城市科学研究会), published the "Scientific Evaluation Standards of Livable Cities" (宜居城市科学评价标准) recognized by the Ministry of Construction. Mr. Gu believes that residents have the clearest ideas as to whether a city is livable or not.

[3] Lian Yuming, *"China Urbanization's Challenges and Opportunities,"* Finance Blog, November 4, 2010, http://blog.caijing.com.cn/expert_article-151296-12925.shtml.

Drawing on descriptions of such cities from the local and international community, we find that a livable city should, from a functional perspective, have the following common characteristics: a stable political environment; fair and just economic opportunities; vibrant centers of activity or local business (this is differentiated from big and trans-regional business powerhouses); affordable housing; efficient and reliable public transportation; a clean natural environment; a fair public investment policy; quality services in health, education and culture; and safety in daily travel. In addition, cities need public spaces and platforms where people can actively interact with each other and the city can maintain vibrancy and listen to and respect the peoples' voices. If we look at the core value of these functions and services, we can see three key aspects: one is accessibility to services; two is fairness in the allocation of services and resources; and three is the participation of the public. These three core values determine a resident's perception of a city's livability and evaluation of its services.

The "Scientific Evaluation Standards of Livable Cities" developed by CSUS is the first-elected among all research studies on the standard of a city's livability.[4] This study was formally filed, in the course of setting up a project, with the Ministry of Construction in December 2005, and was approved in April 2007 (refer to Table 1). This standard evaluates a city's livability from the following aspects: the degree of social civilization, economic prosperity, environmental beauty, resource carrying capacity, convenience of life, and public safety. These aspects can be scored to divide cities into three categories: livable, relatively livable, and livability endangered. In addition, four negative conditions can result in a veto. They are: social contradictions are apparent and a high crime rate relative to the national average; a Gini index of greater than 0.6, and severely polarized social wealth distribution; categorization by the Ministry of Environmental Protection (MEP) as one of the "Top Ten Most Polluted Cities" for the most recent three years; and scarce regional fresh water or severe ecological deterioration.

[4] "Scientific Evaluation Standard of Livable Cities," http://news.xinhuanet.com/politics/2007-05/30/content_6175236_1.htm, or word document http://yjcs.city188.net/Uploadfile/20101793410397.doc.

Table 10.1: Scientific evaluation standards of livable cities (CSUS).

A. Degree of Social civilization (Weight: 0.10) 10/100	Premise: Social civilization is a prerequisite for livability. Four secondary indicators: political civilization; social harmony; community civilization; and public participation
B. Economic prosperity (Weight: 0.10) 10/100	Premise: Economic prosperity is the most important basic condition for livability, and one of the most important determinants as well. Five secondary indicators: GDP per capita; disposable income per capita; fiscal revenue per capita; employment rate; proportion of employment in tertiary industries compared to total employment
C. Environmental beauty (Weight: 0.30) 30/100	Premise: Ecological deterioration is a major problem facing urbanization. Four secondary indicators: ecological health; climate; humanistic environment; urban landscape
D. Resource carrying capacity (Weight: 0.10) 10/100	Premise: A city's resources—particularly water and soil—determine its natural carrying capacity. Five secondary indicators: total fresh water volume per capita; industrial water recycling rate; urban land per capita; food supply safety; other specific negatives and positives
E. Life conveniences (Weight: 0.30) 30/100	Premise: Convenience is the most important and central factor of a livable city. A livable city should provide high quality services and make them convenient and accessible to its residents. Seven secondary indicators: transportation; commercial services; infrastructure; education, culture, and sports facilities; green spaces; housing; public hygiene
F. Public safety (Weight: 0.10) 10/100	Four secondary indicators: lifeline engineering in good order; city government's facilities, mechanisms, and plans for natural disaster prevention and response; city government plans and mechanisms for preventing and dealing with man-made disasters; and the number of successful public safety cases handled by the city government in the most recent three years

The pace of China's urbanization is so fast that the "Scientific Evaluation Standards of Livable Cities" issued four years ago now cannot cover every new development trend. The author thinks that it is necessary to include supplementary notes.

First, the indicator of social civilization is the most significant. This indicator should see "community culture" and "community self-service and self-management" as secondary indicators. The aim is to highlight the special role basic social organizations can play, expand areas for public participation, increase public involvement in the evaluation of the impact of governmental projects on society, the environment, and public health, and to have a feedback mechanism in place to respond to the public concerns.

Second, certain groups have particularly important roles in livable cities, but have living and working conditions that are difficult to reflect. As an example, the urban prosperity index is too general an indicator. The economic conditions of people like migrant rural workers, sanitation workers, and people who scavenge refuse are not reflected. It is reported that there are 300 thousand refuse scavengers in Beijing and that they convert a great deal of waste into productive resources. This is a valuable service, but nevertheless their economic standing and opportunities to participate in policymaking are not properly reflected. Further, an increase in crises resulting from production of household waste and industrial solid waste from 2009 has presented real challenges. Thus, a single indicator of the "household waste treatment rate" is not enough to reflect the real problem.

In addition, green space per capita and the urban greening rate does not reflect the potential use of green space by the public or the participation level of the public in the planning process. Green area allocation and management should not solely depend on the Parks Department's involvement. Experience proves that public participation not only helps decrease administrative costs but also improves or even rebuilds the relationship between the public and the natural environment.

Additionally, the index on public transportation has not been able to correctly reflect the rate of urban traffic congestion over the past two to three years. Low-carbon and low-speed methods of travel like walking and cycling are not included, and the question of road use rights is not considered either. Further, investments in public roads and monitoring of services have been weak. All these require more specific indicators.

III. *Challenges and Opportunities*

Over 100 cities in China aim to develop their livability, but few really use the "Scientific Evaluation Standards of Livable Cities" as a guideline, and many of those that do have not seriously implemented it. In January 2005, the State Council approved the *Beijing City Master Plan* (*2004–2020*) (北京城市总体规划 (2004年–2020年)) and positioned Beijing as not only the national capital, but a global, cultural, and a livable city as well. This represents the first time that the ancient capital has mentioned the concept of the "livable city" and set "building Beijing into a livable city" as one of its goals. Unfortunately, the plan has been dormant for six years, and now Beijing faces increasingly severe challenges to health caused by garbage and sewage, uncertainties brought about by housing prices, and an increasing frustration caused by public transportation and traffic problems.

A. *Surrounded by Garbage*

For the past two years, livability in Beijing has been under the spotlight. Being surrounded by garbage and traffic congestion directly challenges the capital's goals in this regard. Will "a livable Beijing" be just a dream? The increasing volume of garbage production in China's megacities should arouse attention. But how exactly should the garbage production relate to per capita income?

Even though Beijing has proactively set a goal of zero waste, there is a lot more that needs to be done, such as the selection and application of front-end and back-end technology, and dealing with the relationships between various interest groups. Great effort needs to be put into keeping this project open, inclusive, fair, and scientific. Between 2009 and 2010, the Beijing Municipal Government issued and implemented, *An Opinion on Promoting Comprehensive Waste Treatment* (关于全面推进生活垃圾处理工作的意见). This was followed by a certain level of public participation, but still lacked the involvement of civil society organizations in monitoring the 600 waste classification pilot complexes. And Beijing continues to lack a communication channel for the public to be informed of policies on garbage incineration and related projects, as well as their impact on the environment. As a result, there is no channel for public involvement in monitoring how equipment for incineration projects is running. Experiences abroad have shown that it is only through transparent

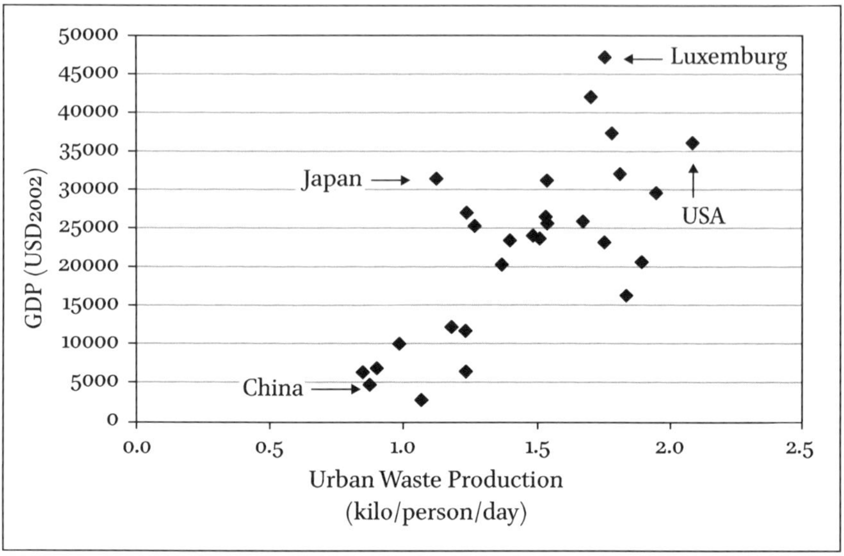

Figure 10.1: Urban waste production vs. GDP.

information disclosure and adequate monitoring that the concerns urban residents have about incineration can be reduced. These measures also help reduce inefficiency and error rates.

B. *Traffic Congestion*

According to the data published by the Traffic Management Bureau of the Ministry of Public Security, in 2009, there were 238,351 traffic accidents nation-wide, with 67,759 people killed, 275,125 injured, and direct property losses of 910 million RMB. In 2009, China owned only 3% of the world's cars, but 16 percent of the world's traffic fatalities. Since the end of the 1980s, when traffic fatalities reached over 50,000, China (excluding Hong Kong, Macao, and Taiwan districts and regions) ranked first in number of traffic fatalities for a consecutive ten years.[5]

Another big problem is that the private car industry has been vigorously developed and that people overly rely on private cars as a preferred trans-

[5] Yan Xiaohong, "China Traffic Fatalities Ranks First in Consecutive Ten Years," *China Xinhua News*, http://news.xinhuanet.com/politics/2008-12/07/content_10467031.htm, 2008-12-7.

portation tool. This results in travel difficulties associated with congestion. Therefore, the 12th Five-Year Plan should focus on maintaining a balance between private car travel and public transportation. As a public property, road resources should be available and accessible to different groups of people. This is an important part of achieving inclusive growth.

But with all of these challenges, also come opportunities. The *Green Beijing Action Plan 2010–2012* (绿色北京行动计划 (2010–2012年)), published at the end of 2009, was a very good example. The Plan set out to implement green transportation travel projects, vigorously develop the public transportation system, actively promote alternative fuel and environment-friendly automobiles, accelerate the construction of low-speed travel systems, and optimize the intelligent transportation system. The Plan also set up the goal of "an initial formation of low carbon and green lifestyle and consumption patterns." Furthermore, it included the goal "to continue to optimize the garbage classification management system." Notably, the action plan also actively solicited opinions from the public. It has been reported as well that the 12th Five-Year Plan submitted for consideration by the NPC also included elements like "inclusive developments."

C. *Low-Carbon Issue*

Apart from national matters, it is also increasingly important that cities consider their impact on the global climate when making decisions. Since the Copenhagen United Nations Climate Change Conference in 2009, low carbon and energy savings have grown as contributing factors for economic development. A livable city also needs to include low-carbon and climate change issues into its evaluation systems. To achieve its energy saving goals, a livable city needs to continuously work on transforming industries and managing technology—in addition to broadening the role of social participation. The principles of public participation and information transparency in climate public policies should be specifically addressed. Understanding and support from the public on energy saving, and effective supervision (to prevent people from finding ways to take advantage of the policies rather than implementing them) are also vital. Urban residents also make up the largest consumer group, so rapid urbanization means that urban consumption will naturally become an important part of China's total carbon emission. Issues such as sustainable consumption and low-carbon consumption are the main energy saving areas in which the public can widely participate.

IV. *Non-Governmental Environmental Protection Organizations Should Focus More on Livable Cities*

The history of Chinese non-governmental environmental organizations is less than twenty years. Friends of Nature (FON)—one of the oldest such organizations—concerns itself with issues of species and habitat protection. FON's course of development is instructive. Their efforts at education on species protection and environmental topics such as biodiversity and climate change have produced a dramatic increase in knowledge of these subjects among the public.

But, the growing number of residents living in cities need to give up the common strand of thought that "environmental protection is somewhere far away." It has become commonplace to use wilderness protection activities to help rural communities develop low-carbon and sustainable lifestyles; but, the issue of unsustainable urban development is often neglected—this despite the fact that a majority of threats to natural resources can be rooted in the demands of urban development. City dwellers' lives, work, and consumption patterns are led by industry, and as such, lack needed self-evaluation and innovation. On the other hand, very few environmental protection organizations take the environmental challenges of cities as a primary objective or mission. They also tend to lack plans to deal with urban environmental problems. Missing too is a systematic guiding conceptual framework for urban environmental protection.

This analysis of both the concept of livable cities and the current situation in China points to the conclusion that the goal of developing livable cities should be adopted by environmental protection organizations and society at large. This needs to become a new base for both community practice and policy promotion. The state of the urban environment needs to be linked with the state of people's livelihood. Disparate organizations can start from their particular interests and integrate them into a livable city framework. Further, more detailed monitoring processes, participation mechanisms, and information disclosure should be developed to make "livability" an important part of urban environmental management. If effectively implemented, these will all contribute to vastly improved quality of life.

THE ENVIRONMENTAL RISKS OF RAPID URBANIZATION: INDICATORS FOR LIVABLE CITIES

Gu Chaolin

Abstract: Rapid urbanization in China has increased the country's environmental risks. Avoiding "Latin-Americanization" and reducing environmental and resource risks in China's urbanization process has become a strategic issue that requires careful consideration. The development of livable cities should take into account the multiple key factors in urban development. At present, the development of indicators and pilots for livable cities has just begun in China. Some indicators are excessively complicated, while other necessary evaluation mechanisms are still missing—unable to satisfy the overall requirements of China's urbanization process while countering environmental deterioration.

Keywords: urbanization, environmental risk, human settlement, indicators of livable cities

China has become the world's second largest economy since its reform and opening up, with its GDP rising from 362.41 billion RMB in 1978, to 39.8 trillion RMB in 2010. As an important part of social modernization, China's urbanization level has also risen from 17.82% to 49.6%. While promoting social modernization and improving production efficiency, as well as quality of life, rapid urbanization is also posing risks to the environment. It consumes huge quantities of material and energy resources, forcing the congregated urban population to face challenges from the uncoordinated development of population, resources, and environment.

I. *Environmental Risks of Rapid Urbanization*

A. *Characteristics of China's "Peri-Urbanization"*

Urbanization serves as an important socio-economic function, providing fundamental solutions to issues concerning agriculture, rural areas, and farmers. It is also used as a powerful lever to promote coordinated regional development. Urbanization creates huge economic demands in habitation, living, education, and employment, and provides strong stimuli and basis for the expansion of domestic demand and investment. It also serves as an impetus for industrial re-structuring, improving independent innovation capability, and building an innovative country. However, the rapid urbanization process has put increasing pressure on China's population, resources, and environment. The problem is heightened as China has lower-than-average land and resources per capita, as well as a unique history and cultural traditions.

"Incomplete urbanization" occurs when urbanization and industrialization do not develop in a synchronized way due to institutional and policy differences. China's urbanization is considered incomplete with distinctive characteristics of "peri-urbanization"—a type of urban sprawl with a high diversity of land use. Certain areas in China suffer from "urban diseases" marked by over-population and over-urbanization. City governments rely excessively on incomes from land and expropriate farming lands in surrounding areas, resulting in the problem of land-lost farmers. Population and industry do not develop proportionately with the large area of urban land consumed, and too much focus on the construction of industry parks, coupled with an insufficient focus on urban function zones, creates pendulum commuter flows between urban and rural areas in some cities. The most prominent problem is that the large number of rural surplus laborers migrating to cities have only transferred geographically and occupationally. Migrant workers living in urban areas cannot become city residents and cannot enjoy equal pay, working hours, or the same rights as those of city residents.

B. *Human Settlement: Environmental Risks of China's Urbanization*

By 2011, China had become the world's second-largest economy, the largest trading country, and an important manufacturing base. There are many lessons to be learned from the unsuccessful cases of urbanization, which are often dubbed the "middle income trap."[1]

Tang Min (汤敏) sums up the characteristics of the "middle income trap" as follows: a large income gap, urbanization at the cost of large-scale slums, fragile financial systems, slow industrial upgrading, and lagging social services. Related to the "middle income trap" is the "Latin American mode of urbanization." Latin America and South Asia are among the most highly urbanized areas after WWII, with 75% of their populations living in cities. They also, however, suffer from large numbers of slums—a consequence of rapid urbanization. This phenomenon can be found in the

---

[1] Gu Chaolin, "China Doesn't Need Latin-Americanized Urbanization," *People's Daily*, April 8, 2010.

world's second and fourth largest cities—Mexico City and Sao Paulo—and Buenos Aires. In Brazil, 80% of the population lives in cities. Meanwhile, in Rio de Janeiro, lower-income areas and slums occupy 80% of the city's 500 square kilometers of urban land. Many factors lie behind the Latin American mode of urbanization. These include the following: farmers are driven to cities, deprived of their lands in the rapid urbanization process; fierce economic competition drives the manufacturing industry away from big cities to remote areas in search of cheaper land and labor; and with no land or jobs in cities, farmers resort to slums for urban residence. The situation is similar to that in Beijing: Xiaoyuehe Village (小月河村), known as the "college graduates village" in the city's Tangjialing (唐家岭) area, holds hundreds of thousands of low-income residents living in extremely crowded quarters. This development trends towards the signs of future slums in China's urbanization process.

Additionally, China's human settlements are facing increasingly serious environmental risks. For example, most cities are beset with the problem of garbage. Burying and burning garbage in large quantities without sorting and reduction poses high risks to public health and environmental resources. As the public transport and low-carbon commuting strategy lags behind the development of the automobile industry, exhaust emission has become a key contributing factor to deteriorating air quality and serious traffic congestion poses another hindrance to urban economic development. Many cities suffer from a shortage of drinkable water, and underground water is over-exploited. Cities depend on large quantities of natural resources, such as water, energy resources, and land. These are being produced, exploited, and consumed, which also causes secondary pollution. This takes a direct toll on the living environment, public health, and quality of life of urban residents as well as the sustainable development of cities. Moreover, the stability of cities' ecosystems is affected by the loss of green lands, rivers, lakes, wetlands, and their ecological functions.

II. *Evaluation Indicators of Livable Cities in China*

It is estimated that over 10 million people from rural populations will move into cities each year over the next 5 to 15 years or a longer. By 2020, approximately one billion Chinese will be living in cities, with 350 million new residents, among which 240 million will be part of the floating population. This makes building livable cities an important plan, in order to avoid "Latin Americanization" and reduce cities' environmental and resource risks.

In a narrow sense, livability refers to suitable living conditions, especially in housing. It refers to a suitable environment for settlement, primarily in living environment. Likewise, a livable city is also defined in two senses. A livable city in a broad sense is a comprehensive concept: a city whose economy, society, culture, and environment are developed in coordination, and whose people feel satisfied with their work, living, and dwelling, and are willing to settle for a long period of time. In a narrow sense, a livable city refers to an area with an amiable climate, harmonious ecological landscape, and is suitable for living.

A livable city is a complex mega-system composed of a natural-material environment and a social-cultural environment. The former consists of three sub-systems, namely, natural environment, artificial environment, and facility environment. The latter consists of the social environment, economic environment, and cultural environment. All subsystems work in an organic and coordinated way to create a healthy, pleasant, and harmonious urban environment for human settlement, forming a livable city system.

A. *Standards for Livable Cities in Other Countries*

Japan is a typical example of a country that had rapid industrialization and urbanization. Water and air pollution resulting from industrial production caused the Four Big Pollution Diseases: the Itai-itai disease (1955), the Minamata disease (1956), the Asthma Event of Yokkaichi (1960), and the Niigata Minamata disease (1965). Japan began to promote its "Amenity Town" plan in 1984. Beginning in 1990, cites with considerable achievements were commended. By 2003, a total of 115 cities and towns were commended.

In 1996, Habitat II (the Second UN Conference on Human Settlements) was held in Istanbul, Turkey, which put forward the concept that cities should be habitable for human settlement. The conference proposed two basic goals in its *Human Settlements Agenda*: adequate shelter for all, and sustainable human settlement development.

The annual international competition of livable cities was launched in 1997 by the U.K., the U.S., Japan, Belgium, and Canada, aimed at improving the urban environment under the theme of livability. The event encourages cities to adopt the most effective means of environmental governance and transformation in an effort to improve urban life quality, promote sustainable development of cities and communities, preserve and carry forward cities' local, historical, and cultural traits, and strengthen the role of city planning as a support for urban environment and culture construction.

The international competition was first termed "Nations in Bloom," with more focus on cities' hardware than software, like the improvement of urban landscaping and urban greening. The focus, however, has begun to shift as the world's cities face similar problems. These include environmental pollution and control, energy shortages, traffic conflicts, production of renewable energy, and energy recycling. The event is now termed "The International Awards for Livable Communities" (IAFLC), and focuses on the reasonable use of urban resources and sustainable development of cities, of which "livability" has become a key criterion.

The criteria for IAFLC fall into five categories. The first is the improvement of a landscape's environment. The focus is on both the hardware and software, including the purpose, effect, and means of landscape improvement, the effectiveness of that improvement in enhancing the urban environment or quality of life, and achievements in its planning, management, maintenance, and financing. The second is the preservation and management of cultural heritages, both tangible and intangible. The focus lies on both the preservation and the effective use of these heritages, such as the means, technical support, financing, and effects of preservation, as well as ways for improvement. The third is on the practices of environmental improvement. This includes the process and results of specific projects, with focus on their achievements in improving the quality of water, air, and land, especially in promoting sustainable development of the environment and its resources, reducing energy consumption and waste production, and an active recycling of resources. The fourth is public participation, with special attention to the method, extent, and effects of such participation. Lastly is city planning. Long-term planning of cities is reviewed in order to see whether a sound basis or guarantee can be provided for environmental transformation, landscape preservation, and the sustainable development of the area and its resources.

B. *Indicators for Development of Livable Cities in China*

Six key issues on the sustainable development of human settlements were put forward in *China's Agenda 21* (中国21世纪议程) in 1994. Focus was on urbanization and management of human settlements, infrastructure construction and enhancement of human settlement functions, improving the human settlement environment, providing adequate shelter for all, promoting sustainable development of the construction industry, and improving the energy efficiency of buildings within human settlements. In 1999, the *Beijing Charter* (北京宪章) further proposed that better and fairer human settlement environments should be built under the world's limited resources to create a more suitable living environment. In January 2005, the State Council asked that Beijing adhere to the people-oriented principle and build a livable city in reply to the city's master plan revision. In July 2005, while attending the national city planning meeting, Vice Premier Zeng Peiyan (曾培炎) stressed that livable city development should be an important part of city planning.

In October 2005, a livable cities research team was established by the Chinese Society for Urban Studies (中国城市科学研究会) to carry out research on the concept, evaluation index system, and city planning practices of livable cities. The team published the *Scientific Evaluation Standards of Livable Cities* (宜居城市科学评价标准) in 2007, in which 30 sub-goals and 83 indicators (including nine bonus and penalty indicators) were proposed under six major goals. These include social civilization, economic prosperity, environmental pleasantness, resource carrying capacity, life convenience, and public security. The team's *Index System Research* (指标体系研究) suggested a "one vote veto" for any cities with severe polarization in wealth distribution, high crime rate, serious environmental pollution, or ecological deterioration. More recent research was carried out by the research center of the Ministry of Housing and Urban-Rural Development, such as the *2011 Livable City Index* (2011年宜春宜居指数) (Table 1). As an examiner of the *Index*, the author believes it has a feasible framework and the participating cities are highly oriented and motivated, although it is still a pilot with a very limited number of participating cities, and has yet to establish itself as a major goal for the development of Chinese cities.

The concept of "green development" was first put forward in China's Twelfth Five-Year Plan, which has green indicators as part of the central government's performance appraisal index system for local officials. Examples include: unit GDP energy consumption, CO2 emission, and forest cover. The concept is also summarized by environmental health, resource conservation, low-carbon development, and ecological livability (Table 2). It can be concluded that China's livable city index and pilots are generally in the preliminary stages. Some indicators are too complex and therefore lack the necessary evaluation mechanism. The index system currently serves more as a self-entertainment club, with a few cities as members, unfit for the requirements of city development affected by urbanization and ecological deterioration.

Table 11.1: 2011 Index system for livable cities and indicator scores.

Goals	Sub-Goals	Indicators	Points	Points Earned
Social Civilization	Political Civilization (3 Points)	Scientific and democratic decision-making	0.9	0.45
		Making government affairs public	0.6	0.30
		Democratic supervision	0.6	0.30
		Administrative efficiency	0.9	0.45
		Government innovation (bonus indicator)	1	0
	Social Harmony (2 Points)	Income gap	0.4	0.40
		Coverage of social security	0.3	0.11
		Social assistance	0.3	0.30
		Crime rate and solution rate of criminal cases	0.6	0.57
		Cultural inclusiveness	0.2	0.18
		Employment service for floating population	0.2	0.20
		Bonus and penalty indicators	1	0
	Community Civilization (2 Points)	Community management	0.5	0.26
		Property management	0.5	0.34
		Community Service	1	0.79
		Penalty indicators	1	0
	Public Participation (3 Points)	Making planning public	1.5	1.50
		Price hearing	1.5	1.50
	Total		10	7.65
Economic Prosperity	GDP per capita		2	1.56
	Disposable income per capita of urban residents		3	1.79
	Fiscal revenue per capita		1	1.00
	Employment rate		2.5	2.46
	Proportion of employed population of the tertiary industry in the total employment population		1.5	1.30
	Total		10	8.11

Table 11.1 (*cont.*)

Goals	Sub-Goals	Indicators	Points	Points Earned
Environmental Pleasantness	Environment (24 Points)	Number of days per year with air quality of or better than Level 2	4.8	4.77
		Water quality compliance rate of centralized drinking water sources	4.8	4.80
		Treatment rate of urban industrial waste water	2.4	1.20
		Rate of harmless disposal of urban domestic waste	2.4	1.20
		Coverage of noise control districts	3.6	3.39
		Disposal rate of industrial solid waste	2.4	0.80
		Public green area per capita	1.2	1.00
		City green area coverage	2.4	2.00
		Bonus indicators	2	2.00
	Climate	Bonus indicators	1	1.00
		Penalty indicators	2	0
	Humanistic Environment (3 Points)	Cultural heritages and their preservation	1.2	1.20
		City characteristics and intentionality	0.6	0.49
		Harmony between traditional and contemporary architecture	0.6	0.48
		Harmony between architecture and environment	0.6	0.49
	Urban Landscape (3 Points)	Landscape in city's downtown	1.2	0.60
		Community landscape	1.2	0.60
		City image	0.6	0.30
	Total		30	26.32
Resource Carrying Capacity	Total fresh water volume per capita		5	5.00
	Industrial water recycling rate		1	0.56
	Urban land per capita		2	2.00
	Food supply safety		2	1.00
	Bonus or penalty indicators	Bonus indicators	1	0
		Penalty indicators	2	0
	Total		10	8.56

Table 11.1 (*cont.*)

Goals	Sub-Goals	Indicators	Points	Points Earned
Life Convenience	Urban Transportation (6 Points)	Residents' satisfaction rate with urban transportation	1.2	0.78
		Road area per capita	0.6	0.60
		Share of public transport	1.2	1.18
		Average commuting time (one-way) of residents	1.2	1.06
		Social parking rate	1.2	0.97
		Accessibility of public transport to other cities, towns, and tourist attractions	0.6	0.60
	Commercial Services (3 Points)	Residents' satisfaction rate with commercial services	1.2	0.91
		Area of commercial facilities per capita	0.6	0.60
		Rate of public facilities in residential areas	0.6	0.45
		Rate of residential areas with supermarkets within 1,000 meters	0.6	0.47
	Infrastructure (6 Points)	Residents' satisfaction rate with infrastructure service quality	2.4	1.94
		Penetration of city gas	0.6	0.40
		Coverage of cable TV	0.6	0.59
		Penetration of internet cable	0.6	0.24
		Normal supply of tap water	0.6	0.60
		Normal supply of electricity	0.6	0.60
		Site visit: reasonable distribution of environmentally-friendly public toilets	0.6	0.30
	Culture and sports facilities (3 Points)	Rate of communities with primary schools within 500 meters	0.6	0.37
		Rate of communities with middle schools within 1,000 meters	0.6	0.42
		Number of public libraries, culture centers, and science museums per 10,000 people	0.6	0.50

Table 11.1 (*cont.*)

Goals	Sub-Goals	Indicators	Points	Points Earned
		Rate of communities with free public sports facilities within 1,000 meters	0.6	0.45
		Residents' satisfaction rate with educational, cultural, and sports facilities	0.6	0.46
	Open green space (3 Points)	Residents' satisfaction rate with distribution of open green spaces	1.2	1.00
		Rate of communities with over 2 square meters of green land per capita	0.9	0.30
		Rate of communities within 500 meters of free public parks	0.9	0.63
	Urban housing (6 Points)	Housing area per capita	1.8	1.80
		Rate of families with less than 8 square meters of housing area	2.4	2.40
		Percentages of commodity housing, low-rent housing, and affordable housing within total housing	1.8	1.80
	Public hygiene (3 points)	Residents' satisfaction rate with public health service	1.2	0.96
		Coverage of community health services	0.9	0.76
		Average life span indicator	0.9	0.90
		Penalty indicators	1	0
	Total		30	25.04
Public Security	Lifeline engineering in good order		4	3.14
	City government's facilities, mechanisms, and plans for natural disaster prevention and response		2	2.00
	City government's plans and mechanisms for preventing and dealing with man-made disasters		2	2.00
	Number of successful public safety cases handled by the city government in the last three years		2	2.00
	Penalty indicators		1	–1.00
	Total		10	8.14
Total				83.82

Table 11.2: Green city index system and weights of indicators.[2]

Index	Index	Indicators
Environmental Health	Environmental load	Industrial COD discharge density Industrial helium-nitrogen discharge density Industrial SO2 discharge density Industrial nitrogen oxides discharge density Industrial smoke discharge density
	Quality of environment	Industrial waste water discharge compliance rate Industrial SO2 discharge compliance rate Industrial smoke discharge compliance rate Industrial dust discharge compliance rate Percentage of number of days with air quality of or better than Level 2
	Management of environment	Treatment rate of urban waste-water Rate of safe disposal of domestic waste Percentage of investment on management of the environment
Resource Conservation	Resource supply	Population density Volume of water resources per capita Penetration of tap water in urban areas
	Efficiency of resource use	Unit GDP energy consumption Unit gross industrial output value water consumption Volume of water consumption per capita Volume of electricity consumption per capita
	Recycle economy	Industrial SO2 discharge intensity Industrial COD discharge intensity Industrial nitrogen oxides discharge intensity Industrial ammonia discharge intensity Comprehensive recycling of industrial solid waste

[2] Shi Minjun and Liu Yanyan, "Green City Index: Evaluation of Environmental Performance in City Development," *National Economic Geography*, 2011:2, 101–107.

Table 11.2 (*cont.*)

Index	Index	Indicators
Low-carbon Development	Low-carbon economy	CO2 emission intensity of industrial enterprises above designated size Unit GDP electricity consumption Low-carbonization: change in carbon emission intensity over the past five years
	Emission-reducing practices	Whether low-carbon city planning is developed Intensity of emission-reducing policies: low-carbon economy pilot cities
Ecological Livability	Green space	Area of green land per capita Greening rate of the built-up area
	Comfort and convenience	Housing area per capita Number of buses per capita Density of urban drainage pipes Number of primary and middle school teachers per capita Number of doctors per capita

III. *Urbanization with Chinese Characteristics and the Building of Livable Cities*

China's conditions for urbanization are complex. On the one hand, the continuous and rapid economic growth and industrialization has an increasingly strong pull on urbanization; on the other hand, the urban-rural migration, and cross-regional flow of hundreds of millions of rural surplus laborers, serves as an impetus for further urbanization. All the while, globalization and information technology are having a large influence on the urbanization of developed coastal areas. China's rapid urbanization is facing a complex set of problems involving resources, the environment, society, and industry. Researchers are further motivated to create a path catering to China's current set of circumstances.

A. *A New Industrialization Model*

China's urbanization has to first take the road of "new industrialization." New industrialization is driven by, and in turn promotes, information technology, with high science and technology content, good economic returns, low resource consumption, less environmental pollution, and full play of China's human resource advantages. Compared with traditional industrialization, new industrialization is marked by three distinctive characteristics. The first is industrialization stimulated by information technology, which enables leapfrog development. With technical advances and innovation as impetus, and focusing on development of technology as well as the quality of laborers, new industrialization can acquire more market shares with products of high quality, and low prices, amid fierce competition. The second is industrialization that strengthens sustainable development. New industrialization can greatly enhance China's capacity for sustainable development and economic stamina. This occurs by emphasizing ecological development, environmental protection, and a well-handled relationship between economic development, population, resources, and the environment. This reduces resource consumption and environmental pollution, providing strong technical support. Lastly is industrialization that fully brings into play the advantages of China's human resources. This is one of the gauging standards for new industrialization, along with high science and technology content, solid economic returns, low resource consumption, and less environmental pollution.

B. *A New Model of Urbanization with Chinese Characteristics*

In its quest for modernization, China is faced with the choice of becoming the world's factory or taking the road of new industrialization. A new model of urbanization with Chinese characteristics must be established through the embracing of ecological and low-carbon development.

The first component is that urbanization must be sustainable and intensive, adhering to the basic principles of environmental protection and conservation. It should emphasize the coordinated development of urbanization, population, and the environment, with reasonable and intensive use of resources, and it should preserve the ecological, historical, and cultural environment.

The market must play an important role in promoting urbanization. Consideration should be taken of the level of economic and social development, market conditions, and the extent of social acceptance. By doing so, the market can realize the effective distribution of various resources in the process of urbanization, attract all necessary production elements to cities, while simultaneously bringing into play macro-control by the government, strengthening and improving its management, guidance, and the regulating of urbanization.

A diversified road of urbanization must be taken, promoting a coordinated development of all levels of cities and townships to form a sound city system, improve the comprehensive capacity of cities and towns, and bring into play their functional roles in a certain region.

Appropriate goals, strategies, and policies for urbanization must be defined according to levels of social and economic development, regional characteristics, resource reserves, and an environmental basis of different cities. Economic connection and coordination between cities should be strengthened to achieve complementary and common development.

Four key issues stand at the center of new urbanization suitable for China. First, the migration of rural populations into cities for employment and settlement should be the core. Second, metropolitan coordinating regions should be the main form. Third, optimizing existing architecture and space structures should be the precondition of the urbanization process. Fourth, system reform, policy improvement, and sound planning should be the focus.[3] With China's current economic restructuring and change in its economic growth pattern, new Chinese urbanization not only helps to solve the problems and conflicts inherent in China's social and economic development, but also fits the requirements posed by the change in impetus for urban-rural development in the economic restructuring. Additionally, it is compatible with the basic requirements of China's conditions and socio-economic development.

Overall, the development of livable cities is actually a shift of emphasis in urbanization from numbers and scale, to quality. From economic development to coordinated social and economic development, and from emphasis on city development to integrated urban-rural development. While maintaining a necessary speed of urbanization, the development of livable cities should enhance the ability of industry congregation to promote population congregation, ensure the healthy development of urban-rural industries, hold a continuous increase of employment opportunities, create a sound social security system, and improve the conditions of living, traffic, education, culture, hygiene, and the environment. The development of livable cities also requires that urban-rural planning, construction, and management levels be upgraded, infrastructure and public service facilities improved, comprehensive and competitive capacity enhanced, institutional hindrances between urban and rural

[3] Yang Weimin, "Four Key Issues of Urbanization with Chinese Characteristics," *Studies on Urban and Regional Planning*, 2008:1(2), 17–25. Beijing: Commercial Press.

areas removed, and the gap in development gradually narrowed. The development of livable cities should realize the basic goals of sustainable human settlement development, as laid out in *China's Agenda* 21, improving human settlement environments, providing adequate shelter for all, promoting the sustainable development of the construction industry, and improving the energy efficiency of buildings.

THE WASTE CRISIS: SEEKING A NEW DIRECTION IN THE DILEMMA

Yang Changjiang

Abstract: 2010 witnessed an escalation in China's ongoing waste crisis. Although waste separation is becoming prioritized, limited progress has been made in the transport of sorted waste, and mixed processing remains a common practice. The government, focusing on terminal processing, emphasizes waste incineration as a means of power generation. The public, meanwhile, values front-end waste separation and collection. Incidents of leachate pollution highlight a deep conflict in waste processing, while a lack of open information on dioxin and heavy metals indicates likely hidden problems. The three-decade-long practice of mixed landfills is largely responsible for today's crisis.

Keywords: waste crisis, dilemma, new direction

On March 5, 2010, Premier Wen Jiabao proposed "Actively promoting sewage and garbage processing in cities and towns and actively utilizing resources from domestic waste." This marked the beginning of the issue being taken seriously by the nation's political and economic powers.[1] Beijing, Guangzhou, and other cities have since begun new waste separation practices, and collected comments and suggestions on local regulations. We have also seen new efforts from environmental organizations and enterprises. But progress on front- and mid-process separation and transportation is slow and difficult, while terminal processes—such as power generation from waste products—have developed rapidly. Current problems with mixed landfills still exist; and an incineration crisis is on the way. What is the crux of the problem?

I. *Waste Separation Caught in a Dilemma*

In 2010, several cities began experimenting with waste separation, addressing both governmental and public concerns. On January 26, Guangzhou launched a domestic waste separation program in part of the city's Yuexiu District (越秀区). Unfortunately the residents' efforts proved useless, as

[1] "Government Report by Wen Jiabao at the Third Session of the Eleventh National People's Congress," *People's Daily*, March 6, 2010.

separated and non-separated wastes were mixed in compactors by workers. In addressing the issue, Zhang Jianguo (张建国), associate director of the Guangzhou Municipal Administration Committee (广州市城管委), announced plans to introduce new garbage trucks; but he did not announce a detailed timetable. The enthusiastic public soon found their efforts going nowhere.

In 1957, Beijing was the first city in China to propose that: "Waste should be classified before collection." Half a century has passed, yet waste separation is still at the experimental stage. On April 11, the Beijing Municipal Office for Building a Civilized Society (北京市文明办) and the City Appearance Committee (市政市容委) formally launched a program under the headline "Waste reduction and waste separation start with me!" Six hundred residential communities and 30% of party and government offices carried out trials aimed at improving standards. All kinds of facilities for food waste classification, collection, transportation, and treatment achieved their targets on waste separation by the end of the year. But reporters who made secret visits to these so-called "model residential communities" found that they were "nothing but fake." They found that, while visitors would be shown great efforts at addressing the problem during tours, such actions ceased as the tours finished. One reporter, Chen Yuanyuan (陈媛媛), described the situation as thus:

> What the public faces, after all their enthusiasm was aroused, is a disorganized collection and transportation system for domestic waste separation. Mixed waste collection and transportation are still the main practice. The waste energetically and enthusiastically separated by the public is mixed again in one vehicle for further processing. Waste separation and recollection have become a showpiece directed by a number of government departments.[2]

Mixed waste transportation produces great difficulties for classified collection. Zhou Hongchun (周宏春), a researcher from the State Council Development and Research Center (国务院发展研究中心) expressed just such doubts: "Can separation solve the municipal waste problem? The answer is negative."[3] And Xu Haiyun (徐海云), chief engineer at the Chinese Architecture Research Center, (中国建筑研究院) held the view that, "Classified collection, in particular sorted collection of waste from

---

[2] Chen Yuanyuan: "Waste Separation Should Not Be Just a 'Show'," *China Environmental News*, May 5, 2010.

[3] Zhou Hongchun, "Is Waste a Resource Put in the Wrong Place?" *China Economic Times*, April 8, 2010.

direct collection, can only be comparatively limited.[4] The domestic waste in China is not the same as the domestic waste abroad. In our country, domestic waste is classified the same as 'residue waste' in Germany. The re-collection value is low, despite the fact that it contains plastic bags and waste paper."[5] There *have* been attempts at addressing this, though. The Beijing Green Space Biology Science and Technology Corporation (北京绿色空间生物科技公司) produced high-speed fermentation food waste processing machines, and turned food waste from the Beijing Yuying School (北京育英学校) and Beijing West Hill Garden Community (北京西山庭院小区) into green organic fertilizer. The Beijing Zhihuiyihe Environmental Protection Company (北京智汇谊合环保公司) has also produced food waste processing machines—300 of which were purchased by a Hong Kong company. And the Beijing Hongxiang Taihe Company (北京鸿翔太和公司) made use of bits of wood and plastic waste to produce high strength model-board for building construction. But among the waste processed by the Beijing Asuwei Waste Comprehensive Processing Factory (北京阿苏卫垃圾综合处理厂), only 10% is suitable for waste incineration.

And there might be an even more profound reason for the slow progress in waste separation: government policies that favor terminal treatment. One report explained it thusly, "For the waste collection companies, if the standard is more detailed and results in better practice among people and more waste being recycled, companies will get fewer subsidies. This is the primary reason why waste separation failed to be strongly promoted in the past few years."[6] Over the 30 years since environmental protection reform started in 1979 (when responsibility for environmental protection was shifted from the Ministry of Health to the Ministry of Housing and Urban-Rural Development), classified and separated collection, transportation, and processing of waste has seen little progress despite a worsening crisis.

---

[4] Xu Haiyun, "How Should we Promote Domestic Waste Separation?" *China Solid Waste Web*, http://news.solidwaste.com.cn/k/2010–11/20101127911229752.shtml, October 27, 2010.

[5] Nie Yongfeng, "Our Country's Domestic Waste Should Not Be Compared with that of Abroad," *China Solid Waste Web*, http://news.solidwaste.com.cn/k/2010-8/20108271354478103.shtml, August 27, 2010.

[6] "Waste Supports Cities," *Teng Xun News*, http://news.qq.com/photon/tpyk/laji.htm.

II. *Expert Opinions on Waste Incineration and the Government Sets the Tone*

Solving the waste crisis is now a matter of great urgency. Experts on incineration and government officials alike are calling for accelerated promotion and implementation.

On February 11, Xu Haiyun delivered a lecture to over 1,200 people in positions of responsibility for Guangzhou city and district administration bureaus, as well as people in charge of local communities, towns, and villages. He talked about the advantages and necessity of waste incineration. On August 18, he gave a lecture on waste processing to around 450 people from the Dongguan Municipal Administration Bureau (东莞市城管局) and environmental protection departments. He pointed out that as the land needed for construction shrinks, and landfill methods remain out of date, waste incineration provides a reasonable alternative. "Waste incineration is better than landfills because not only can it solve the urban crisis, but the heat produced from incineration can also generate electricity." Professor Nie Yongfeng (聂永丰) from the Environmental Engineering Department of Tsinghua University expressed the view that "Land resources in big cities are limited, as there are more people and less land. Waste incineration should necessarily play an important part in waste processing and reduction. Whether you like it or not, the country's reality has determined this." He added that, "Waste is a kind of resource put in the wrong place. By using waste incineration to generate electricity this kind of resource can be utilized to the utmost extent." Wang Jiuhua (王久华), CFO of the Sino-German Group (中德集团) adds that, "In developed countries, 70% of total waste is processed through waste incineration to generate electricity. But in China, the percentage is only seven."[7] He believes that waste incineration is the only practical method for dealing with the problem at present in China—even if doubts and questions remain.

Official media has continued the push. A February 26 meeting of experts looking at Guangzhou's domestic waste treatment and headed by Nie Yongfeng and Xu Haiyun concluded that the city was suitable for a domestic waste treatment model mainly focused on waste incineration supplemented by landfills. They urged speeding-up the construction of a

---

[7] Xu Pei, "The Bundled Up Prometheus," *Beijing Weekly Web* http://www.beijingreview.com.cn/2009news/guonei/huanbao/2010-08/09/content_289544.htm, August 9, 2010.

modern domestic incineration plant. Only one expert of the 32 participants expressed doubts about focusing on incineration—pointing-out that there were risks in current waste incineration technology and saying that Guangzhou should not blindly choose the method. The *Yangcheng Evening News* carried an eye-catching headline "31:1! Supporters for Waste Incineration Win!"

Meanwhile, government departments launched intense publicity campaigns. Zhang Lijun (张力军), vice minister of the Ministry of Environmental Protection, stated on March 10 during the period of the "Two Meetings" (两会) (the People's Congress and the Chinese People's Political Consultative Conference): "We support waste incineration factories that have met standards for pollution discharge. The problem on opacity of dioxin can be solved." On March 11, a seminar on comprehensive treatment for local and domestic pollution was held in Shenzhen. Wu Xuelong (吴学龙), a member of the Shenzhen Municipal Administration Bureau and Shenzhen Environmental Protection Bureau suggested:

> The Ministry of Housing and Urban-Rural Development should publicize the *Guide on Domestic Waste Treatment Technology* (生活垃圾处理技术指南) to provide guidance for the nationwide construction of waste processing facilities, so as to turn around the poor situation of 'practicing waste separation with no waste processing facilities established.' Provincial environmental protection departments should further clarify criteria for waste incineration regarding environmental impact assessment in order to fully address the difficult situation in environmental impact assessment.

On March 12, Vice Minister Qiu Baoxing (仇保兴) of the Ministry of Housing and Urban-Rural Development said in an interview with China Central Television (CCTV), that the country had a large population with insufficient land. This, he said, was particularly true in areas near big cities, where land resources are sharply limited. He concluded that every inch of land needed to be utilized, and thus, waste incineration was a must. This means incineration will become the mainstream in waste processing. In defending this he said, "Some people are crying that the waste processing factories will cause a negative impact on the local environment. They are misleading the public!"[8]

On April 22, the Ministry of Housing and Urban-Rural Development, the National Development and Reform Committee, and the Ministry of

---

[8] Ling Qinghui, et al., "Housing And Construction Bureau Saying that Waste Incineration Is The Direction For China Waste Treatment," *Central TV Web*, http://news.cctv.com/china/20100402/104360.shtml, April 12, 2010.

Environmental Protection, jointly issued the *Guide on Domestic Waste Processing Technology* (生活垃圾处理技术指南), which states that "waste incineration technology should be used in areas with limited land resources, which require the heat energy from burning domestic waste." One netizen with the name of "Lüshidaner" (驴屎蛋儿) wrote online: "The signal flare for incineration has been launched!" Another netizen going by "Lie Feng" (烈风) said: "This guide is equivalent to law." The *Guide* is the legal basis provided for all levels of government departments to promote waste incineration. On June 18, the Ministry of Environmental Protection issued the draft of *Opinions on Enhancing Comprehensive Treatment of Domestic Waste and Pollution* (关于加强生活垃圾处理和污染综合治理工作的意见(征求意见稿)) by the three ministries mentioned above seeking suggestions from the public. Friends of Nature, an environmental NGO, pointed out that the opinions still focused on "terminal treatment" and that the meaning of "clean incineration" was not clear. This leads to doubts about practical operation and the utilization of such systems.

Government policy forged ahead in spite of the concerns. On October 30, CCTV's news channel broadcast a special feature, "Incineration—Searching for a Way Out of Waste." This was seen by most as the official and final tone on waste incineration.

III. *Waste Incineration Strongly Promoted and Implemented on a Large Scale*

Both official opinions and informal signals indicate that waste incineration plans represent a profound shift for China. During the period of the Twelfth Five-Year Plan the nation invested an unprecedented 800 billion RMB in the solid waste industry. There's no doubt that this was viewed as a powerful action. According to Li Yangdan (李阳丹):

> With the nation's strong support for the strategic development of newly emerging industries, the solid waste industry entered into a golden age. The market capitalization for the municipal solid waste processing industry has the highest prospects among these industries. In the next decade, China's waste incineration will increase from the present 2%–3% of total waste treatment to 10%–20%. The annual investment volume for electricity generation from waste will reach 80 billion RMB.[9]

[9] Li Danyang, "City Solid Waste Treatment Industry Welcomes the Rapid Development Period," "The Investment in Solid Waste in the Twelfth Five-Year Plan Reaches 800 Billion RMB," *China Security News*, March 14, 2010.

It is clear that the nation has attached never before seen importance to the industry, and that the market is responding.

On May 10, the Beijing Nangong Domestic Waste Incineration Factory (北京市南宫生活垃圾焚烧厂), with a daily processing capacity of 1,000 tons, won the design bidding process and began construction.

On July 31, in his "Progress Report on Domestic Waste Treatment" (关于生活垃圾处理工作进展情况的报告) to the Beijing People's Congress, Chen Yongxiang (陈永向), director of the Beijing Municipal Administration and City Appearance Committee, emphasized the urgency of waste processing: "Waste will be unable to be processed if Beijing does not establish a waste incineration plant. The development of waste incineration cannot afford to be delayed by controversy or other obstacles. It must start immediately without any delay." His advice was heeded. On October 23, the construction of the Lujiashan Waste Separation, Incineration, and Power Generating Plant (鲁家山垃圾分类处理焚烧发电项目), with a daily processing capacity of 3,000 tons—the biggest in Asia—got under way. Total investment in the project reached 2.16 billion RMB, while investments in environmental protection were 420 million RMB. The planned operational period is 30 years. On October 16, Beijing's Haidian District (海淀区) announced a 1.26 billion RMB investment in a waste-processing factory at Sujiatuo (苏家坨) with a daily incineration capacity of 1,800 tons.

Netizen "Lushidaner" was surprised—expressing opposition to the project: "China is not yet ready to take a great leap forward in setting up such a waste incineration project. There remain unresolved questions over technology, waste output, flawed regulations, and a lack of supervision." Zhao Zhangyuan (赵章元) also lent his voice to the criticism, saying, "In a few years, the Beijing city environment will be completely destroyed by the nine waste incineration facilities. Environmental destruction will inevitably occur in the capital, Beijing."[10]

Despite such protests, there was a strong push for power generation from waste at the beginning of 2011, and surprising growth in the time since.

According to the "List of Newly Developed Projects on China's Energy Resources in 2010," 41 projects have been proposed on the construction of waste power generation stations in 16 provinces, cities, and autonomous

---

[10] Zhao Zhangyuan, "The Nine Waste Incineration Stoves Will Deteriorate the Living Environment In the Capital," *People's Web*, http://env.people.com.cn/GB/12370308.html, August 6, 2010.

regions.[11] In Dongguan, Zhong Yaoxiang (钟耀祥) from the Municipal Comprehensive Administration Bureau (城市综合管理局) stressed the urgency of solving problems in waste processing. "The problem concerning waste processing is becoming extremely urgent and must be solved quickly." At that time, the sites for four new waste processing factories in Humen (虎门), Qingxi (清溪), Mayong (麻涌), and Changping (常平) had been chosen, and the construction of new waste incineration factories would soon go ahead. On October 24, the *Proposal for Accomplishing the 'Four Great Developments' through the Implementation of The Essentials in the Reform and Development Plan for the Pearl River Delta (2008–2020)* (实施<珠江三角洲地区改革发展规划纲要(2008–2020年)>实现"四年大发展"工作方案) was formally released. Eighty-four important projects were included in the four years of development, including 15 waste incineration power generation factories to be built by 2012 in Guangzhou's Likeng (李坑), Foshan's Nanhai (佛山南海), and other locations with a total investment of 20 billion RMB.

How many waste power generation factories does China need? On March 25, Cui Yan (崔焱), vice general manager from the Sinochem International (中化国际招标有限公司) estimated that according to current city waste production volumes (roughly 1,000 tons per day) the nation would need 600 to 700 waste incineration plants with a capacity of 1,000 tons each. Only 70 factories were in operation at the time. The large shortfall suggests great business opportunities. On November 6, a special feature called "Fight against Waste for Everyone" on CCTV's News Weekly said that, "By the end of last year (2009), 1,000 projects for large scale incineration plants had registered with the National Development and Reform Committee."

A netizen named "Bai Fuqin" (百扶勤) estimated that 1,000 waste incineration factories would require a total investment of 6–8 trillion RMB, based on estimates for plants with capacities of 1,000 tons. "It seems that the country could regard the waste incineration industry as one of the most important industries for expanding domestic market demand. This equals the real estate industry, automobile industry, or the express rail industry. This scheme is too crazy! Who has made this mistake, the National Development and Reform Committee or the Ministry

[11] Liang Zhongrong, "41 to Start Operation In 2010: Waste Power Generation 'On Interlinked Ring'," *21st Century Economic News*, January 30, 2010.

of Housing and Urban-Rural Development?"[12] He then made a dire prediction: "Another poisonous bubble in the national economy will emerge. The State Council should put a halt to this crazy scheme to establish 1,000 large scale waste incineration plants!"

IV. *Waste Pollution Leading to Incidents and Mass Protests*

In 2010, there were an increasing number of incidents and mass protests caused by pollution from waste incineration and landfills. On the morning of January 7, a cold-water pipeline at the Likeng Domestic Waste Incineration Power Plant (李坑生活垃圾焚烧发电厂) in Guangzhou suddenly exploded, leading to five workers being scalded and injured—two gravely. On April 17, the Guangzhou Likeng Waste Incineration Power Plant was opened to the public for the first time. The villagers in Likeng were dissatisfied with the "staged visits"—besieging the plant for over three hours. Han Zhipeng (韩志鹏), a member of the Guangzhou People's Political Consultative Committee appealed for support on the Internet. In the end, the "visitors" left under the protection of the police. On the morning of May 16, over 500 villagers from Shixin Community in Dongwan (东莞市石新社区) held a walking protest in the town's four main crossings. This resulted in massive traffic jams which spread to Huangjiang (黄江), Tangxia (塘厦), and Changpin. On the same day, over 100 villagers dissatisfied with the site chosen for the construction of another waste incineration plant in the Rome Community of Qingxi (清溪罗马社区) held a "Group Walk" at the bus station in town.

On June 17, leachate from the waste incineration plant at the Environmental Resource Processing Center (环卫资源处理中心) in Chao'an County (潮安县), Guangdong flooded farmer Xiao Zhang's (小张) fishpond in the town of Caitang (彩塘镇). In one day, this killed over 5,000 kilos of Grass carps raised over the course of three years. Adding in the 19 *mu* of vegetables that were destroyed in 2009, Xiao Zhang asked for a compensation of 150,000 RMB. The waste incineration plant responded, saying that they should not be held responsible for the losses and should not be required to compensate the farmer.

On July 3, around 800 people from Beishan (北山), Shaotian (绍田), and other villages in Lingchuan County (灵川县) in Guangxi's city of Guilin

[12] Bai Fuqin, "The Plan Is Too Crazy! Who Is Wrong: The Development and Reform Committee or the Accommodation and Construction Ministry? Or Bai Yansong?" http://www.rg-gd.net/viewthread.php?tid=229879, October 9, 2010.

went to visit the Lingchuan County government to protest a decision to place a domestic waste landfill in Beishan. Five hundred people were persuaded to return before reaching their destination, but the rest continued and by noon, over 300 had arrived at the offices of the Lingchuan County government to protest. Later, over 200 people continued on to the Guilin government offices for further protests. On July 24 and 25, hundreds of people from Ziyuan Village (紫园村) in the city of Liu'an (六安市) in Anhui Province twice visited the township government to protest landfill pollution. The group disrupted the National Highway 206 and disturbed the work of the township government. On August 9, villagers from Qutan (衢潭村), Tongshan Town (通山县) in Hubei Province prevented the environmental protection department from dumping waste in Liangtingling (凉亭岭). Over a hundred tons of daily waste in Tongshan Town could not be cleared and transported. The villagers suspected that waste pollution had produced a dramatic spike in diseases, as 37 of the villagers there had died of cancer since 1994. Villagers in Houlong (后龙村), in the city of Ningde (宁德市) in Fujian Province who could no longer stand the waste landfill pollution, made visits and lodged complaints to the government departments concerned. They also traveled to the plant on a number of occasions, trying to prevent construction and halt waste from being dumped on the spot. On October 18, Zhang Changquan (张长全) and another five villagers went to visit the Position Letters Bureau of the Fujian Provincial Government (福建省政府信访局), but were stopped on the way by people sent from Pingcheng Town's (屏城乡) government. On October 19, they were arrested by police in Pingnan Town (屏南县) as criminal offenders under the charge of "gathering to disturb social order."

People were generally left unsatisfied with the responses to their actions and demands for compensation. Further, the courts did not support their lawsuits. Zhao Lei (赵蕾), a resident of Beijing, took legal proceedings against the Gao'antun Medical Waste Incineration Plant (高安屯医疗垃圾焚烧场) charging violations of human rights for health and clean air. The Chaoyang District Court of Justice maintained that causes of disease were many and that other causes apart from those produced by the plant should not be excluded. The ruling held that Zhao Lei, who had chosen to purchase a nearby apartment, should bear the obligation to tolerate the existence of the waste landfill. His appeal was turned down.[13] Netizens

[13] Wang Dianxue, "Residents Sue the Waste Spot Stinking And Causing Disease, the Court Says that Residents 'Have the Responsibility' to Tolerate It," *Xin Jing News*, May 22, 2010.

commented: "In asking the residents to tolerate poisonous gas, the verdict actually deprived the residents of their rights to health and life." Yang Zi (杨子), who could not tolerate the offensive smell from the same plant, sued the Beijing Environmental Protection Bureau. Yang questioned whether the bureau had illegally approved the construction of the waste incineration plant. The first verdict from Haidian District Court was that Yang Zi, who lived 2.5 kilometers away from the waste incineration plant—a distance beyond the 800 meters set in the *Regulations on Medical Waste Treatment* (医疗废物管理条例)—was not qualified to launch the proceedings. Thus, the court turned down that appeal as well.

Disappointed with the courts' rulings, both refused to take their cases to the higher courts. Ultimately, though, the two separate cases against the Gao'antun Medical Waste Incineration Plant were turned down. The difficulty in ensuring citizens' environmental rights results both from partiality in the government's enforcement of laws and the weaknesses in the legal justice system.[14] Inaction by administrative bodies and unjust legal applications have posed great challenges for citizens who wish to maintain their environmental rights. This has pushed the increase in riots and other threats to social harmony. As people's faith in government is drained, society is left with hidden dangers that pose a consistent, but uncertain threat to social order.

V. *Without Truth, What Will Be the Future of Waste Treatment?*

Zhang Lijun, vice minister of the Ministry of Environmental Protection, stressed that "The problem of opacity on dioxin can be resolved," and Qiu Baoxing, the vice minister of the Ministry of Housing and Urban-Rural Development said that the waste processing plants "are completely up to national or even international standards." But reports from the Chinese Academy of Sciences (CAS) have revealed the truth. The Dalian Chemical and Physical Research Academy (大连化学物理研究所) of CAS tested and analyzed the emissions for dioxin opacity from the domestic waste incineration plants in 19 cities. They found that the average opacity of dioxin emitted from 19 enterprises reached 0.423ng TEQ/m3, four times higher than the European Union's standards of 0.1ng TEQ/m3. The emissions

[14] Fu Dalin, "Why The Court of the Environmental Protection Bureau 'Takes Care Of' the Waste Incineration Factories," *People's Web*, http://env.people.com.cn/GB/11668225.html, June 26, 2010.

data released from the Gao'antun Waste Incineration Plant did not include opacity of dioxin and heavy metal.[15] Residents were skeptical and asked, "How do we know that the data are true?" On October 19, nine ministries including the Ministry of Environmental Protection issued the *Guiding Principles on Strengthening the Protection against the Opacity of Dioxin Pollution* (关于加强二恶英污染防治的指导意见). This was looked upon as a new policy to control the opacity of dioxin pollution. But how can we talk about "control" if no true information regarding the polluting plants is disclosed? Which of the 19 waste incineration plants disobeyed the emission standard? "I won't tell anyone" was the answer from Chen Jiping (陈吉平), the person responsible for the project. Mao Da (毛达), a PhD candidate in Environmental History at Beijing Normal University said: "In the emission control over opacity of dioxin, the problems lie in the lack of integrity of the enterprises and the lack of public faith in the supervision of the environmental protection departments. The participation of the public is the only final motivation for the protection against opacity of dioxin emissions."[16] At present, though, public supervision is out of the question.

In September 2010, media in Beijing and Shenzhen repeatedly disclosed the covered up emission of large amounts of leachate from waste landfills and waste incineration plants. The Guozhong Biology and Technology Corporation (国中生物科技有限公司), responsible for waste separation at the Beijing Asuwei Waste Processing Plant, directly dumped large amounts of daily-produced leachate into city wells.[17] The corporation paid the No.4 Clearance Yard (清洁四场) 70 RMB per ton in "transportation fees" to handle the leaked leachate. They were thus paid 20,000 RMB per day in such costs—a draw of around 6 to 7 million RMB per year. In Shenzhen, waste leachate from the Nanshan Waste Incineration Power Generation Plant (南山垃圾焚烧发电厂) and the Bao'an Laohukeng Waste Incineration Power Generation Plant (宝安老虎坑垃圾焚烧发电厂) was dumped directly into rivers and the sea without being treated. This led to severe pollution of the environment.[18] And a large number of waste incineration power generation plants in Shenzhen were not

---

[15] Xu Zhiguo, "Arguments On Waste Incineration," *Science News*, Nov. 8, 2010.

[16] Mao Da, "Answer To The True Problem—The Fourth Comment On the Guide on Opacity of Dioxin," Blog, http://elephantmao.blog.sohu.com, October 9, 2010.

[17] Zhang Yongsheng, "Why The North River Turned Into a Black River," *Beijing Morning*, August 5, 2010.

[18] Cheng Xi, "The Sewage From The Waste Power Generation Factories Was Secretly Dumped Into The River and Sea," *South City News*, September 13, 2010.

even equipped with waste leachate treatment facilities before beginning operations.

Such leachate does great harm. It produces at least one cancer-inducing virus and five supplementary cancer-inducing viruses. If it is leaked into underground water discharge systems, it can cause very wide scale contamination. If it is leaked into the ground, it can cause pollution of a scale and nature that takes many years to clean up. Disturbingly, today it is unknown how many waste landfills and incineration plants are secretly dumping their leachate. Mao Da once asked the Beijing Environmental Protection Bureau to publish the "Risk Assessment on the Pollution from Beijing Domestic Waste Landfill Plants." He got the following answer: "The information is categorized as confidential, thus it is not allowed to be disclosed" because the information "needs further investigation and analysis."[19] It is unknown right now whether these documents hold key evidence to a hidden environmental crisis.

In 2002, at the invitation of the Beijing Municipal Management and Administration Bureau, the Chinese Geophysical Society (中国地球物理学会) tested leaking pollution in the three biggest waste landfill sites and found that the problem of leakage was very common in Beijing's waste landfill sites. The Beijing Municipal Management and Administration Bureau held a positive view towards the result and expressed that they would continuously work on it to find a quick solution.[20] But Wang Weiping (王维平), senior engineer of the Beijing Municipal Management and Administration Bureau, stridently argued at a later news conference that: "The result from the assessment has confirmed that not one of Beijing's waste landfill sites will cause any underground water pollution, neither now, nor in the future."[21] The serious environmental problem in Bejing has been covered up for a long time now, and thus the ideal window to solve the capital's problem with underground contamination has been missed. From incidents like pollution from opacity of dioxin, heavy metal, and leachate, how much has been covered up in the Beijing waste landfill and processing industry? How much further can an industry proceed if the truth is covered up?

---

[19] Mao Da, "Why It Is So Difficult to Publicize Environmental Information," *Southern Weekly*, June 30, 2010.

[20] Zhao Zhangyuan, "Questioning the City Administration Committee" Blog, http://zhaozhangyuan.blog.sohu.com/144523264.html, February 20, 2010.

[21] Huang Jianhua, "Beijing Underground Water Is Polluted By Waste," Beijing Youth, March 12, 2010.

The year 2010 was a line of demarcation, indicating that the country had entered into a period of dual crises of mixed landfills and incineration. Since 1979, the administration of environmental sanitary work has been handed from the sanitary departments to construction departments. In the three decades since opening reforms, accumulated problems have resulted in a crisis impacting everywhere from cities to villages. An even deeper crisis on a much larger scale might have just started. Waste incineration has reached a level of "being practiced everywhere, as crazy as in the U.S. previously, and perhaps even worse. But the difference is that the U.S. has come down from its madness, while we are still in the middle of the craze."[22] If indeed the plan to construct 1,000 new waste incineration plants goes through, it is hard to imagine how China can possibly control its pollution.

Since 2010, the choices of solutions to the waste crisis have differed greatly between the government and the people. It may be that in the future, hope might be brought about by the unremitting efforts of those enthusiastic residents who have persisted in waste separation. But we might also look elsewhere—to creative waste recycling processing enterprises that have yet to catch social attention, or NGOs that continue to push for the font-end waste reduction.

[22] Mao Da, "The Craziness Of Waste Incineration In American History," *Interdynamic Encyclopaedia*, http://www.hudong.com/e61d7265dc4c47c2b13ebb8222b3b51d.html, February 1, 2010.

2010: A RESTLESS YEAR FOR ENVIRONMENTAL POLICIES ON PACKAGING

Mao Da and Yang Weihe

Abstract: 2010 was a restless year for environmental policies on packaging in China. The government departments responsible for the *Plastic Bag Restriction Policy* were making efforts to fully implement it, while civil society groups tracked and reported its effectiveness. In the late summer, debate was sparked on whether the nine-year-old ban on polystyrene disposable food containers should be lifted. Different opinions were expressed via media channels, providing useful references for policy makers. Regulations against over-packaging were issued. However, according to some civil society groups, amendment and follow-up policies are still needed. In summary, with the positive progress made, a proposal and discussion of an integrated environmental policy on packaging should not be far away.

Keywords: packaging, environmental policy, *Plastic Bag Restriction Policy*, polystyrene food container; over-packaging

Packaging includes the inner and outer wrappings of common merchandise, shopping bags, and disposable food containers. In 2010, three policies responding to the environmental problems caused by packaging generated certain social effects. These policies were the *Plastic Bag Restriction Policy* (塑料购物袋限制政策) that began on June 1, 2008, the nine-year-old ban on polystyrene (PS) disposable food containers, and the regulations against over-packaging. A review of the paths of these policies can help us better understand the present and future of China's environmental policies on packaging.

I. *The Struggle over the Plastic Bag Ban*

In June 2008, a policy approved by the State Council to restrict the production and use of plastic shopping bags came into effect. The policy enforced a national ban on the production, sale, and usage of plastic shopping bags thinner than 0.025 millimeters (referred to as "ultra-thin bags" hereafter), and forbade all retailers from supermarkets and shops to farmers' markets to offer free plastic shopping bags to customers. Now, after more than two years of its implementation, the objectives, measures, and targets of the ban remain the same, with no adjustments made to address the issues that appeared in practice.

Some regulations of the *Plastic Bag Restriction Policy* encountered difficulties when put into practice. The two prominent problems were the wide usage of ultra-thin bags and the free supply of plastic shopping bags by farmers' markets and individual vendors. At the beginning of 2009, the government began to take measures to solve these problems, such as strengthening inspection and supervision. For example, the commercial and industrial bureaus on local levels sent law enforcement staff to inspect the performance of the ban, and to correct and punish wrongdoing.

However, no official information has been released since August 2009 on the performance and effects of the policy. As a result, the public could only figure out its influence from their daily experiences. Faced with the lack of authorized examination and reporting, some individuals and non-governmental organizations did independent investigations and released feedback to the policy makers as well as the public.

In the summer of 2010, Envirofriends Institute of Environmental Science and Technology (环友科学技术研究中心) (Envirofriends), an NGO dedicated to environmental protection, and three other NGOs conducted a joint survey, which showed the changes and problems that emerged since the implementation of the ban.[1] First, most of the supermarket chains charged fees for plastic bags. (More than 60% of the supermarkets on average in Hangzhou, Zhengzhou, and Harbin complied with the policy). However, the overall implementation rate was reduced year by year. Second, individual vendors still tended to offer free plastic bags to their customers, both due to their neglect of the policy and the lack of supervision. Scattered in various places, the individual retailers were hard to organize. The low cost of plastic bags also contributed to the situation. However, there have been positive changes in shopping behavior in the last two years, with four out of ten consumers bringing their own shopping bags to the markets. Compared with the other retailers, supermarket chains show more uniform indications that customers are using their own shopping bags.

---

[1] The three other NGOs are Hangzhou Ecological Culture Association, Green Longjiang (in Harbin, Heilongjiang Province), and Friends of Nature Henan group (in Zhengzhou, Henan Province).

Although the communication between the responsible governmental bodies and the public is far from adequate, the former are making efforts to improve the effects of the policy. On June 1, 2010, the National Development and Reform Committee and three other ministries or commissions jointly issued a notice on the self-examination of the restrictions on the production, marketing, and usage of plastic shopping bags, requiring the governmental bodies at local levels to reinforce the implementation of the *Plastic Bag Restriction Policy*. Later, with the support of policy makers, China Central Television (CCTV) conducted a series of open and secret investigations into the production and circulation of the non-standard plastic bags, especially the ultra-thin ones. By doing so, CCTV exposed illegal producers in its program, "Weekly Quality Report," not only in an effort to help the public understand the harm of unqualified plastic bags, but also to create favorable public opinions for a new round of enforcement of the restrictions. Meanwhile, the industrial and commercial bureaus at local levels attempted to improve the situation in the weaker links, such as the farmers' markets, by rewarding those complying with the regulations and punishing the violators. For example, the Bureau of Commodity Prices in Liuzhou (柳州市物价部门), in the Guangxi Zhuang Autonomous Region included the implementation of the *Plastic Bag Restriction Policy* in an evaluation of commercial integrity.[2] The Industrial and Commercial Bureau of Chaoyang District in Beijing and the organizers of the farmers' markets took various measures to reinforce the ban, in the form of signing an agreement of liability, appealing to the public, providing free recyclable bags, and setting up counters for plastic bag sales. The bureau put those markets refusing to comply with the policy or correct their behaviors on record, and gave punishments accordingly.[3]

The non-governmental organizations paying close attention to the *Plastic Bag Restriction Policy* also initiated some activities. For example, on June 3, 2010, the second anniversary of the implementation of the policy, they invited experts on environmental protection from Taiwan to share their experiences restricting the use of plastic bags. Some other activities included holding a "Q&A on the Plastic Bag Ban" to make the policy

[2] "Plastic Bag Ban in Liuzhou Integrated into Long-Term Management of Authentic Goods Guarantee on Model Street and Integrity in Pricing," *Liuzhou Daily*, October 25, 2010, via http://pic.gxnews.com.cn/staticpages/20101025/newgx4cc53e3b-3355628.shtml.

[3] Lishuang Lin, "Violation of Plastic Bag Ban in Farmer's Market to Face Legal Record and Punishment," *Beijing Youth Daily*, October 22, 2010, via http://bjyouth.ynet.com/article .jsp?oid=70290826.

known to the public and the release of research results, showing the benefits as well as shortcomings of the policy.

II. *Controversy around the Return of Disposable PS Food Containers*

Compared with the *Plastic Bag Restriction Policy*, which aroused few discussions in the public, the controversy over whether or not to lift the ban on disposable PS food containers in the summer of 2010 drew much more public attention.

On August 24, 2010, in a press conference jointly held by the China Industry University Research Institute Collaboration Association (中国产学研合作促进会) (CIUR) and China Plastic Processing Industry Association (中国塑料加工工业协会) (CPPIA), the two associations called on the public to update their information about disposable PS food containers, and claimed it to be a "typical green product." According to the experts attending the event, disposable food containers were both clean and recyclable. Therefore, it was a pity to ban the production, sale, and usage of this type of food container in 2001, when the government called for a stop to its circulation in the food and beverage industry.

This statement met with public outcry. Disposable PS food containers and plastic bags had long been considered the enemy of environmental protection, the embodiment of "plastic pollution," and thus separate from what we usually consider a "typical green product."

Tencent (腾讯网), a well-known media network in China, soon produced three feature reports presenting the views of both the advocates and the opponents of the ban in the form of a "Q&A." The opinions of the advocates of disposable PS food containers were listed in the first report as follows: First, the material, polystyrene, was safe under 70° Celsius, and the temperature of take-away food was usually lower. Second, PS food containers could be recycled and would not generate dioxin when burned. In contrast, the substitute degradable food containers were not environmentally friendly at all. Third, there were already several successful cases recycling the disposable PS food containers, such as the "three fen" (三分钱) recycling project in Shanghai, in which the plastic tableware makers must submit three fen (about 0.005 US dollars) for the management, transport, and recycling of PS food containers. It proved that recycling was feasible. Fourth, the low cost of the PS food container made it a competitive product in the market. As a matter of fact, it was a common choice among customers. All these considered, the ban issued by the government almost a decade ago seemed too hasty.[4]

In making the second report, Tencent invited Dong Jinshi (董金狮), an expert on the packaging industry, as a spokesman for the opponents of lifting the ban. According to Dong, the return of the PS food container to the market would be an environmental disaster. When interviewed, Dong acknowledged certain points raised by the advocates for the PS food container, such as the physical and chemical nature of the product, the significance of the "three fen" recycling project in Shanghai, and the haste in carrying out the ban on the PS food container back in 1999. However, he was very concerned about whether it was justified to lift the ban on the PS food container, since the advocates had not taken into full consideration the risks to environment and health, whereas the other types of products at present had already covered the needs for tableware. Talking about the potential harm to the environment, Dong said that once the garbage materials had nowhere to go, it would mean more pollution. As for harm to health, the extensive existence of low-quality materials, the residue of styrene monomer, and the leaching of toxicants under high temperatures could all be threats to consumers. Besides, in his view, the current market for tableware was already a mature one, much improved compared with the time when the ban first came out. More importantly, almost a decade's practice had proven that the ban did greatly reduce the amount of garbage caused by the PS food containers, thus easing the "plastic pollution." Lifting the ban without a sufficient reason would shatter the authority of the government and its public policy. Moreover, Dong strongly protested against the advocates who claimed the PS food container as a "green product," because it was not "green" by any standard of "green products" in China.[5]

The third report mainly reflected the opinions from the retrieving industry of plastic tableware and environmental NGOs. The recyclers confirmed Dong's concern, while the NGOs criticized the appeal to lift the ban from the perspectives of the common environmental awareness of society as well as the public participation. According to them, as a disposable product, the usage of PS food containers should be restricted rather than encouraged—even if they were safe—from the point of view of saving

---

[4] " 'Safe and Green,' Trade Association Calls for Return of PS Food Container," *Green View*, No. 4, http://news.qq.com/zt2010/whitetrash/index.htm.

[5] Dong Jinshi, "The Return of PS Food Containers Will Be an Environmental Disaster," *Green View*, No.5, http://news.qq.com/zt2010/whitetrash2/.

resources. They argued that the kind of "green campaign" launched by the advocates of lifting the ban differed greatly from the environmental awareness of the public. Besides, the opinion of some individual trade associations or certain experts should not be the reason to ban or lift the ban on a product. The government should guarantee the consumer's right to know and involve public participation and supervision.[6] At the same time, 14 NGOs, such as Envirofriends, Friends of Nature, and the Da'erwen Institute of Environmental Protection (达尔问自然求知社), jointly released an open letter to the media calling for the restriction of all disposable tableware.[7]

Although different parties held different views on the issue of PS food containers, they all acknowledged that the ban issued in 2001 lacked feasible management. The ban denied the legitimacy of the PS food container in society, but the goal of "zero tolerance" was far beyond the reality of society. This gap not only led to extensive violations of the policy, but also shattered the enthusiasm of the qualified producers and recyclers of PS food containers. As a result, the problem of pollution was left unattended. In 2010, the NDRC crossed out the PS food container from the list of obsolete products in its draft of the *Catalogue for the Guidance of Industrial Structure Adjustment* (产业结构调整指导目录), which showed signs of the government's intention to amend the policy. However, without regard to the changed reality of the tableware market or the trend of going green in the food and beverage industry, to lift the ban on PS food containers would cause as much negative impact on society and the environment as the ban did in 2001. With the widespread existence of "illegal products" in the market, to redefine their "legitimacy" was a remedy worth further discussion. More importantly, the responsible governmental bodies should have an overall plan on the management of garbage in the food and beverage industry, including the reduction and recycling of kitchen waste, disposable tableware, and other packages. They should also be able to find the point of balance between the reasonable demand from the food and beverage industry and health and environmental protection.

---

[6] "Collective Response from Environmental NGOs: To Lift the Ban on Plastic Food Container is a Farce," *Green View*, No.6, via http://news.qq.com/zt2010/whitetrash3/index.htm.

[7] "Collective Response from Environmental NGOs against Lifting Ban on Disposable PS Tableware," Green QQ, via http://news.qq.com/a/20100902/001775.htm.

III. *"Excessive Packaging" under Legislative Management*

Following the plastic bag and PS food container bans, excessive packaging was also taken into account in the environmental resource legislation of 2010. On April 1, another national standard (GB23350–2009) approved and issued by the Administration for Quality Supervision and Inspection and Quarantine (国家质量监督检验检疫总局) (AQSIQ) and the Standardization Administration of China (国家标准委) (SAC) came into effect. It was named: *Requirements for Restricting the Excessive Packaging of Food and Cosmetics* (GB23350–2009) (限制商品过度包装要求一食品和化妆品) (hereafter referred to as the *Standard*). The *Standard* set restrictions on the commercial packaging of all foods and cosmetics. For example, the packing should not exceed three layers, the volume of packaging should take no more than 60%, and the total cost of the packaging for a product—except for the initial packaging—should not exceed 20% of its retail price, with adjustments made for different types of goods. According to the *Metrology of Testing for Food and Cosmetics Packaging* (JJF 1244–2010) (食品和化妆品包装计量检验规则) launched at the same time, violation of any of the three restrictive indices listed above was defined as "excessive packaging."

The issuance of the *Standard* was of great significance. It had long been considered a problem that the packaging of many goods took too much space, consumed too many materials, cost too much, and had extravagant designs. On the one hand, it is a waste of resources and damage to the environment. On the other hand, it laid financial burdens on the consumer's shoulders, disturbed competition in the market, and even misled or cheated consumers. Therefore, it was indeed necessary to address the issue. Before the release of the *Standard*, AQSIQ announced a standard on the packaging of moon cakes (GB19855-2005) back in 2005, putting restrictions on the cost and volume of packaging on this particular food for the first time. In comparison, the new *Standard* in 2010 was more comprehensive in the sense it not only covered all foods and cosmetics, but also applied more indices defining "excessive packaging."

However, as many environmentalists observed, the first strike in the government's effort to restrict excessive packaging was not effective. Above all, the *Standard* itself had obvious defects. First, it lacked requirements on the material such as quality and structure of the packaging. In reality many packages with fewer materials could still be hard to recycle, because of the type of materials they were made from as well as the complexity of their structures. Second, regulations on the proportion of packaging costs of certain products to its retail price were impractical, as it was unlikely for either the consumers or law enforcement officials to estimate the cost of a product's packaging.[8] Yet another problem was that the *Regulation on the Restriction of Excessive Packaging* (限制商品过度包装条例) had not been issued. As a result, there was no legal ground to correct excessive packaging. Despite such defects, the announcement of the Standard meant the country started legislation and management of excessive packaging. It could promote accompanying rules and regulations, such as the *Regulation on the Restriction of Excessive Packaging*, and provide basis for public participation in the management of excessive packaging. Before the Mid-Autumn Festival in 2010, Envirofriends conducted an investigation on the packaging of moon cakes in Beijing by the *Standard*, and offered some suggestions on the improvement of packaging policy.

During the investigation, the researchers collected more than 1,000 moon cake packages from two schools, then further selected a total of 300 samples for examination in strict accordance with the requirement of the *Standard*. By rigorous measuring and calculation, they deduced the following: First, the number of moon cake packages from six professional and time-honored pastry brands, such as Daoxiangcun (稻香村) and Gongyifu (宫颐府), took up 48% of the total. It showed that in the management of excessive packaging, priority should be given to the big brands to achieve twice as much with half the effort. Second, the disposable plastic forks and knifes attached to the packages were seldom used, which implied that they were not a necessity in moon cake packaging, but should be offered in accordance with the actual needs of customers. Third, all the samples had no more than three layers of packaging. On the one hand, it showed most of the moon cake packages followed the national standard. On the other band, it also showed that the *Standard* could be stricter. Fourth, the volume of 40% of the samples exceeded 60%, resulting in excessive packaging. Fifth, the average weight of the packaging materials was more than 1.5 times that of the moon cakes, among which more than 95% of the packaging materials were deemed as unnecessary, a serious waste of natural resources. Sixth, the mixture of various materials in moon cake packages made the recyclable waste collectors unwilling to retrieve them.

[8] Duo Zhang, "Environmentalists Warn against Hidden Trouble of Moon Cake Over-Packaging," *Xinhua News Agency*, September 19, 2010, via http://www.gov.cn/fwxx/sh/2010-09/19/content_1705574.htm.

It demonstrated the necessity to add regulations on packaging materials in the *Standard*.

The survey not only revealed the extent of excessive packaging against the *Standard*, but also reflected how the policy was applied in real situations. While providing factual basis for the further improvement of the *Standard*, the NGO researchers also called for timely release of the corresponding regulations, otherwise the Standard alone would lose its purpose. The discourse analysis of the public opinion in recent years shows the issue of packaging received far less social attention than the issue of garbage disposal. However, more and more people are beginning to realize that the management of packaging is an important link in the process of solving the problem of garbage disposal. A sound environmental policy on packaging with accompanying measures is a way to manage the garbage disposal in its earlier stages. The obstruction against the enforcement of the *Plastic Bag Restriction Policy* and the ban on disposable tableware shows a gap between the policy goal and reality. It also demonstrates the necessity to regard all packages and packaging materials as an integrated entity by applying a unified strategy in their management. Besides, the gradual invalidation of the bans meant that prohibition and restriction were only part of packaging management, and needed to be taken into account with policies on the recycling of waste. Although there is mention of recycling in the current policy on excessive packaging, it is not connected with the policies on the promotion of recycling, such as the *Circular Economy Promotion Law* (循环经济促进法). Apart from this, no clear assignment of responsibility to the producers and suppliers of packaging in the existing policies are targeted at excessive packaging or certain kind of packages. In summation, although the 2010 environmental policies on packaging were still quite scattered, isolated, and ineffective, their development presented the need for deepening reforms. It should not be long before the emergence of proposals and discussions on a comprehensive environmental policy on packaging.

RESOLVING THE "GARBAGE SIEGE" BY DEVELOPING A RESOURCE REUTILIZATION INDUSTRY

Du Huanzheng, Zhu Yucheng, and Hu Mengjie

Abstract: Domestic municipal waste has increased drastically after China began urbanizing in the 21st century. Out-of-date garbage treatment technologies and deficient facilities have resulted in a "garbage siege." To solve this problem, we need to vigorously develop a recycling economy, perfect resource reutilization systems, establish garbage classification regulations, promote resource utilization and decontamination of municipal waste, and improve classified garbage recovery, airtight transportation, and centralized processing.

Keywords: resource reutilization industry, "garbage siege"

I. *The Serious Problem of Domestic Municipal Waste in China*

Coinciding with social and economic development and rapid urbanization, China has produced an astonishing amount of domestic municipal waste. According to the *Research Report on a Market Analysis of China's Municipal Waste Disposal and Investment Prospects* (2011–2015 年中国城市生活垃圾处理行业市场分析与投资前景研究报告), domestic municipal waste has increased at a rate of more than 10% per year. This has yielded nearly 0.15 billion tons of domestic municipal waste per year, while the storage of domestic municipal waste has reached 7 billion tons. This issue has become a focus of concern for China and the world.

In China, municipal domestic waste treatment technologies are lagging and facilities are deficient. By the end of 2009, there were 567 domestic waste treatment plants in China, out of which 447 were hygienic burial, occupying 78.8% of the total; 16 were composting plants, occupying 2.8%; and 93 were incineration plants, taking up 16.4%. On a national level, 29% of Chinese cities lack domestic treatment facilities. Quite a few domestic treatment facilities and technologies are imported directly from abroad.

Cities across China are facing a domestic waste crisis. Over one-third of cities have been besieged by domestic waste. With the rapid development of the national economy and the rapid increase of the urban population, it is difficult to stop the increase of domestic waste (See Figure 14.1).

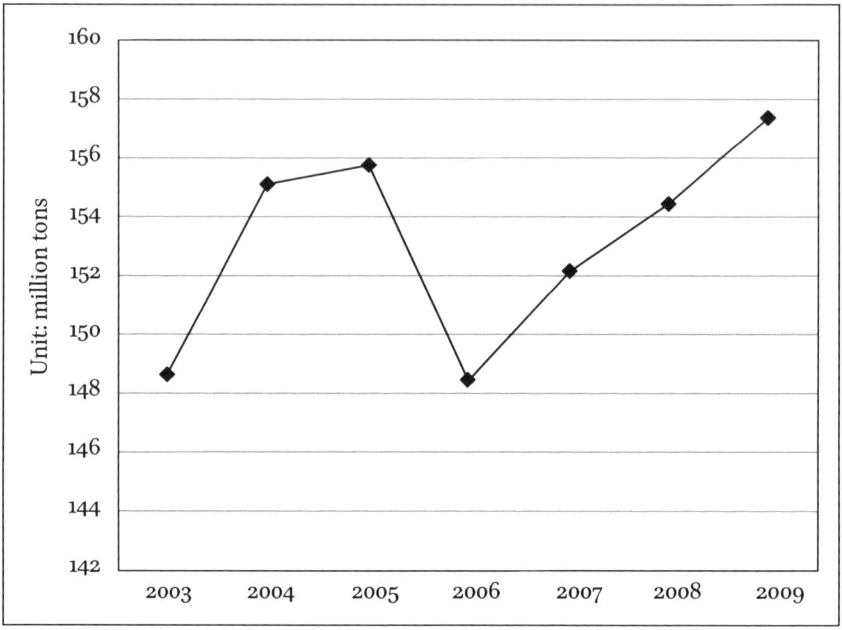

Figure 14.1: Fluctuations in domestic waste treatment.

II. *The Origin of Domestic Waste*

A. *Accelerating Process of Urbanization*

With rapid urbanization and the building of towns, construction waste has increased every year. It is estimated that as many as 500 to 600 tons of construction waste are produced per 10,000 square kilometers of construction. Moreover, demolishing 10,000 square kilometers of old construction produces 7,000 to 12,000 tons of waste. Construction waste has amounted to 30–40% of China's total domestic waste. Urbanization has caused a large number of enterprises to congregate and brought about the arrival of enterprise parks or industrial parks, creating a large amount of production waste. In 2009, industrial solid waste in China reached 2.039 billion tons, an increase of 138 million tons from 2008. Urbanization has also caused population centralization and produced a large amount of domestic waste. In 2009, the amount of municipal waste treatment was 157 million tons, as compared with 154 million tons in 2008. The traditional method of dumping and burying waste has not only occupied vast amounts of land and polluted the urban environment, but also polluted land water and deep water to varying degrees.

B. *Rise in Living Standards*

Rapid economic development has produced a dramatic income increase for urban people. The increase of expenditures per capita has thus produced a municipal waste increase. In 2009, the average manageable income per capita was 17,174.65 RMB, as compared with 10,493.03 RMB in 2005—an increase of 63.68%. In 2009, the average expenditure per capita for urban people was 12,264.55 RMB. In 2005 it was 7,942.88 RMB in 2005, an increase of 54.41%. The increase of income and expenditure has caused a total production of domestic waste of over 150 million tons in recent years.

C. *The Popularity of a Consumerist Lifestyle*

The rise of incomes has produced a consumerist ideology and way of life. Excessive and extravagant consumption is prevalent. According to the newly-published *World Luxury Association Official Release, 2010–2011* (世界奢侈品协会2010–2011年度官方报告), by the end of 2010 mainland China's total expenditures in the luxury market amounted to 10.7 billion USD, up from 9.4 billion in 2009. Data published by Goldman Sachs shows that within the next five years, the population of people willing to spend on luxuries will increase from 40 to 160 million. It is predicted that by 2014, China will become the largest luxury market in the world, occupying around 23% of the total.

D. *Waste of Resources due to Traditional and Extensive Mode of Production*

China's traditional and extensive mode of production has led to a serious waste of energy resources. According to statistics by the International Food Packaging Association, out of the yearly production of nearly 3 million tons of domestic waste, different kinds of commodity packaging materials amount to 830,000 tons. Out of this, 600,000 tons are caused by over-packaging. One-third of domestic waste comes from packaging, out of which one-half is from luxury packaging. The cost of discarded packaging materials amounts to 400 billion RMB. Raw materials used in packaging, such as paper, rubber, glass, steel, iron, and plastic, come from wood, petroleum, iron, and steel, all of which are scarce resources in China. If over-packaging and luxurious use of these materials without recycling are not controlled, they will produce a large amount of waste.

III. *Measures for Resolving the Problem of Domestic Waste*

A. *Municipal Waste Classification*

By establishing an enforced and voluntary waste recycling system, municipal waste can be collected and classified in the origin of its production with many parties being involved and measures being taken to increase the domestic waste recycling rate, facilitate the separate treatment of hazardous discarded substances, and reduce and recycle garbage. In 2011, Guangzhou published the *Temporary Regulations on Municipal Waste Classification and Management of Guangzhou* (广州市城市生活垃圾分类管理暂行规定). It stipulated that the principle of garbage classification should first be implemented "from easy to difficult, then steadily improved, and finally enforced step-by-step." It requires the citizens to separate the "dry" from "wet" garbage at home. Sanitary workers will separate the dry garbage as recyclable garbage and then sort it again at the recycling station.

B. *Promoting the Reduction of Municipal Waste*

Cities need to turn from the end treatment of garbage to promoting the reduction of municipal waste from its source. In Taiwan, when dumping garbage, the pre-treated waste must be classified. The daily municipal waste production per capita has been reduced from 1.143 kg at the end of the 1980s to 0.35 kg in 2011, which is far below the 1 kg international daily average waste-treatment per capita. In 2011, Guangzhou implemented the *Scaled End-Measuring Fees, Rewards, and Fines in Guangzhou's End Treatment of Municipal Waste* (广州市生活垃圾处理终端阶梯式终端计量收费及奖惩办法) to collect scaled fees from citizens. The more municipal waste is produced, the more people must pay. This not only reduces the amount of municipal waste but also promotes living a low-carbon and environmental lifestyle.

C. *Set Up a Waste Recycling System*

China must promote and perfect the system of "from the recycling station to the classification center" and adopt various forms of recycling. In 2011, the Suzhou Industrial Park published its *Execution Plan of Establishing and Managing the Network System of Recycling Renewable Resources* (苏州工业园区再生资源回收利用网络体系建设和管理实施方案). It aims to establish and standardize the community recycling station and domestic waste classification as the foundation, use the preliminary division and classification center as the vehicle, and set up the recycling network system as the goal. By referring to international recycling modes, such as the German DSD, Austrian ARA, and French ECO, one can combine the established recycling with a mobile recycling form and rely on different types of waste recycling to purchase the domestic waste, such as through telephone, internet, or contracts.

D. *Developing a Resource Reutilization Industry*

1. *The "Three-Intensives and One-High" Characteristic of the Resource Reutilization Industry*

As an emerging industry, resource reutilization includes the classifications of artistic works, products (second-hand products), spare parts (remanufacturing), and materials. It is different from the simple waste-recycling industry which the "farmer guerrillas" (农民游击队) used to be engaged in. It possesses the "Three-Intensives and One-High" (三密一高) characteristic, meaning it is policy-intensive, technology-intensive, labor-intensive, and has a high added-value.

a. *Policy-Intensive*

The resource reutilization industry is a policy-oriented industry. The country's laws, regulations, and policies play a leading role in its development. The producer- and the customer-responsible system and green purchasing system promoted by the developing resource reutilization industry all need to be regulated and limited. For many years, China's recycling of waste TV sets has not made much progress. In 2011, the *Regulations and Rules for Recycling Household TV Sets and Electrical Appliances* (废旧家电及电子产品回收处理管理条例) were officially enforced. These clearly regulate recycling and disassembling a large number of waste electrical and electronic products on the market in order to promote the sizable and industrial development of recycling and disassembling household electrical appliances.

b. *Technology-Intensive*

The resource utilization industry is a reverse direction manufacturing process. It includes new and advanced technologies such as restoration technology (including remanufacturing technology), surface modification technology, waste-separation technology, waste reutilization equipment manufacturing, and safe waste disposal technology. Technology that can separate a paper milk carton from its aluminum will solve the problem of 1 million tons of discarded paper drink cartons in the country. Equipment that can handle used household appliances has the potential to create a market worth 680 billion RMB. Applying information technology to resource utilization will also become a hot topic. Consequently, China needs to unceasingly push forward innovation in this field, to provide both circular and economic technology.

c. *Labor-Intensive*

The resource reutilization industry relies more on the excessive use of labor than on technology and equipment. Discarded electrical and electronic equipment, such as automobiles, cell phones, computers, and air conditioners require an excessive amount of manual labor in the preliminary stage of disassembling to take out the spare parts from automobiles and air conditioners and dismantle computers and cell phones for components. During the first stage of such disassembling, only manual labor is the most efficient method. It is estimated that, on the national level, 20 million people are participating in the work related to recycling, disassembling, and reusing.

d. *High Added-Value*

The added value of the new material made by applying new technology to change the characteristic of the original material is high. Take artistic products for example. By making artistic products out of waste materials and selling them, something the price of a few *yuan* is turned into a few thousand *yuan*. The production cost of remanufacturing a car is only 50% that of a new product and it saves energy and materials by 60% and 70% respectively. These industries not only embrace the idea of environmental protection but also produce a higher economic value.

2. *Multi-Parties Promote the Development of the Resource Reutilization Industry*

The resource reutilization industry covers the public, enterprises, communities, and society. In order to collaboratively establish a conservation society and ecological economy, it is required that enterprises have a productive cycle, industries have a planning cycle, and society has a consumption cycle. Developing the resource reutilization industry is a comprehensive reform of the traditional mode of economy and mode of production.

a. *Government Guidance*
The resource reutilization industry is a burgeoning industry. It needs the government to stipulate a complete set of legal policies to safeguard it. From a life cycle perspective, the resource reutilization industry requires a large investment, has a long cycle, and has a high risk during its initial development.

b. *Enterprise Dominance*
Enterprises play the main part in the development of the resource reutilization industry. The principle of “reduce, reuse, recycle” (3Rs) is mainly carried out by enterprises. The operation of enterprises is directly related to the effectiveness of the development of the resource reutilization industry. The Qihe Tiandi Metal Products Company (台州路桥齐合天地金属有限公司) in Taizhou, Zhejiang Province is mainly engaged in the disassembling, deep processing, and comprehensive use of waste metal products. In 2010, the company imported and disassembled 480,000 tons of waste products. The overall recovery rate was above 98.5%. In the past 10 years, it has produced 250,000 tons of recycled copper, 180,000 tons of recycled silver, 1,500,000 tons of recycled steel, and 30,000 tons of plastics. Comparatively, it saved from being mined 30,750,000 tons of copper ore, 1,080,000 tons of silver ore, and 2,400,000 tons of iron ore and promoted the minimization and cyclic utilization of resources.

c. *Public Participation*
Public participation is the main form of modern democracy. According to a national opinion poll of public awareness of environmental protection conducted by the Ministry of Environmental Protection and Ministry of Education, only 25% of people think about environmental protection while shopping, 35% are willing to accept a comparatively higher price for environmental protection, and 30% act in accordance with environmental ethical standards when handling waste. So there is a need for publicity and education in order to: raise people’s level of understanding and awareness of participation; introduce the public to training themselves to develop a good habit of green consumption; and steadily raise the awareness of public participation and supervision, decision-making, and responsibility.

d. *International Cooperation*
The policy, technology, and rationale of developed countries in the safe treatment of waste and development of a recycling economy are far more

advanced than in China. China needs to continuously expand the scope of international cooperation, learn the technology and rationale of developed countries, and using a combination of China's experience developing a recycling economy and that of developed countries, form China's unique mode of a recycling economy.

Establishing a system of international cooperation is needed to promote the Asian 3R strategy and start Asia's low-carbon mode. In Taiwan, the renewable resources industry has a developed to a high degree, but the resource-product market has a low degree of acceptance. South Korea established a unique "Resource-Reclaiming Commune," while Japan changed into a recycling society due to a severe lack of waste burial land. It has been a trend that Asian recycling societies have become more and more related to a low-carbon society.

It is necessary to deepen the exchanges and cooperation of official and non-official organizations and platforms, such as the United Nations Environment Program, the International Union for Conservation of Nature and Natural Resources, and the International Committee of China Resources Recycling Association. By depending on the international platforms of the Chinese-German Forum for Economic and Technological Cooperation, the Asia 3R Regional Forum, and the China-Japan Energy Saving and Environmental Protection Comprehensive Forum, China needs to encourage its resource recycling enterprises to start developing international resources and investment.

IV. *Recent Developments of the Resource Reutilization Industry in China*

As clearly indicated in China's Twelfth Five-Year Plan Outline, it is highly necessary to "put in great efforts to develop the recycling economy, complete a circular resource recycling and recovering system, establish and complete a waste-classification recovering system, perfect the recovery classification, airtight transportation, and central processing system, and promote the resource utilization and safe treatment of waste." China is now putting great efforts into improving waste-classifying recycling technology, setting up classification recycling standards, promoting a waste classification recycling rationale, enforcing waste-classification recycling measurers, and easing the environmental conflicts between national and social resources.

A. *Implementation of the "City Minerals Resources" Action Plan*

In 2010, the National Development and Reform Commission (NDRC) and Ministry of Finance organized and developed a model base of "City Mineral Resources" (城市矿产) with a plan to establish 30 model bases within five years. "City Mineral Resources" refers to recyclable iron and steel, nonferrous metals, rare metals, plastics, rubber that are generated and stored in discarded mechanical and electronic equipment, electric wires and cables, communication devices, automobiles, household electric appliances, electronics, metal and plastic packaging materials, and other discarded materials produced during the process of industrialization and urbanization. The recyclable amount is equal to that of the raw mineral resources. By October 2011, the NDRC and Ministry of Finance had designated two groups of 22 national-level model bases of "City Mineral Resources" and pushed forward the sizable development and recycling of discarded mechanical and electrical equipment, automobiles, lead-acid batteries, plastics, and rubber.

B. *Promotion of the Pilot Sites of the Recycling System of Renewable Resources*

In 2006, the Ministry of Commerce started establishing pilot sites in the renewable resources recycling system. In the last two years, the Ministry of Commerce and Ministry of Finance supported the establishment of 33,075 pilot networks in 44 cities of 181 sorting centers, 22 collection and distribution markets, and 36 regional collection and distribution markets of renewable resources. By 2011, the Ministry of Commerce had initiated two groups of construction pilot sites for the renewable resources recycling system, and decided on 55 pilot cities and 11 regional recycling bases, out of which the pilot cities have initially formed the recycling development mode of community recycling networks, sorting and processing centers, and collection and distribution markets.

C. *Determination of the Resource Reutilization Industry as a Strategic Emerging Industry*

In October 2010, the State Council released its *Decision on Accelerating the Fostering and Development of Strategic Emerging Industries* (关于加快培育和发展战略性新兴产业的决定). It clearly points to developing and fostering the seven major strategic emerging industries. These are industries that involve: energy-saving environmental protection, new-generation information technology, biology, high-end equipment manufacturing, alternative energy, new materials, and alternative energy vehicles. The resource reutilization industry is one of the three major sub-industries of energy-saving environmental protection.

D. *Structuring of the System of Resource Reutilization Laws and Regulations*

In 2006, the Ministry of Commerce passed the *Administrative Measurers of Renewable Resources Recovery* (再生资源回收管理办法) to regulate the development of the renewable resources recycling industry. In 2008, the fourth meeting of the Eleventh National People's Congress Standing Committee passed the *Circular Economy Promotion Law* (循环经济促进法), which is the most important, complete, and secure basic law in the resource recycling field. Then, in 2008, the State Council passed the *Regulations on Recovery Processing of Electrical and Electronic Product Waste* (废弃电器电子产品回收处理管理条例), stipulating the recovery processing system and measurers for electrical and electronic product waste. In addition, more than 10 laws and regulations are being drafted. A series of laws, regulations, and department rules have steadily formed a legal system for the resource reutilization industry and provided a secure legal system to promote the development of the industry and resolve the "garbage siege."

HOW CAN CITIES BE FREE FROM URBAN FLOODING?

Yan Weiqi*

Abstract: Urban flooding has become a major threat to Chinese cities, receiving particular attention beginning in 2010. Local governments have typically focused on construction and ground-level infrastructure rather than planning and underground work. As a result, urban drainage systems have not kept up with expansion. This has resulted in ever-growing coverage by non-permeable surfaces less able to absorb rainwater—putting cities at greater risk of flooding. This has made urban flood protection a challenging, long-term task.

Keywords: city, flood protection, urban flooding

2010 was a particularly bad year. Numerous instances of urban flooding severely disrupted a number of major cities, resulting in heavy property damage and even deaths. This has clearly become an issue demanding close attention and scientific approaches to management if cities want to achieve effective protection.

I. *Frequent Floods Sweep Cities*

A. *269 Cities at the County or Higher Levels Experienced Flooding*

Let us review some of the major incidents. On the evening of May 6, 2010, Guangzhou found itself hit with unusually heavy rain. As a result, flooding struck 102 localities impacting some 32,000 people and killing six. One hundred and nine houses collapsed. More than 17,000 hectares of farmland were flooded, and 118 low-lying areas in the central urban parts of the city were inundated. Automobiles, too, saw particular damage. Some 1,400 vehicles in parking lots were flooded. All this amounted to financial losses of nearly 550 million RMB, according to preliminary statistics.[1]

---

* He Ping and Li Jinjie, both reporters from *Gaungming Daily*, also contributed to the writing of this article.

[1] Zeng Ni and Zhu Dan, "Rainstorm on May 7 Flooded 102 Towns in Guangzhou and Caused a Financial Loss of 540 Million RMB," *Nanfang Daily*, May 12, 2010.

Just over a month later the city of Wuzhou (梧州) in Guangxi Zhuang Autonomous Region was hit with over 100 millimeters of rain within three hours. The result was the most severe flooding in four years. Eleven low-lying urban roads were flooded, leading to traffic jams of up to two hours. In some areas, flooding forced drivers and passengers to abandon their cars and swim to safety, while the local police had to use boats to rescue others. Two workers drowned in an underground parking lot under construction.[2]

On July 1 of the same year, Kunming, the capital of Yunnan Province, was hit with 113.6 millimeters of rainfall. The waters in more than 30 areas across the city reached a depth of more than a meter. Traffic ground to a halt on main roads, and Kunming Airport was forced to close as terminals and runways became inundated by overflowing rivers. More than 100 flights were delayed and over 5,000 passengers were affected.[3]

Three days later, Beijing's Chongwenmen (崇文门) subway station closed after being inundated. A day later, heavy rain left the city of Cangzhou (沧州) in Hebei Province with knee-deep water. A sudden rainstorm on July 19 left parts of Jinan, the capital of Shandong Province, flooded. On July 27, numerous low-lying parts of cities in Jilin Province, including Changchun, Yanbian (延边), and Jilin (吉林), were severely hit. A 50-year rainstorm caused heavy losses across Hainan Province on September 30, impacting 16 counties and cities. A second round of rainfall on October 15 brought even further woes.

Urban flooding generally impacts cities along rivers or coasts. But nowadays, inland cities are also finding themselves subjected. Across China the impacts are characterized by broad swaths of affected areas, long durations, and particular challenges faced in fighting them. This has begun to draw widespread attention. Chen Lei (陈雷), the head of the Ministry of Water Resources (MWR), said at a national flood and drought protection meeting on December 7, 2010: "Floods in 2010 covered the widest areas, lasted the longest, and caused the heaviest losses and casualties across China since 1998." A total of 269 cities at the county or higher levels were flooded.

---

[2] Zhang Zhoulai and Xie Daming, "Rainstorm Hits Wuzhou, Leads to Severe Flooding and Kills Two," www.xinhuanet.com, June 11, 2010.

[3] Chen Peng and Guan Guifeng, "Spring City Becomes Flooded Overnight—Behind the Flood Caused by the Rainstorm in Kunming on July 2, 2010," www.xinhuanet.com, July 4, 2010.

B. *Urban Flooding: A Stubborn Problem*

In fact, the problem is neither new nor rare. Experts point out that 100 millimeters of rain over a short period in an urban area will usually produce flooding.[4] The harms are also well known. One study has shown that such flooding typically lasts one to five hours, leading to significant increases in fractures, leg and foot injuries, fevers, colds, and digestive disorders. It severely impacts road transportation systems, increasing traffic accidents and claims for vehicle repairs eight to tenfold. Other negatives include failures in utilities, communications equipment, underground cables, as well as flooded markets, casualties, and social disorder.[5]

Some cities that are subject to frequent rainstorms have increased investment in urban drainage systems and flood protection infrastructure in recent years, but have not seen significant effects.

The series of floods in Guangzhou in 2010 not only highlighted the typical characteristics of urban flooding but also produced greater awareness of the challenges of flood protection. The city had already spent 900 million RMB modifying vulnerable streets in 2009. One local authority had even announced that "this year will undoubtedly be better than previous years," ahead of the 2010 rainy season. Obviously this prediction turned out not to be true.

The city of Nanning provides another telling example. In 2007, the local government spent more than one billion yuan on water-management improvements, including work on eighteen rivers. In 2009, it planned to invest more than 1.6 billion RMB dredging waterways, improving the sewer network, and modifying areas vulnerable to floods. It also developed and implemented the *Comprehensive Work Plan for Flood Protection in Nanning Urban Areas (2009–2010)* (南宁市市区内涝综合整治工作方案(2009–2010年)), promising to spend two years working on local rivers, main drainage networks, and main areas vulnerable to flooding. Once again, however, the city was severely disrupted by the floods of 2010.[6]

In an interview, Cheng Xiaotao (程晓陶), the deputy chief engineer of the China Institute of Water Resources and Hydropower Research (IWHR) and the executive director of the Research Center of Flood and Drought

[4] Zhang Xinxin, "Why Is South China So Hurt by Rainstorms?" www.xinhuanet.com, June 22, 2010.

[5] Zhang Zhiguo, "Reflection on Urban Flooding in Cities along the Lower Reach of the Yangtze River and Solutions," *Yangtze River*, November 2009.

[6] Xie Yang, "Why Can't Nanning Stop Flooding Despite All Its Efforts?" *China Youth Daily*, June 7, 2010.

Disaster Reduction at the MWR, proclaimed that efforts to solve the urban flooding problem could no longer be delayed.

China is in a phase of rapid urbanization, as the urban population now represents more than 45% of the total population compared with less than 20% in 1980. Led by rapid expansion of cities along rivers and coasts, the urban population has seen a net increase of about 400 million persons over the past three decades. Cheng Xiaotao reminded us that urban water safety assurance has become more complex and challenging due to a combination of more severe floods, shortages of water resources, and deteriorating aquatic environments.

Rain-related disasters can, however, be mitigated with increasing awareness of urban flood protection, enhanced infrastructure development and management, and the creation of an effective disaster protection and reduction system. In the event that flooding is unavoidable, we should be able to minimize its duration and area and, accordingly, property losses and casualties.

II. *Causes of Frequent Urban Flooding*

There are three main causes of floods in cities across China:

A. Old Sewer Networks, Low Standards, and Inefficient Maintenance due to Management by Multiple Organizations

A sewer system known as the Fushougou (福寿沟) designed and built in the Song Dynasty in Ganzhou (赣州), Jiangxi Province drew much public attention after a series of rainstorms in 2010. Despite its thousand-year age, the system remains an effective protection. Qingdao of Shandong Province also found itself adequately protected in 2010 thanks to a German-designed sewer system more than a century old. This has caused many to question the more modern efforts of other Chinese cities.

"A good urban drainage system serves as the foundation of a sustainable city," emphasized Kong Yanhong (孔彦鸿), director of the Urban Aquatic System Planning and Design Institute at the Chinese Academy of Urban Planning and Design (CAUPD) when I interviewed him. Unfortunately, this obvious truth has failed to receive sufficient attention in the current wave of urbanization.

When talking with journalists, Wang Mingshu (王明书), director of the General Affairs Department at the Kunming Municipal Flood Protection Office, confessed that the sewer networks for more than 80% of the flooded areas across the city in 2010 had been decades or even a century old and overburdened in rainstorms. Those networks were also found to be incompatible with each other due to different sewer diameters and depths. Cui Shenghui (崔胜辉), an associate researcher at the Institute of Urban Environments at the Chinese Academy of Sciences (CAS), believes that authorities underestimate the impact of extreme rainfall when designing urban drainage systems.

This is a problem impacting the whole of China. The old urban areas of the city of Jinan still use sewer systems decades or more old, according to Kong Yanhong. Precipitation levels have dropped since a flooding in 1987, and as a result, the local government has not paid sufficient attention to the urban drainage system. There are only 810 kilometers of storm sewers and eleven rainwater pump stations across the city—which is far from sufficient. Main problems include a lack of improvements, sewers systems that are used to carry rainwater and sewage together in old urban areas, storm sewers prone to blockage, and stretches of road unsupported by rainwater collection and discharge systems. Sluiceways around the city also need improvements. At present, there are many structures above watercourses in the urban areas that block discharges, adding to the pressure.

This is not just a question of infrastructure. The lack of a unified system of management and coordination for urban drainage systems, as well as poor maintenance, poses a major challenge according to Yang Baojun (杨保军), deputy director of CAUPD. He believes that flood protection relies more on management than on planning. "Take pumping stations for example. They may be subject to insufficient maintenance if there is a lack of flooding for two or three years. It is not uncommon for pumps to be found to have been rusted when turned on during a sudden rainstorm. Another common problem is an increasing severity of floods due to blocked or damaged sewer systems."

Underlying all this is the fact of management by multiple organizations. The efficiency of urban drainage management is directly impacted by the lack of coordination among relevant organizations, according to Ye Wenhu (叶文虎), director of the China Center for Sustainable Development Research at Peking University. Urban drainage infrastructure is typically owned by multiple organizations like drainage companies, urban construction bureaus, water resources bureaus, and infrastructure developers. Functions such as flood protection guidance, infrastructure development, and administrative law enforcement are performed by different organizations among which there is no clear functional division.

This makes it difficult to create a unified drainage system and to realize unified coordination in the event of a sudden rainstorm.

B. *Rapid Increases in Urban Areas and Hardened Ground Surfaces Add to the Drainage Load*

Chinese cities have been growing rapidly since the 1980s. The land area occupied by urban development is increasing quickly—bringing with it great changes to the terrain in and around cities: original vegetation has been removed, soil has been altered, and many ground surfaces have been hardened into non-permeable ones. Accordingly, water bodies and permeable surfaces such as ponds, fields, and farmland that absorb surface runoff continue to decrease. All this leads to higher runoff and flow rates, and more frequent peak outflows.

This assumption has been borne out through measurements. Kong Yanhong points out that Shanghai had an urban area of 91.6 square kilometers in 1947 and a local surface runoff coefficient of only 0.29. As urban areas grew to 364 square kilometers in 1997, the coefficient rose to 0.67.

The hardened ground surfaces in cities have added to the workload of the local drainage systems, according to Li Xiaoquan (李小泉), chief meteorologist at Weather China. Natural drainage systems such as ponds and vegetation-covered land in cities have been replaced by concrete buildings or cement/asphalt road surfaces. As a result, water flows into low-lying areas.

Despite this, most cities have no rainwater collection plans to guide drainage system development. According to Kong Yanhong, they simply lay sewers where they build roads; storm sewer designs are focused on drainage for roads; and design margins are considered only based on the scope of road designs and designers' experience. Cities, then, are left with ineffective systems and a lack of necessary consideration for long-term use.

C. *Local Governments Focus on Short-Term Benefits and Pay More Attention to Ground-Level Infrastructure Than Underground*

Cheng Xiaotao believes that problems with urban drainage system development are compounded by a lack of basic information on sewer systems. Further to this, underground spaces are unsupported by basic flood protections. Common problems include: destruction of water bodies without proper consideration; watercourses being turned into roads; and groundwater overuse leading to subsidence.

The root cause of this lies in local governments' ideas about administrative performance and urban development. Urban expansion is given a top priority, which leads to cities whose appearance belies big problems with underground infrastructure. Rapid expansion comes at the cost of needed infrastructure development, urban planning and management, and ecological balance.

Urban development should be based on long-term planning, said Yang Baojun during an interview with journalists. Frequent urban flooding reveals focus on the short term in urban development. Local governments pay excessive attention to what is *on* the ground without respect for the underlying realities of growth.

Fortunately, some cities have begun to take note, in particular Harbin, Jinan, Tengzhou, and Nanning. Yang Baojun reminds us that urban safety should be given the highest priority during periods of development, and that without this achieving more livable cities and higher quality of life will be impossible. This requires city officials to be educated on urban development and administrative performance.

III. *Urban Flood Protection: A Long-Term Task*

This will be a long-term task all across China. More than 600 cities are now on board, including 57 in coastal areas facing the toughest challenges from events like typhoons and tidal flows; 288 cities in plains, typically along lower and middle reaches of main rivers; and 297 in hilly and mountainous areas that are subject to waters running from peaks. One aspect of this worth noting is that Chinese cities vary widely in terms of economic development and levels of management. Thus, flood protection measures need to consider not only natural factors, but also socioeconomic conditions in their quest for human-oriented, comprehensive management and unified, scientific planning.

A. *Accelerating Sewer System Modification to Create an Effective Urban Aquatic System*

Given the current poor state of water management across numerous cities, Chinese authorities should play a leading role by increasing investment in this area. Experts recommend that local governments enhance sewer cleaning activities and push the construction of embankments, watercourses, sewer systems, and pumping stations. They should also have rainwater collection and diversion systems built at railroad crossings prone to floods so as to reduce pumping station workloads. Additionally, they should enhance flood protection infrastructure management and maintenance, and dredge watercourses while working harder to identify and remove hindrances to water flow, e.g. obstacles in watercourses and illegally built structures. There is also now an urgent need to launch reforms aimed at water service integration in order to bring about management by a single authority. This would facilitate collaboration and increase efficiency. Such reforms have seen early results in Beijing, Shanghai, and Guangzhou.

A city works like an organism. Leveraging rivers and lakes to create healthy aquatic environments is also an effective approach. Chinese government authorities should enhance water management strategies to restore and increase the storage capacities of urban rivers and lakes. Adjustments to urban water systems should be made compliant with flood protection plans and strictly regulated.

B. *Valuing Local Rainwater Storage and Utilization*

Traditional methods of management require the construction of high-capacity drainage systems—leading to greater peak flows downstream. According to Kong Yanhong, this creates additional flooding. Even with city-wide drainage systems, Chinese cities still face high flood risks due to insufficient floodwater storage capacity. Moreover, traditional systems seldom consider storing rainwater for later use, instead focusing solely on discharging these waters. This occurs even in cities that regularly suffer from water shortages.

Xiang Liyun (向立云), the deputy director of the Department of Water Hazard Research, agrees, encouraging the development of systems to store water for later use. Resources available to this end include rivers, lakes, wetlands, ponds, and decentralized rainwater storage facilities spread across cities. Such systems have enormous potential and deserve greater attention from authorities. Properly designed, such systems can be neatly integrated into urban landscapes and environments—thereby boosting rainwater utilization, reducing drainage system workloads, and ultimately mitigating the effects of flooding.[7]

Expanding natural bodies of water or creating man-made lakes in cities is a common method used worldwide, according to Xiang Liyun. This is

[7] Lü Xianru, "Xiang Liyun, Deputy Director, the Disaster Reduction Institute, the IWHR, Talks about Flood Protection Methods," *Guangming Daily*, May 28, 2010.

less applicable for highly developed urban areas, however, due to land use issues. Thus new developments will likely be the primary focus.

Decentralized systems can be significant. Outdoor urban spaces such as parks and fields may serve as temporary rainwater storage facilities through controllable drainage equipment, and sunken urban green spaces can also store large quantities of rainwater while supplementing groundwater, especially in northern China. Additionally, urban ground surface permeability can be increased by building permeable boxes, ditches, tanks, and for hardened surfaces, making use of permeable pavement.

C. *Enhancing Land Planning and Assessing Rainstorm Flood Risks*

Effective urban planning and management—in particular following through on plans—should be given higher priority. Development should not occur in areas prone to geological disasters such as mountain torrents, landslides, and debris flows. On the other hand, bodies of water, wetlands, and other rainwater storage resources should be made part of urban land use plans, as they are necessary features of flood protection and help maintain good urban ecosystems and landscapes.

China's flood protection law specifies that flood impact assessment and administrative approval mechanisms regarding floodwater storage and flood plains be launched for urban development projects. Assessment is also required for projects that may significantly impact the course and likelihood of flooding. Analysis of potential interactions between urban development and flooding may help reduce flooding on the whole and reduce its impact when it does occur.

The current problem with urban development is a lack of respect for planning, rather than a lack of ability. Relevant organizations tend to brush aside advice when given. Experts recommend that laws and regulations specify that no development project may increase urban flood risks. Risk transfer, that is actions that increase flooding in areas outside development, should also be avoided.

D. *Becoming More Aware of Disaster Prevention and Enhancing the Capability of Emergency Response*

One further area relevant to these questions is a city's capacity for emergency response. Experts recommend that local governments develop urban flood risk maps based on the flood prevention and drainage capacities of watercourses, lakes, and sewer systems. They should also inform relevant organizations and the public of flood risks from rainstorms at different magnitudes as a means of helping citizens take necessary precautions. Governments should also conduct disaster prevention exercises and drills to help the public better deal with urban flooding.

Governments should also improve contingency plans and develop measures to address flooding when it does occur to minimize casualties and property losses. Creating or improving warning systems, launching contingency plans, and implementing disaster prevention and reduction measures in advance will all help achieve these ends.

PART FOUR

SUSTAINABLE CONSUMPTION

AN OVERVIEW OF CHINA'S UNSUSTAINABLE CONSUMPTION IN 2010

Zhang Boju and Zheng Yisheng

Abstract: In 2010, the absence of sustainable consumption continued both in terms of awareness and practice. Many social and environmental problems caused by unsustainable consumption drew the attention of the public and the government. Though it is still very difficult to advocate for sustainable consumption, we are convinced that continuing to push for it is the only way forward for both the government and all sectors of society. Furthermore, we firmly believe that resisting and preventing the spread of consumerism is a prerequisite for China's sustainable development.

Keywords: sustainable consumption, consumerism, difficulties in advocating sustainable consumption, resource-efficient society

In the *China Environment Yearbook, Volume 5*, we highlighted the importance of China's consumption patterns for sustainable development, as well as the key role environmental organizations have to play in this regard. One year later, we must regrettably acknowledge the unsatisfactory status quo in advocating sustainable consumption.

I. *Unsustainable Consumption and Environmental/Social Problems Reinforce Each Other*

Although 2010 witnessed China's economic recovery, it also saw more pressing environmental and social problems. One of the root causes was a pitiful lack of awareness and practice of sustainable consumption. The country saw more rampant conspicuous consumption while unsustainable consumption was not curbed. A key phenomenon in 2010 was the close connection between unsustainable consumption and environmental and social problems. In retrospect, the following events deserve our special attention:

A. *Conspicuous Water Consumption Getting Worse*

In June 2010, the environmental NGO Friends of Nature issued its *Report on the Awareness and Consumption of Water Resources of Beijing*

Residents,[1] (首都市民水资源状况认知及利用调研报告) pointing out how water resources available in Beijing are continuing to fall short of the city's rising population. Seventy percent of Beijing residents have realized that the city's water shortage is very severe, and the vast majority of them think that such a problem should be addressed as soon as possible.

Ironically, a number of water-intensive industries are booming in the thirsty city. One such example that came under the spotlight was the spa industry, which "consumes the water equivalent to 41 Kunming Lakes in a year."[2] According to statistics, there are over 3,000 bathing centers, spas, and hot spring resorts in Beijing, and flagship bathing centers are often as large as 10,000 or 100,000 square meters. The annual water consumption of this industry alone is 81.6 million tons. It is obvious that a city like Beijing cannot afford such a "fashion."[3] Thanks to sustained appeals from environmental organizations and the media, at the end of September 2010, the Beijing municipal government issued the *Water Saving Measures of Beijing* (*Draft for Review*) (北京市节约用水办法(修订草案送审稿)) by the Beijing Water Authority (水务局) for public review. The *Measures* forbid the opening of new high-end public bathing facilities and will impose on these special industries water saving measures such as a particular billing system and strict cap on water consumption.[4]

The large number of public bathing facilities, however, is just the tip of the iceberg. Golf courses, artificial snow ski resorts, and car washes are some other examples of water-intensive industries in the city. In the meantime, as the Chinese demand for meat and poultry products continues to grow, such water-intensive industries as feed grain cultivation, feed grain production, and livestock breeding, slaughtering, and processing are also booming. This is adding more pressure to the scarce freshwater resources in China. Although the promulgation of Beijing's new policy is commendable, it is far from enough.

---

[1] Chen Yuanyuan, "Beijing Residents' Understanding of Water Resources: Pressing Crisis, Immediate Attention." *China Environment News*, July 21, 2010.

[2] Li Shasha, "Beijing's Bathing Industry Consumes Water Equivalent to 41 Kunming Lakes in a Year," *Legal Evening News*, March 22, 2010.

[3] Hu Kanping, "Severe Water Shortage vs. Conspicuous Water Consumption in Beijing," *The China Environment Yearbook 2010*, Social Sciences Academic Press, March 2010.

[4] Liu Ke, "Strictest Water Use Measures in Beijing in Draft," *Beijing Daily*, January 17, 2011.

B. *Over-Packaging and the Municipal Garbage Crisis*

Garbage has become yet another major environmental crisis in China following water and air pollution. The year 2010 saw more garbage pollution and disposal problems, and the country's garbage production was still on the rise. Packaging accounts for a large proportion of municipal solid waste. According to Dong Jinshi (董金狮), vice chairman of the International Food Packaging Association (国际食品包装协会), the average recycling rate of the entire packaging industry is as low as 20%. The monetary value of the financial loss from packaging waste and the unrecycled recyclable resources is 30 billion RMB annually.[5] Although the *Requirements for Restricting Excessive Packaging—Foods and Cosmetics* (限制商品过度包装要求——食品和化妆品) jointly approved and promulgated by the General Administration of Quality Supervision, Inspection and Quarantine of China (国家质检总局) and the Standardization Administration of China (国家标准委) became effective on April 1, 2010, the regulations' effectiveness has proven to be very limited. According to a nationwide survey conducted in September 2010 by the environmental NGO the All-China Environment Federation (中华环保联合会), 45% of the respondents believed that over half of packaging could be deemed as over-packaging, while 45% thought the greatest danger of over-packaging was pollution and the waste of resources. In terms of its cause, 42% believed that over-competition in the market has led to over-packaging, 34% thought the lack of government supervision was another major reason, 13% attributed it to the unhealthy social tendency of flaunting consumption and the craze to vie with each other, and 10% thought that manufacturers were to blame by raising prices, misleading consumers with excessive packaging, and depriving consumers of their right to know and independent choice.[6]

C. *"Haste Makes Waste:" Automobile Consumption and Tragedy of the Commons*

The year saw dramatic growth in China's automobile production and sales. According to statistics from the China Association of Automobile Manufacturers (中国汽车工业协会), the two totals were 182,647 and 180,619

---

[5] Guo Jia, "Some Chinese Rich into Conspicuous Consumption, CPPCC Member Suggests Luxury Tax," *People's Daily*, November 24, 2010.

[6] Hu Lijuan, "Majority of Consumers against Over-packaging, Lack of Supervision Leading to Waste," *Science and Technology Daily*, December 9, 2010.

respectively, or a year-on-year growth of 32.44% and 32.37%.[7] China's automobile production and sales volume topped the world again in 2010 after it first reached number one in 2009. What are the consequences of such an automobile craze?

The year was also a landmark year for traffic congestion in China. In August, the great gridlock of the Beijing, Hebei, and Inner Mongolia sections of China's National Highway 110 earned a welter of global publicity and the *New York Times* called it one of the worst traffic jams in world history. On September 17, a small rainfall in Beijing caused a dead halt in the entire city. Over 140 roads were heavily congested during the evening rush hours, breaking the congestion "record" that had been set at the beginning of the year.[8]

Even worse, traffic congestion was not only found in such mega-cities as Beijing. Many other medium- and large-sized cities, such as Xi'an, Hefei, Zhengzhou, Urumqi, and Lhasa, have also been trapped in traffic jams. According to the Traffic Administration Bureau of the Ministry of Public Security (据公安部交管局), some two-thirds of the cities among the 667 in China witnessed traffic jams during rush hours in 2010.[9] The research of Professor Niu Wenyuan (牛文元), the chief scientist at the Research Group in Sustainable Development Strategy of the Chinese Academy of Sciences (中国科学院可持续发展战略研究组), shows that the daily financial loss resulting from traffic congestion and poor administration of 15 Chinese cities could be nearly 1 billion RMB.[10] Furthermore, health problems due to tail gas pollution have become a major public health crisis that is comparable to smoking.

According to Dong Yang (董扬), the secretary general of the China Association of Automobile Manufacturers, the biggest challenge for the auto industry is petroleum. By the end of the year, China's reliance on imported crude oil was as high as 53.7%, hitting a historical high. As the international oil price fluctuates, such a heavy reliance on imported oil may well intensify the fragility of China's economy.

---

[7] Tang Jieqiong, "China's Auto Market Tops the World in yet Another Year." *Xiaoxiang Morning News*, November 17, 2011.

[8] Guo Chao, "140 Jammed Roads during Evening Rush Hours in Beijing Sets a New Record," *Beijing News*, September 18, 2010.

[9] Jiang Ye, "Second- and Third-tier Cities in China Experiencing Traffic Jams—More Jams, Greater Auto Sales," Xinhua News Agency, http://news.sohu.com/20101202/n278057924.shtml, December 2, 2010.

[10] Chen Ying and Wei Ning, "Traffic Jams in China: Seeking a Way Out and Looking for a Role Model," CRI Online, December 21, 2010.

D. *Life of Wanton Extravagance: Increasing Luxury and Luxurious Consumption*

A report by the Chinese Academy of Social Sciences indicates that by the beginning of the year, the total volume of China's luxury consumption had reached 9.4 billion USD, accounting for 27.5% of the global total.[11] The country also surpassed the U.S. for the first time and became the world's second largest luxury goods market. Furthermore, over the past few years, luxury brands have expanded from first-tier cities to many second- and third-tier cities. It is expected that in the next five years, China's luxury goods market will reach 14.6 billion USD, when it will surpass Japan and become the world's largest luxury goods market. This certainly contradicts China's intention to position itself as a developing country. To a large extent, this serves as one more piece of hard evidence for China's growing gap between the rich and the poor.

II. *Advocating Sustainable Consumption Faces Multiple Challenges*

It is quite obvious that in the foreseeable future, we will face multiple difficulties and challenges in advocating for sustainable consumption in China.

A. *The Government is Half-Hearted at Curbing Conspicuous Consumption and the Relevant Policies Merely Address the "Symptoms" Rather Than the Underlying Cause*

It is the Chinese government's intention to maintain a high GDP growth, control the unemployment rate, and even help stabilize the global economy through stimulating domestic demand. Therefore, it has to adopt an ambiguous attitude toward consumption problems. It is thus quite natural for the authorities to deal with overconsumption-led environmental and social problems on an ad hoc basis. Both the *Requirements for Restricting Excessive Packaging—Foods and Cosmetics*, a national policy, and Beijing municipal government's commitment to turn the city into a "traffic

[11] Li Xiaoyi and Lin Jinghua, "Growth of China's Luxury Goods Market," in *Annual Report on China's Commercial Sector (2009–2010)*, Social Sciences Academic Press, April 2010.

jam-free city in the next five years,"[12] a local policy, are such examples. This perfectly illustrates the zero-sum relationship between environmental protection and economic growth. It also means that the positive role the government could play in advocating sustainable consumption might be limited, and its negative role might be bigger than expected.

B. *Mainstream Consumer Culture Runs Counter to the Spirit of Environmental Protection*

It is a sad fact that "flaunting consumption" has been gradually accepted by mainstream culture and the general public. The media has to be partly blamed for this. Such media products as "You Are the One," a popular TV dating show, are filled with the approval and endorsement of living a materialistic life. It seems as if the airwaves promote consumerism wherever there are people, words, sounds, and images.

The most vibrant offspring from such fertile land is flaunting consumption, which means wasteful, luxurious, and extravagant consumption that is beyond the practical utilization of the product and daily necessity. Such consumption is meant to showcase one's wealth and social status as well as corresponding fame and prestige. In short, flaunting consumption is a reflection of one's vanity in materialistic consumption. The opposite of flaunting consumption is practical consumption. It is naturally true that everyone has some vanity. However, we do have to be vigilant when materialistic consumption meant to fill one's vanity has become a part of mainstream culture, and when such behavior has been wholeheartedly embraced rather than criticized by the majority of the people. This is because such a trend not only will lead to major environmental consequences, but also social instability.

C. *Power of Advertising Cannot Be Underestimated*

Contemporary marketing, or "advertising," has become a major means to stimulate consumption. However, the difference between contemporary marketing and traditional advertising is that the former features more and more sophisticated study of consumer psychology and more skillful application of this research. As a result, consumers have been reduced to mere puppets of manufacturers. Whether they cater to happiness, hope, van-

---

[12] Yao Yi, "Four Megacities Drafting Five-Year Plans, No Traffic Congestion a Major Goal," CPC News Website, http://news.qq.com/a/20101208/000675.htm, December 8, 2010.

ity, jealousy, depression, loss, or loneliness, salespersons can easily cajole consumers into buying almost anything. Faced with constantly updated sales pitches, few can resist the temptation. Furthermore, in a "paralyzed" world where no philosophy is powerful enough to compete with the idea of "materialism" on an equal footing, the impact of advertising will be even bigger, more difficult to resist, and will lead to more worrying damage to the sanity of the people.[13]

D. *An Irreversible Trend: Popularization of Credit Cards and Online Shopping*

Even as the Chinese government stimulates domestic demand and materialist consumption becomes more fashionable, consumers have become weaker at resisting the temptations of a materialistic world. If no convenient shopping methods are available, the frequency of consumption will be lower. However, online shopping and credit cards have made shopping much simpler, easier, and more convenient. Thanks to the Internet, consumers can shop around the world without stepping out of their living rooms. Credit cards, moreover, have fueled impulsive consumption, premature consumption, and overdraft consumption.

According to the *Analysis of Chinese Credit Card Holders' Behavior in 2010* (2010中国信用卡用戶使用行为分析报告) published by Tencent.com (腾讯网), "Credit card overdrafts are unsecured micro loans ... Compared to cash, convenient and fast bank card payment may better stimulate consumption as well as increase the efficiency and lower the cost of monetary circulation, which will contribute directly to the growth in consumption."[14] Many other studies done both at home and abroad have shown that credit card holders tend to shop more. The same analysis by Tencent.com also acknowledged that "One quarter of the respondents admitted that they shopped more by over 20% as a result of credit cards." This clearly illustrates the impact of credit cards on consumption.

The total number of credit cards issued by Chinese banks topped 190 million by 2010. In the long run, more and more consumers are likely to be lured by a wide range of credit card services, including so-called "premium services" as bonus point reward programs, flight insurance, roadside service,

[13] Lü Xiantao, "Study on the Marketing Strategies for Irrational Consumer Behavior," *Science Technology and Industry*, Issue No. 1, Vol. 10, January 2010.

[14] "*Analysis on Chinese Credit Card Holders' Behavior—A Study Based on the Survey of Tencent.com Users.*" DATA100 Market Research, July 2010.

and VIP medical treatment—unwittingly joining the ranks of credit card holders. The trend of using credit cards and the rapid growth of online shopping in China will surely reinforce each other. If we do not learn any lessons from other countries in this regard, China's financial wealth and cultural heritage may well be encroached on by impulsive consumption, premature consumption, and overdraft consumption. Needless to say, it is extremely difficult to advocate sustainable consumption against this background.

III. *Consumerism Is Exerting a Growing Negative Influence*

The impact on China's environment and society brought about by the emerging consumer culture and consumerism has gained the attention of more and more people.

The transformation in the pattern of economic growth is being interpreted from different angles. Many economists believe the only way to sustain China's economic growth is through stimulating domestic demand. Some scholars, though, have challenged the idea—do we really have the resources necessary to support consumption once the people's desire to consume is kindled? If not, what will happen then? Surely, since China's per capita resources are too scarce, we can make use of the resources of the world in this globalized economy. However, we have witnessed the fact that China's growing demand for resources has led to a fear on the international resource market as well as its dramatic growth. "The price of anything China buys will rise," so the saying goes. A major difference between stimulating domestic demand by Western countries and by China is that the former controls access to world resources while the latter does not. Over the past few years, China has suffered great losses due to the rising price of resources. Much of its economic growth has been offset because of the monopoly of the international raw materials market by the West. In the long run, the reckless stimulation of domestic demand may establish China's consumption patterns as well as technological and economic development on an outdated path—a path that does not comply with China's unique situation. Take the government's policies to encourage private cars as an example. They, in effect, will drive up the demand for petroleum, which will inevitably lead to a rise of oil prices on the international market. In turn, more and more Chinese will not be able to afford driving their cars even if they are car owners. In the absence of monetary and resource independence, if China's domestic demand has been driven up and if it cannot have access to the resources necessary on the international market, China may well end up in galloping inflation because it cannot export such inflation to the rest of the world like the U.S. with the U.S. dollar as an international currency.[15]

Furthermore, the negative influence on society brought about by consumerism—in particular, the excessive pursuit of commodities and money—has been well recognized. People have been controlled, or even been enslaved by, commodities. While paying more attention to making money, they tend to neglect the nurturing of morality, character, and philosophical wisdom that guides the overall development of human beings. Eventually, social morality will deteriorate, human wisdom will degenerate because of the sole pursuit of material wealth, and the overall performance of human beings will become worse. Thanks to consumerism, mass-produced cultural products will make cultural communication more and more shallow and simplistic, while elite and high-brow culture may be gradually watered down.[16] Some scholars have even gone as far as asserting that the age of consumerism is an age without classics.[17] In China, consumers who are too into fashion and who do not have the proper information or awareness are likely to become a frightening force in destroying the environment.

In his speech delivered in December 2003 at Harvard University, Premier Wen Jiabao said, "A large population and underdevelopment are the two facts China has to face. Since China has 1.3 billion people, any small individual shortage, multiplied by 1.3 billion, becomes a big, big problem. And any considerable amount of financial and material resources, divided by 1.3 billion, becomes a very low per capita level. This is a reality the Chinese leaders have to keep firmly in mind at all times." It is very clear to us that given such a large population, the shift in the consumption patterns of a rapidly growing China carries special significance for the country and the world. Fortunately, the decision-makers have reached a consensus to build a conservation-minded society. If the strategy can be successfully implemented, it will make an invaluable contribution to the entire human race in the long run.

---

15 Zhang Jie, "China's Stimulation of Domestic Demand is a Trap Set by the West," *Green Leaves*, Issue No. 11, 2009.

16 Wang Zhijun and Wu Weihua, "Information Production in a Consumer Society," *Jiangxi Social Sciences*, 2005 (1).

17 Yao Dengquan, "Post-modern Culture and Consumerism," *Qiusuo*, 2004 (1).

In the meantime, the shift in China's consumption pattern has drawn more and more international attention. For instance, Karl Gerth, a scholar of modern Asian history at Oxford University, argues that China's attempt to embrace American-style consumerism is an important development deserving international attention as it will affect all aspects of human life and even change the world. A small change in Chinese consumption patterns will lead to greater and more significant changes to the world than military budgets, carbon emissions, and trade disputes.[18]

We are also keenly aware of the fact that while a small number of Chinese are leading a luxurious life, many others have just shaken off poverty, and that the gap of per capita resources between a small number of developed countries and the vast majority of developing countries has not been reduced. Despite such a complex and difficult background, it is still our conviction that advocating sustainable consumption must be the only choice of the Chinese government and public, and we must resist and prevent the spread of consumerism to achieve sustainable development of the country.

IV. *Prospects of Advocating Sustainable Consumption and Some Recommendations*

At the end of 2010, the Central Committee of the Communist Party of China adopted an important document—*CPC Central Committee's Proposal for Formulating the 12th Five-Year Program for China's Economic and Social Development (2011–2015)* (中共中央关于制定国民经济和社会发展第十二个五年规划的建议). The proposal carries major significance for advocating sustainable consumption. Chapter Six of the document stipulates: "In the face of greater environmental restraints, we must have a sense of crisis and advocate a green and low-carbon lifestyle. The importance of energy-saving and emission-reduction should be stressed and a sound incentive and restraint mechanism should be set up. We should accelerate the steps in building a resource-saving and environment-friendly pattern of production and consumption to achieve sustainable development." Such a statement in this important document is a mixed blessing for advocating the pattern of sustainable consumption. Even though the section contains some encouraging key words, it has not

[18] Karl Gerth, *As China Goes, So Goes the World—How Chinese Consumers Are Transforming Everything*, Hill and Wang, 2010.

clearly defined the path and method for fostering such a pattern. We have every reason to question if only lip service will be paid.

Tang Dingding (唐丁丁), the director general of the Environmental Development Center of the Ministry of Environmental Protection (环境保护部环境发展中心), made the following statement in public. China has achieved some remarkable progress in setting up the legal framework and economic policy framework to support sustainable consumption. And, the policy framework for sustainable consumption is largely in place. First, in terms of the legal framework, China has set up an extensive manufacturer responsibility regime, clean production system, environmental labeling system, and government green purchasing system. Secondly, in terms of economic policies, China has designed such favorable polices as green financing, green credit, and green investment.[19] Unfortunately, despite such a sophisticated policy framework, we have yet to see evident progress in sustainable consumption, which is another indicator of the difficulty in advocating it. Faced with all these challenges, it will be hard to achieve tangible results with only some legal documents and government action. While having to acknowledge the long-term nature and urgency of advocating sustainable consumption, we have to admit that the process calls for the joint efforts from all sectors in the country. Therefore, in the 12th five-year program period, we have the following recommendations:

A. *Set Up Role Models and Properly Guide the Public to Raise Awareness of Sustainable Consumption*

The government should set up role models and properly guide the public, extensively and continuously educate the public about proper consumption patterns, and cultivate their awareness of sustainable consumption. It should engage in targeted, well-planned, and organized education campaigns to spread the right information, advocate the right philosophy, and raise consumer literacy. Such campaigns should be incorporated into and become a major component of the system of education for all. The government should also encourage consumer education campaigns by NGOs, social organizations (such as consumer associations), and the private sector.

---

[19] Tang Dingding, "Proactively Advocating Sustainable Consumption," *Wenhui Newspaper*, October 18, 2010.

B. *Advocate Green Consumption*

We should encourage government bodies, social organizations, and businesses to initiate green consumption awareness campaigns to raise customers' awareness of conservation, environmental protection, and sustainable development.

C. *Further Improve Relevant Policies and Regulations and Shape Consumer Behavior through Economic Tools*

The government should draft and implement a law on the protection of natural resources, and any consumer behavior that contradicts the concept of sustainable consumption should be subject to legal liabilities. It should impose sales taxes on specific purchases to curb unsustainable consumer behavior. It should formalize and legalize the "end-control" on consumption—while setting up a waste recycling and comprehensive utilization system. The government should also step up the control and punishment of consumption waste. Actions that are detrimental to the ecological and consumer environment should be subject to legal liabilities.

D. *Develop a Green Logistics Channel*

While paying attention to sustainable production and sustainable consumption, the government should also pay attention to green logistics and sustainable trade. In addition to performing regular logistics functions, green logistics may also perform such specific roles as engaging in clean production, dealing in green products, advocating green consumption, and waste recycling, which may help address such problems as overpackaging, packing waste pollution, and urban environmental pollution.

E. *Resist a Consumerist Lifestyle and Encourage Sustainable Consumption*

The active participation of consumers is the prerequisite for sustainable consumption. Any consumer behavior is ultimately performed by the consumer, and consumption and production are inter-related. As a result, only when every consumer has realized the dangers of consumerism and is willing to reduce his/her own unnecessary consumption can sustainable consumption be achieved. All consumers should take joint efforts to make purchases in a sustainable manner.

In sum, in order to set up a model of sustainable consumption that complies with China's unique situation, the government should formulate policies to guide and encourage sustainable consumption by the public;

enterprises should operate within the industrial framework outlined by the government and shoulder due corporate responsibilities instead of solely and recklessly seeking profits; social organizations should actively promote the right concept and behavior at the community level and look for innovative methods to offer more solutions for society; and consumers should develop their independent thinking ability and be aware of misinformation from the media and businesses. Only when innovation and improvement are achieved in all of the above interrelated sectors can we welcome the dawn of sustainable consumption.

A NEW ROUND OF LUXURY WATER CONSUMPTION: ARTIFICIAL SNOW PARKS[1]

Hu Kanping

Abstract: Snow was not seen until very late in the winter of 2010, worsening existing water scarcity. However, ski resorts and snow parks continued to thrive and prosper unaffected in Beijing, using artificial snow. Is it sustainable to enjoy recreation by man-made means? Through field investigation and public opinion tracking, this paper reviews the prosperity of the ski industry that disregards the snowless winters, analyzes the possible price the capital city might have to pay in water resources and ecology, provides suggestions for government agencies to step up administration and supervision of the ski industry, and reminds the public to care more about the speechless natural environment.

Keywords: artificial snow, water resources, environmental problems, sustainable consumption, luxury water consumption, investigation and study

Beijing's water crisis did not improve at all in 2010, with a 50-year low rainfall volume in the high-water rainy season and a winter without any effective waterfall (the first time in the last 60 years). In spite of this snowless winter, ski resorts in suburban Beijing and snow parks in the city prospered. Related water consumption and environmental problems drew extensive attention from the media and the public. Unless its water crisis is solved effectively, Beijing's development will not be sustainable, let alone will it achieve its grand goal of building a "World City."

I. *A Dry Capital with a Snowless Winter*

As a mega-city with a serious water shortage, Beijing is constrained in its efforts for greater development by its limited water resources. According to official statistics, in the first decade of the new millennium, Beijing's per capita water resource volume was halved. As specified in *Explanation On Beijing Municipal Water Conservation Measures* (*For Approval*) (关于北京市节约用水办法(修订草案送审稿)): "With rapid social and economic development and a deteriorating drought, Beijing's per capita

[1] This paper is part of the Green Book 2010 special research project, funded by Delta Environmental & Educational Foundation.

water resource volume has dropped to about 150 cubic meters, from less than 300 cubic meters at the end of the last century."[2]

Since 1999, Beijing has been in a period with insufficient rainfall, worsening the existing water shortage. 2010 witnessed a historic low point of rainfall in Beijing, with a quarterly average rainfall of only 463.1 millimeters in the first three quarters, 44.8 millimeters lower than the quarterly average of the previous decade and 31 millimeters lower compared with that of the corresponding period in 2009. Generally speaking, rainfall in the main high-water seasons accounts for 1/3 or more of Beijing's total annual rainfall. However, the 2010 average rainfall during the high-water season was 25% lower compared with that of the same period on average, and the lowest point since 1960. Droughts of varying levels were found across Beijing.

According to a *Beijing Evening News* report on January 25, 2011, the latest coming of first snowfall in Beijing over the last 60 years came on January 29, 1984 and "we may have a new record this year." A meteorologist from the Beijing Meteorological Bureau explained that snow did not fall because with a very low relative humidity, the atmosphere was too dry for its formation.

With no major rivers, Beijing depends mostly on rainfall, underground water, and water diverted from neighboring provinces (or from the south of China in the future). According to the statistics provided by the Beijing Water Authority, the city's annual water demand reaches up to 3.79 billion cubic meters, with an annual shortage of 1.79 billion cubic meters. Zhang Tong (张彤), chief engineer of the Beijing Institute of Water Resources Planning and Design (北京市水利规划设计研究院), is reported as saying, "Water shortage has become a normal issue in Beijing," a problem made more serious by difficulties in water diversion, limited potential reserves of underground water, and inadequate rain and snow in 2010.

II. *Unprecedented Prosperity of Ski Resorts Dependent on Man-Made Snow*

The booming trend of man-made snow parks is absolutely startling in Beijing, a city with very limited water resources, particularly in a dry winter.

[2] *Explanation On Beijing Municipal Water Conservation Measures (For Approval)*, the Legal Affairs Office of the Beijing Municipal Government, http://210.73.64.50/advice/user/content_new.asp?UNID=317 (accessed October 10, 2010).

Figure 17.1: Distribution of registered ski resorts and snow parks in Beijing.

No accurate calculation has been made as to exactly how many ski resorts Beijing has. One press conference of the Beijing Municipal Bureau of Sports reported 17 ski resorts with new facilities, while many industry insiders disagreed with this figure. In addition, many of the large ski resorts in Beijing, such as Shijinglong (石京龙), Badaling (八达岭), Jundushan (军都山), Huaibei (怀北), Yuyang (渔阳), Lianhuashan (莲花山), Yunju (云居), and Nanshan (南山), are operating as usual this snowless winter, while many snow parks, such as those at the Bird's Nest (鸟巢), Longtanhu Park (龙潭湖), Yuyuantan Park (玉渊潭), Taoranting Park (陶然亭), Haidian Racing Park (海淀狂飙乐园), and Chaoyang Park (朝阳公园), are attracting more enthusiasts, with their customer volume reaching a record high this year. *Beijing Evening News* reported that during the three-day, 2011 New Year holiday, the 12 ski resorts monitored by the relevant department entertained 57,000 visitors, an increase of 9.7%.[3]

[3] Fu Yang, "Packed Hot Springs and Ski Resorts in the New Year Holidays," *Beijing Evening News*, January 4, 2011.

Where does the snow used by these ski resorts and snow parks come from? A common scene at these venues helps us answer this question: snowmakers are busy spraying pressured water into the air, leaving behind shiny ski slopes.

Man-made snow is often used to recreate winter scenes in movies and TV productions as well as the living conditions for polar animals in zoos. Now, man-made snow is widely used by ski resorts around the world. Even in France and America where the awareness of water conservation is relatively high, man-made snow is used to satisfy the demand of skiers. In February 1980, Lake Placid in the U.S. was hit by a winter with little snowfall and it was man-made snow that eventually ensured the success of the 13th Winter Olympics. However, Beijing's normal reliance on man-made snow for ski resorts is exceptional.

III. *Is Skiing a Sustainable Recreation?*

Skiing generally is a special kind of sport and recreation in an environment with natural snow. However, a new definition of skiing has developed in Beijing. Due to the general lack of snowfall, many ski resorts have resorted to man-made snow to meet the basic requirements of a ski slope. As a result, skiing is no longer a chance for people to get closer to nature but one to immerse people in unnatural snow lands, a luxurious trend that consumes exorbitant amounts of water.

On December 22, 2010 (the winter solstice), Yuanming Park (圆明园) launched the "Royal Snow Festival," claiming to recreate the scene when the royal families entertained themselves on snow-covered grounds. However, with no snowfall that winter, the park had to use snowmakers to create the 200 meter long ski slope covered with 20 centimeters deep man-made snow. According to workers on site, in order to create the necessary scenes on the 60,000 square meters snow ground, they had to operate dozens of snowmakers nightly.

This ski slope with man-made snow has kept the public and the media concerned about the waste of water and the negative impacts on the water system in the park. The Park Administration Commission explained that the water used by snowmakers came from the irrigation system and would not affect the water system and that when melted, would go directly underground. Relevant officials also explained that snowmakers in some of these parks used the existing water reserves in these parks and melted water would be recycled. Meanwhile the Bird's Nest and other snow parks, with no natural water reserves, used "reclaimed water."

Source: Website promotional image.

Figure 17.2: A snowmaker at work.

Many ski resorts in the Beijing suburbs claim to be using "reclaimed water" in their snowmakers. For example, the Badaling Ski Resort claims that 70% of the water used by their snowmakers comes from "reclaimed water" generated in neighboring villages and 70% of melted water is recycled.[4]

However, the validity of such claims is seriously questioned by the media and relevant administrations because "reclaimed water" cannot be used in ways involving direct human contact.

In an article written in 2010, Feng Yongfeng (冯永锋) pointed out, "While man-made snow parks are entertaining people, the environmental cost of such recreation has been overlooked." Feng refuted the claim of recycled man-made snow, saying, "Man-made snow consumes an appalling amount of water, with one snowmaker using up to 16 tons of water an hour. Most man-made snow will vaporize or go underground, unable to be recycled. The luxury use of water in Beijing, a dry city with a paralyzed water system, is absolutely irrational."

[4] Zhang Hao, "Reclaimed Water Used to Make Snow at the Beijing Badaling Ski Resort," *Technology of Water Treatment*, No. 1, 2008.

Photo taken by Hu Kanping.

Figure 17.3: Appalling damage done by ski resorts to the local environment and landform.

IV. *Dispute over Water Consumption*

Evidence shows that most suburban ski resorts use underground water in their snowmakers. The fierce debate on ski resorts' consumption of underground water over the years has involved China Central Television (CCTV), the Xinhua News Agency, and many other mainstream media outlets in China. Reporting based on the figures provided by one expert, the 13 ski resorts then consumed a total of 3.8 million cubic meters of water a year, enough to provide for 42,000 residents for a year.[5] This report made many "relevant departments" restless, attempting to defend the ski resorts. Case in point, the Chinese Skiing Association claimed that there were only 10 ski resorts in Beijing, with a certified annual water consumption of 416,300 cubic meters, of which only 297,300 cubic meters was underground water.

[5] Liu Puquan, "13 Ski Resorts in Beijing Use Up to the Annual Water Consumption of 42,000 Residents," *Beijing Youth Daily*, July 18, 2005.

The fact is that ski resorts have been taking advantage of the bull management of the ski industry for years. "Relevant departments" do not have any accurate figures concerning water consumption. According to the standard set by the *Beijing Municipal Regulation on Ski Resorts Water Use Management* (北京市滑雪场用水管理要求) (which mandates that the annual water consumption of a ski slope must not exceed 0.48 cubic meters per square meter and that of the green land of a ski resort no more than 0.3 cubic meters per square meter) and the total skiing surface area in Beijing, a simple calculation shows that snowmakers consume a total of at least 1 million tons of water every year, enough to provide for 8,300 families annually.

V. *Is Skiing on Man-Made Snow Luxury Consumption?*

The Beijing Municipal Bureau of Sports has established the following guideline concerning the ski industry: "To realize the harmonious development of an ecological civilization and the sports industry." This highlights the major challenge facing the ski industry in Beijing. At the heart of the sustainability of this industry is not how much water it consumes, but whether its exclusive reliance on man-made snow corresponds with such ideals as "Green Beijing," "Ecological civilization," "Low-Carbon Economy," "Environmentally Friendly," and "Resource Conservation" If not, what is the rational limit to the development of this industry?

In practice, to build an ecological civilization means to build a resource-saving and an environmentally friendly society. So, is Beijing's ski industry compatible with such a goal? Continuous public debates on this question have been heard in recent years.

The power consumption of ski resorts is also controversial. The power consumed by a snowmaker averages 15–20 kW, that is, up to 480 kWh a day if the snowmaker works on a 24 hour basis. Cable cars also use a significant amount of power. Moreover, many ski resorts use high-power lighting equipment at night. According to the calculation in a newspaper report, the "carbon footprint" of snowmakers in Beijing adds up to 287 tons of carbon. "Those who go skiing," said the newspaper, "quickly increase their carbon deficit." Therefore, skiers, aware of skiing-related carbon emission, must pay more attention to saving energy in their daily life so as to offset the deficit.[6]

[6] Gao Peng, "Ski Resorts with Snow Makers Consume a Large Amount of Power," *Liaoshen Evening*, December 28, 2010.

During a telephone interview with *China Comment* on April 15, 2010, the publicity director of the Beijing Water Authority questioned the validity of the notion of "luxury consumption," a concept put forward by the author in the *China Environment Yearbook, Volume 5*. "With the rising level of living standards, ordinary people now want to ski. How can we say it is a luxury?" said the director. "In the past, a ton of water could create 1 RMB in GDP, but the bathing industry can now generate 100 RMB in GDP with the same water. Taking into account the greatly enhanced water efficiency and its contribution to GDP growth, this is not a luxury." Such "GDP over water resource protection" beliefs are not rare, but it is rare for a government official to make such a comment.

Liang Fan (梁帆) proposes a criterion to decide whether a consumption pattern is luxurious or not from an environmental ethics perspective; that is, if something cannot satisfy one person's need and another person's desire at the same time, the ethical way is to give it to the previous. Or put in another way, the satisfaction of need triumphs over that of desire. Therefore, the question now is whether the water resources available in Beijing can satisfy the daily need of ordinary citizens and the special recreational desires of some people at the same time? If not, is it not ethically wise to put a limit on the latter?

Article 29 (on industries using water for special purposes) in the draft of the *Beijing Municipal Water Conservation Measures* (*For Approval*) (北京市节约用水办法(修订草案送审稿)) publicized from September 20 to October 18, 2010, stipulated, "New businesses in the following high water consumption industries are prohibited in Beijing: high-end bathing centers, production factories with water as raw material, ski resorts, and golf courses and parks with a monthly water consumption of more than 5,000 cubic meters. The existing above-mentioned businesses must apply the categorized measurement, control their water consumption, install data transmission equipment, and report their water consumption monthly to the water conservation office." Later, *Notes on Revision* (修订说明) of the measures reads, "To standardize the water consumption of special industries and special purpose water consumption, including purified water production, auto care, bathing, ski resorts, golf courses, real-time water purifying and vending machines, and engineering and heating."

If ski resorts were, as claimed by some, sustainable with no water waste, what is the rationale for setting a limit?

In fact, the government has always emphasized the protection and rational use of water in the development of ski resorts. Article 14 on man-made snow in the *Chinese Ski Resort Management Standard* (*Trial*) (中国滑雪

场所管理规范(试行)) jointly issued by the National Winter Sports Management Center and the Chinese Skiing Association stipulates, "Improve the cost efficiency of snow making to 65–70% based on scientific analysis of temperature, humidity, wind, and water temperature. In the absence of necessary conditions, snow-making is prohibited to avoid energy and water waste. Monitor the amount of snow made closely to avoid waste. It is encouraged to use surface water stored during the year in snowmakers and recycle melted water."[7]

VI. *Watch Out for Ecological Damage in the Construction and Operation of Ski Resorts*

Ski resorts can cause more ecological problems than just water waste. Ecologists have pointed to the environmental damage to the neighboring environment.

Most ski resorts sit on sloping hills, which used to be home to a complete ecosystem of trees, bushes, and grass. Because of the special requirements of ski slopes, these trees and bushes are removed. When the snow melts, the lane surface, originally covered by snow becomes bald and fragile to wind and rain. Moreover, the piling waste left behind by skiers also adds to the ecological pressure in the area. Therefore, it is urgent and important to emphasize ecological protection in the development of ski resorts.

Article 19 on man-made snow in the *Chinese Ski Resort Management Standard* (*Trial*) has set the following environmental and sanitation requirements: "Pay special attention to environmental protection during the construction and operation of ski resorts by minimizing the damage to existing foliage. Ski slopes shall be built with minimum change to the original landscape. If earthwork cannot be avoided, the construction site must be reforested within two years." Article 22 in the *Chinese National Tourism and Ski Resort Management Standard* (*Trial*) (全国旅游滑雪场管理条例(试行)) has the same provision. However, these regulations are intentionally ignored during the construction and operation of ski resorts. Many experts have been calling on ski resorts to take measures to reforest the barren land so as to prevent wind-borne dust and soil erosion and advocate controlling the scale and number of ski resorts. Unfortunately,

7 *Chinese Ski Resort Management Standard* (*Trial*), http://focus.lwcj.com/FocusReport/focus0090811140002_7.htm.

Photo taken by Hu Kanping.

Figure 17.4: An abandoned ski slope looks barren even with new grass planted.

no humanly possible efforts can quickly restore the damaged mountain ecosystem to replace the original one, as emphasized Guo Geng (郭耕), deputy director of the Beijing Biodiversity Protection and Research Center (北京生物多样性保护研究中心). The development of sports and recreation must not jeopardize the health of nature, the precondition to the health of all living things.

VII. *Suggestions on Strengthening the Management of Man-made Snow Parks*

In the *Twelfth Five-Year Plan Outline of Beijing* (*Draft*) (十二五 规划纲要 (草案)) publicized in January, 2011, the Beijing Municipal Government is preparing to tighten water resource management, by raising the water-conservation threshold for industries and sifting out high water consumption industries, so as to promote the construction of a water-conservation society. Accordingly, the author proposes the following measures to enhance the management of man-made snow parks:

Photo taken by Hu Kanping.

Figure 17.5: Foliage at the same location at different times with photo below showing the normal state of foliage on the slope.

1) Increase water and energy conservation and environmental protection education, popularize the ideal of ecological civilization, and promote the principle of "prioritizing necessary water consumption in production and living while limiting luxury water consumption."
2) In accordance with the suggestions in the draft of the *Beijing Municipal Water Conservation Measures* (*For Approval*), revise, finalize, and issue the new measures as soon as possible.
3) On the basis of in-depth study, formulate the guidelines and the exit mechanism of the ski industry. Specify the threshold of the industry and the responsibilities of relevant governing departments. Increase the monitoring of water consumption and environmental conditions at man-made snow parks. Renovate and eliminate ski resorts that exceed water consumption quotas, fail to report the actual water consumption, or have caused serious ecological damages, and hold the responsible person(s) accountable.
4) Enhance the transparency of administrative affairs and disclosure of the annual water and power consumption data at man-made snow parks on a regular basis.

GOLF COURSES IN BEIJING AND OSTENTATIOUS WATER USE

Hu Kanping

Abstract: There are about 60–70 golf courses in Beijing. With a total annual water consumption of as much as 40 million cubic meters, these golf courses have raised pressure on the Beijing's water supply and ecological safety. This paper, on the basis of fieldwork, makes an in-depth analysis of the relationship between the number of golf courses and water consumption volume in Beijing, and discusses the serious impact of ostentatious water use on the ecological safety of Beijing. Suggestions are also proposed to limit the development of golf courses while enhancing supervision according to the requirements of building a conservation culture and water-saving society.

Keywords: ostentatious water use, golf, conservation culture, Beijing

In 2011, eleven ministries, including the National Development and Reform Commission, the Ministry of Supervision, the Ministry of Land and Resources, and the Ministry of Environmental Protection, jointly issued the *Notice on the Comprehensive National Campaign to Clean up and Rectify Golf Courses* (NDRC Society [2011] No. 741) (关于开展全国高尔夫球场综合清理整治工作的通知). Illegal land use and construction of golf courses have drawn criticism from the government and the public. In addition, ostentatious water use by golf courses, a major pressure on Beijing's water supply, has constantly come under the media spotlight, as well as scrutiny from public and civil organizations.

In 2011, the author, together with volunteers, conducted fieldwork amongst several golf courses in Beijing. Also using relevant media data, this paper studies water use by golf courses and its impact upon Beijing's water supply and ecological safety.

I. *Beijing's "Disastrous Water Shortage"*

The "No. 1 Document" from the CCP Central Committee and the State Council in 2011 ranks prompt reform of water conservation as an issue of China's "economic, ecological, and national security." The National Water Conservancy Work Conference, held in July, was attended by many high-ranking Party and government officials, which was unprecedented. The

Central Government defined the themes of this year's water conservancy work; that is, to "most strictly enforce the water resource management system, determinedly prevent wasteful water use, and effectively incorporate water conservancy into the whole process of socio-economic development and people's life and work."

According to official data, China, ranking sixth worldwide with its total water resources amounting to 2.84 trillion cubic meters, is only 88th in terms of per capita water resources. According to data disclosed by the Ministry of Water Resources, at present, the amount of water available per person in China is 2,100 cubic meters, only 28% of the world average.[1]

As for Beijing, water shortages have become a major constraint to its development. According to the *Report on China's Sustainable Development Strategy* (中国可持续发展战略报告) published in February 2002, Beijing ranked third to last in terms of per capita water resources (325m^3/person). However, in 2010, the Beijing Municipal Government stated clearly in the introduction to the *Beijing Municipal Water Conservation Measures* (*Draft for Approval*) (北京市节约用水办法(修订草案送审稿)) that Beijing's per capita amount of water available has dropped to about 150 cubic meters.

In fact, not much change has been recorded in 2011 as compared to other years in terms of total water resources. According to a report by the Beijing Water Authority on November 27, the total storage volume of medium- and large-sized reservoirs (with a gross storage capacity of 9.20 billion cubic meters), Beijing reached 1.47 billion cubic meters, or 60.81 million cubic meters more than that of the same period of the previous year. However, the ever-growing population and many other factors have contributed to the shrinking of per capita water resources in Beijing. According to Bi Xiaogang (毕小刚), vice director and spokesman for the Beijing Water Authority, the per capita water resources for Beijing's 19.62 million residents is calculated to be merely 100 cubic meters, that is about 1/20 the national average in China and 1/80 the world average.[2] During an interview with a China Central Television (CCTV) reporter, a

1 Man Zhaoxu, "China to Implement Strictest Water Management Measures to Control Water Consumption," CNBN.cn, accessed on July 8, 2011 at http://www.cnr.cn/china/news/201107/t20110708_508198244.html.

2 Author's Note: Permanent residents include registered residents and unregistered residents who have lived in the city for over half a year. However, the water-using population of Beijing is definitely much larger than its permanent residents, since short-term residents are not taken into account in the calculation of per capita water consumption. Experts estimate that Beijing's water-using population is far greater than 20 million.

Beijing Water Authority official pointed out that Beijing's per capita water resources account for only 1/3 that of people in Israel.

Among the world's mega cities with a population of over 10 million, Beijing is the driest, with a "catastrophic water shortage" according to UN standards. With reference to the UN standards, China has developed its own water shortage grading system; that is, mild water shortage with a per capita water availability of less than 3,000 cubic meters, serious water shortage with a per capita availability of 500 to 1,000 cubic meters, and extreme water shortage of less than 500 cubic meters. According to the Chinese system, 300 cubic meters is the minimum requirement for survival, but Beijing has only a third of the minimum now.

Though there is no denying that great achievements have been made with all the efforts in water saving and comprehensive water use, Beijing is still facing an increasingly extreme water crisis with its rapid urbanization and population growth. However, as specified in the *Report on the State of the Environment in China* (中国环境年度报告), entertainment facilities with ostentatious water use such as bathing centers, man-made hot spring resorts, and artificial snow ski resorts are not rare in Beijing, running counter to the strategic goal of establishing a water-saving and environmentally-friendly city. Among these facilities, golf courses use water most ostentatiously.

II. *Over 60 Golf Courses in the Water-Starved Capital City*

In January 2004, the State Council issued the *Notice on Suspending the Construction of New Golf Courses* (关于暂停新建高尔夫球场的通知), banning the approval of new golf courses until further notice and mandating local governments conduct a comprehensive survey of golf courses in operation and under construction. Since December 2006, golf courses have appeared on the *Inventory of Banned Land Use Projects* (禁止用地项目目录). Up to now, the Chinese government has issued 10 injunctions against the construction of golf courses.

The question now is: How many golf courses are there in China? Though official statistics from the Ministry of Land and Resources are not available, the media has been keeping track of their denied information disclosure requests.[3] It is estimated that by the deadline of 2004, there were

[3] On August 21, 2011, People.com.cn and GMW.cn both reproduced Dong Peiwen's article from *Workers Daily*, which says an official in charge of land registration of the Ministry of Land and Resources hesitated to answer a reporter's questions on the matter.

170 golf courses (standard 18 holes). According to the *Industry Orientation White Paper: A Report on China's Golf Industry* (朝向白皮书：中国高尔夫行业报告) published in 2010, China has 490 golf courses in operation. Another widely quoted figure is 595, reportedly given by leaders of the Small Ball Game Administration of the General Administration of Sports of China and China Golf Association during interviews. More dramatically, the Gold Education and Research Center of the Beijing Forestry University estimates that by May 2011, China had about 600 golf courses.

However, the development of golf courses is restricted by the government mostly for the purpose of regulating land use. In the first half of 2011, over 34 cases of illegal golf course land use and construction projects were reported to the Illegal Behavior Reporting Center of the Ministry of Land and Resources. That is an increase of 31% year on year. Many land law enforcement officers have found law enforcement difficult because golf courses are typically constructed by local governments to enhance their regional image and attract investment. In fact, the administration of golf courses remains unspecified, making it difficult for any government agency.

No government statistics have ever been disclosed on the number of golf courses in Beijing, except a Xinhua News Agency report in 2004 which reported 38 golf courses in the city. The author once logged onto Qunar.com to search "Golf Course" and "Golf Driving Range" and found 111 different results. Figures reported vary greatly. For example, a media report says there are at least 75 golf courses in Beijing, covering a total area of 132,257.75 *mu*.[4] Another one reports, "There are altogether 132 golf courses, clubs, and driving ranges in Beijing, with the land area available for 103 of them covering a total area of 135,682.1 *mu*. Within Beijing, Chaoyang District (朝阳区) hosts 41 golf courses of various kinds, followed by Haidian District's (海淀区) 19." However, none of these reports specify its sources.

It was not until early 2011 that a report, with figures provided by a Ministry of Land and Resources insider, said: "A recent satellite survey by the Ministry confirms the serious violation of laws and regulations banning golf courses and the existence of 170 golf courses (including driving ranges) in and around Beijing." If this figure, not yet the largest among those reported, is true, Beijing's golf courses account for a third of the national total. The author has searched www.golf72.cn, an authoritative

4 1 mu≈666.67 m^2

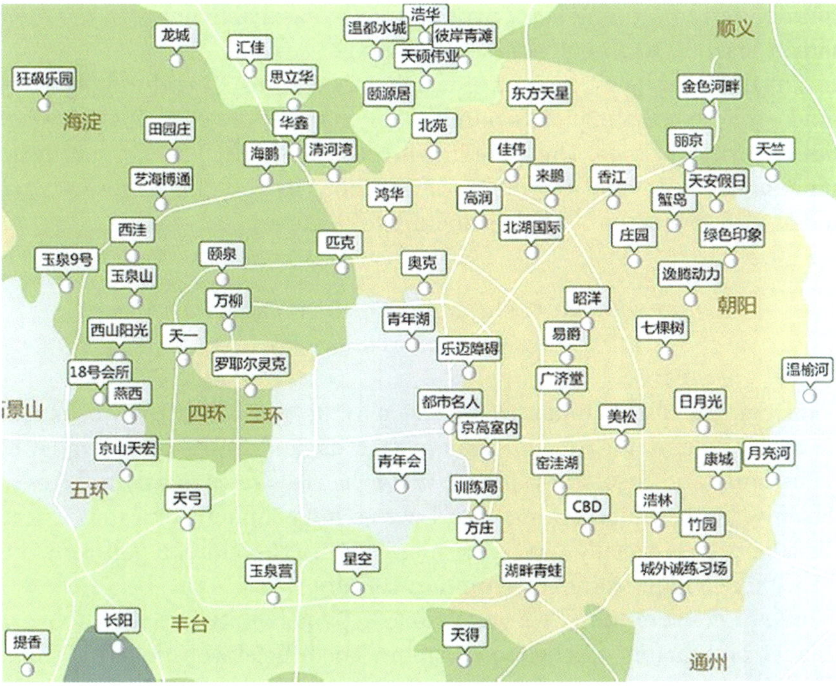

Figure 18.1: Map of golf courses around Beijing.

golf industry website and found 73 golf courses (including 5 in neighboring Hebei Province), 81 driving ranges, and 55 golf-themed real estate projects adding up to over 200.[5,6]

So how many standard golf courses are there in Beijing? According to Cui Zhiqiang (崔志强), former vice chairman of the China Golf Association (中国高尔夫球协会) and president of the Beijing Pan-China Sports Company (北京泛华新兴体育发展有限公司), there are about 60 standard golf courses in Beijing, confirming the figure of registered golf courses in the 114 Directory Query Service. Among the 170 golf courses shown in the satellite photo mentioned above, "at least 70 are involved in serious illegal occupation of farming land." Since driving ranges and real estate projects generally cannot be classified as "land use projects of serious illegal

[5] http://www.golf72.cn/custom/listByProvince.do?province =北京.

[6] Author's Note: Golf courses, driving ranges, and golf real estate projects are greatly different in terms of function, land area, and water consumption.

occupation of farming land," would it be a reasonable guess to say there are at least 60 to 70 golf courses in Beijing?

During interviews, many insiders say that 60 to 70 is the right figure and agree that Beijing ranks number one in terms of golf course numbers and density in China. The Yongding River Basin (永定河畔) in southwest Beijing has been crowned the "Golf Corridor," by golf fans with seven golf clubs and seventeen 18–hole standard golf courses.

III. *Annual Water Consumption of 40 Million Cubic Meters by Golf Courses in Beijing*

Golf courses' ostentatious water use may be attributed to two factors: 1) a golf course must include water hazards; and 2) the greens must be constantly and regularly watered. Water hazards refer mostly to lakes or shallow pools, which are sustained with a large amount of water. Irrigating the greens usually uses even more water. Therefore, a golf course is inevitably a high water-consumption venture.

Water consumption by a golf course depends on its size, area, types of plants, and quality of irrigation systems. The official website of the China Golf Association says that a standard 18-hole golf course generally covers an area of 750 to 1,500 *mu*. According to a golf course greens management expert, a standard 18–hole golf course uses 2,000 to 2,500 cubic meters of water on average every day. The *Industry Orientation White Paper: A Report on China's Golf Industry* reported the annual total irrigation water consumption of 51 golf courses. Based on the figures, the annual irrigation water consumption of an 18-hole golf course is calculated to have reached about 276,000 cubic meters, though the *White Paper* acknowledges, "Owing to geographic, environmental, and climatic differences, the annual amount of water used in irrigation may vary, from less than 150,000 tons to more than 600,000 tons."[7]

So how much water do China's golf courses consume annually? No definite answer has been given. According to Su Derong (苏德荣), professor at the Beijing Forestry University and secretary general of the Course Management Committee of the China Golf Association, "In Beijing, an 18-hole golf course uses 2,000 to 2,500 cubic meters of water every day on average, throughout the year except for three months in winter, which

[7] For fresh water, "ton" and "m^3" are interchangeable. Here, the original uses "ton" unchanged here.

means an annual total of 400,000 cubic meters. Since many golf courses in Beijing may have 27, 36, or even 54 holes, their water consumption is much higher."[8]

During field investigation, a golf course designer told the author that during the operation period, in case of zero or limited rainfall, the lawns must be watered with five millimeters of water at least once a day, a protocol confirmed by golf course workers.

Based on this, a simple calculation may produce the total annual water consumption of golf courses in Beijing: 5mm/day×568,000m^2/golf course[9] (national average per standard golf course)[10] ×60 golf courses×240 days=40.896 million cubic meters. If the number of golf courses is set to be 70, the total annual will reach 47.712 million cubic meters.

This calculation approximates the 40 million cubic meters or so specified in a report prepared by the Beijing Municipal Committee of the CPPCC. In August 2011, the "Economic 30 Minutes" program on CCTV disclosed the figure of 40 million cubic meters, which led to dissatisfaction in the general public. Although no official response has been issued, this figure remains the most reliable projection.

Though the above-mentioned figures remain to be confirmed, the author will base his analysis below on these figures.

According to the *Urban Resident Domestic Water Consumption Standard* (GB/T 50331-2002) (城市居民生活用水量标准), daily water consumption for Beijing's urban residents ranges from 85 to 140 liters per person; that is, 31 to 51 cubic meters per year per person, with a mean value of 40 cubic meters. This means the 40 million cubic meters of water is enough to provide for 1 million people for a year. In fact, the actual domestic water consumption of Beijing residents is generally at the lower end of the spectrum; that is, 31.7 cubic meters in 1995, about 30 cubic meters from 1995 to 2004 and 31 cubic meters in 2004, according to the Beijing Bureau of Statistics. In this case, the annual water consumption of golf courses exceeds, by a large margin, the total standard water consumption of a million residents.

---

8 "Economic 30 Minutes," CCTV, August 8, 2011. http://video.sina.com.cn/p/news/c/v/2011–08–08/221061438005.html.

9 Data quoted from the *Industry Orientation White Paper 2010*. 568,000 m^2≈850 *mu*.

10 A calculation based on samples show an average lawn area of 568,000 m^2 for each 18-hole golf course. This paper assumes that Beijing's average is the same as the national average.

The *Report on Main Statistics of the 6th National Census in 2010* (2010年第六次全国人口普查主要数据公报) shows that the permanent resident population of the two new downtown districts of Beijing, Dongcheng District (东城区) (the old Dongcheng District and the former Chongwen District) and Xicheng District (西城区) (the old Xicheng District and the former Xuanwu District) is 919,000 and 1,243,300 respectively. Then, the golf industry consumes the total standard water resources for about half the population of downtown Beijing.

IV. *Excessive Water Consumption as a Result of Extensive Management*

Why do golf courses consume so much water? A standard 18-hole golf course, covering no less than 50 hectares and nearly 60 to 90 hectares of land, consists of such landscape as woods, lawns, lakes, and sandpits, with the lawns taking up over 80% of the area. The lawn is divided into service areas, fairways, greens, and tall grass areas. Under most circumstances, a golf course's design highlights its scenic effects and usability, with little emphasis on maintenance in the future. Therefore, to ensure good drainage, sand is used extensively, so that the golf course may be used on rainy days and after watering. However, because of the land's poor water-retaining property, regular irrigation is mandatory.

Another reason for the high level of water consumption is that golf courses in China generally tend to be big. According to the *Industry Orientation White Paper*, the lawn area of a standard golf course reaches 568,000 square meters on average in China, 43% larger than the average of 396,000 square meters in the United States. The central area (including the green, the service areas, the fairways, and the tall grass areas) in China on average is 30.8% bigger than that of the U.S. The gardening area of a golf course in China on average is 5.7 times that in the U.S.

An insider's story has it that although Australia, a similarly dry country, has many more golf courses than China, the Australian government has played a much more prominent role in regulating the construction and operation of golf courses. A feature worth noting is that in dry seasons, many golf courses are closed temporarily to reduce water consumption. In the United States, golf courses are blamed for its ostentatious water use and excessive use of pesticide, with an average annual water consumption of 50 million gallons, or 189,000 cubic meters. However, according to a report in *The New York Times* on August 5, 2009, great progress has been made after the management of golf courses in the U.S. started using recycled water, local lawn grass, and closely monitored irrigation systems.

V. *Grave Risks in Excessive Withdrawal of Groundwater*

The *Notice on Water Management at Golf Courses* (关于加强高尔夫球场用水管理工作的通知) issued by the Beijing Water Authority in 2006 set strict rules concerning water use standards, water use measurement, and the use of water-saving facilities. For example, any golf course within range of a reclaimed water system has to use reclaimed water for irrigation purposes. If applicable, a golf course must build reclaimed water treatment facilities. Man-made lakes must be seepage-proof. Sidewalks, cart paths, parking lots, and roofs of buildings must be designed to facilitate the collection and utilization of rainwater. Water meters of various purposes must be installed at different locations accordingly. Surface water and groundwater can be used directly only after its water use feasibility study is approved. If the existing wells are originally drilled for farming (forestry) purposes, relevant purpose alternation formalities must be first completed at the local water authority to verify the amount of water to be used and to specify the new purpose-based charging standard.

However, these rules have not been effectively enforced. Fieldwork shows an epidemic of unapproved wells and pumping stations.

It is said that many of Beijing's golf courses have nothing to do with the development of golf itself. If we type in the official names of these golf courses at the official website of the Beijing Administration for Industry and Commerce, the search results confirm this. Many golf courses have been built in sports parks, eco-parks, leisure parks, or even green lands, many of which had not been approved. As a result, the local water authority does not have a record of the status of water consumption at these courses. For example, the Reignwood Pine Valley (华彬庄园), the Lafeite (拉斐特), and the Bi'anqingtan (彼岸青滩) golf courses in Changping District (昌平区), do not have the necessary approval to build their own self-provided wells. The other three surveyed golf courses do have approvals, but two of them are authorized to withdraw 20,000 cubic meters and 45,000 cubic meters respectively, which is far less than their actual water consumption.

According to the Beijing Water Authority, the total water consumption of Beijing reached 3.5 billion cubic meters in 2010, 2.3 billion cubic meters of which comes from rainfall. For the other 1.2 billion cubic meters, the

Beijing Municipal Government encourages the use of reclaimed water and water diverted from Hebei Province. Groundwater is generally the last choice. However, statistics show that in 2010, an additional 500 million cubic meters more groundwater was withdrawn than planned.

We are overdrawing groundwater, without planning or much consideration for the future. Environmentalists call this "the tragedy of the commons."

Not only does excessive withdrawal of groundwater endanger strategic reserves and causes eco-crises, but also land subsidence may pay dearly. According to China Geology Survey's five-year study called the *Comprehensive Survey of Land Subsidence in North China Plain* (华北平原地面沉降调查与监测综合研究), land subsidence has caused a direct economic loss of 40.45 billion RMB and an indirect loss of 292.39 billion RMB, adding up to 332.828 billion RMB. Another study by the Institute of Hydrology, Environment, and Geology shows that since the 1980s, a total of 600 million cubic meters of groundwater has been over-drawn annually in the Beijing, Tianjin, and Tangshan (京津唐地区) area. The over-drawing of groundwater in the North China plain ranks number one in China, with an accumulative over-draw of over 100 billion cubic meters. The Beijing Municipal Government and China Geology Survey have funded their respective land subsidence research projects. In the paper *An Analysis of the Status and Future of Land Subsidence in the Plain Area of Beijing* (北京平原区地面沉降现状及发展趋势分析), Professor Yang Yan (杨艳) at the Beijing Institute of Science and Technology found that by the end of 2009, the maximum annual land subsidence in Beijing had reached 137.51 millimeters and the maximum accumulative subsidence reached 1163 millimeters.

In 1999, about 989 square kilometers of plain area showed an accumulative land subsidence of more than 100 millimeters and the figure reached 3385 square kilometers in 2009 according to a survey of the Beijing Municipal Commission of the CPPCC. The percentage of land subsidence increased from 15% to 53% in 10 years. For example, the annual land subsidence of the Jinzhan (金盏地区) area of Chaoyang District reached 137 millimeters in 2009, ranking first in China. Ironically, in August 2011, a large golf course was under construction in the Jinzhan Village section of the Wenyu River (温榆河).

An expert at the Beijing Institute of Water says that in 1960, the groundwater was only 3 meters deep on average in Beijing and that in 2010, the average depth of groundwater reached 24.9 meters.

Experts say that land subsidence as a result of the over-drawing of groundwater may undermine city security in the following ways: Firstly, municipal facilities are endangered. At present, land subsidence has greatly affected buildings, roads and infrastructure in Beijing, with the underground pipeline system under the greatest threat. Over a third of running water pipe fracture accidents since 2000 are caused by land subsidence and these accidents are most common in Chaoyang and Dongcheng Districts, which are most severely troubled by land subsidence. Gas pipe fractures and roadbed collapses are also partly caused by land subsidence. Secondly, Beijing's rail transportation system is under constant threat. Among Beijing's subway and light rail lines, Line 13, Line 10, Daxing Light Rail, and Yizhuang Light Rail all go through or by land subsidence areas and thus constantly face the potential danger. Moreover, Beijing is also the national railway system hub. The Beijing-Tianjin High-Speed Railway, the Beijing-Shenyang High-Speed Railway, and the Beijing-Tangshan High-Speed Railway, among others, all pass areas of serious land subsidence. And fourthly, land subsidence may also endanger the South-to-North Water Diversion Project. For example, the Hebei Province section of the East Line of the project is seriously affected by land subsidence with its mild slope greatly reducing the flow of water. If land subsidence is not effectively dealt with, dams will be endangered.

VI. *The Future of Eco-Friendly Golf Courses*

Water is a strategic resource for public interest, indispensable for the development of Beijing. The government and society have been attaching unprecedented importance to the prevention of ostentatious water use and the protection and rational use of water resources.

In 2011, the Minister of Water Resources Chen Lei (陈雷) said, "We should strengthen and standardize the management of water use in service industries, investigate water use behaviors of high-water-consuming industries, and fight determinedly against law-violating behaviors."[11] Later, the Ministry of Water Resources issued the *Notice on the Investigation of Water Use Behaviors of High-Water-Consuming Industries* (关于开展高耗水服务业用水专项检查行动的通知), launching a national campaign to fight against illegal and wasteful use of water and to standardize water

[11] Author's Note: "Ostentatious water consumption" and "water consumption of high-water-consuming industries" refer to the same phenomenon from two different perspectives.

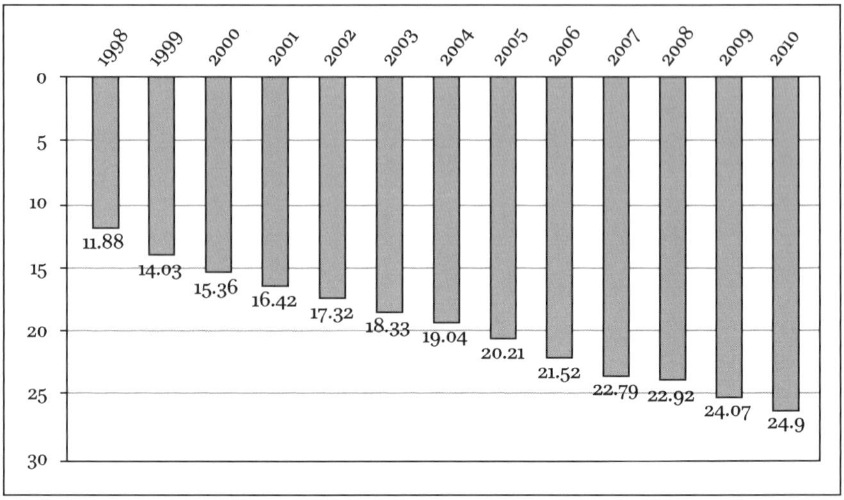

Figure 18.2: Changes in the depth of water in Beijing (1998–2010).

use behavior. At the National Water-Saving City Conference, Xie Zhenhua (解振华), vice director of the National Development and Reform Commission, pointed out that wasteful use of water, such as bathing, car washing, skiing, and golfing, had become a serious problem in China's social and economic development, which called for prompt and well-targeted efforts. In a policy recommendation document from the Beijing Municipal Government, it is recommended that further development of bathing centers and golf courses be banned and that a reward system be developed with respect to the reporting of illegal and wasteful water withdrawal and use.

To prevent ostentatious water use, we must follow the principles of an eco-civilization, so as to develop an energy and resource-saving and environmentally-friendly industrial structure, growth model, and consumption pattern. In the long run, solving the water crisis in Beijing means controlling population growth and changing the illogical production mode and lifestyle.

As mandated by the Beijing Water Authority according to the newest water conservancy measures, no new high-water-consumption businesses will be approved, such as high-end bathing centers, water-based production enterprises, ski resorts, golf courses, and playgrounds with a monthly water consumption exceeding 5,000 cubic meters. In case of a water supply breakdown or the daily water consumption reaching 90% of the daily water supply capacity, water supply for production purposes at golf courses, high-end bathing centers, and other high-water-consumption businesses may be cut with the approval of the municipal government.

The development of golf courses in Beijing must be based on Beijing's water supply conditions. In addition, the monitoring, management, and renovation of existing golf courses must be strengthened. According to the *Beijing Municipal Water Conservation Measures*, "Reclaimed water should be used for civil services, scenic water areas, golf courses, ski resorts, and car washing when the reclaimed water supply system is within reach." Other solutions include increasing the use of reclaimed water, applying more effective water-saving facilities, and replacing natural lawns with artificial ones if the conditions are fit. In Beijing's campaign to build an energy-saving and environmentally-friendly eco-civilization, the golf industry has a lot to contribute.

INVESTIGATION INTO THE POLLUTION IN APPLE INC.'S SUPPLY CHAIN AND A CALL FOR GREEN CONSUMPTION

Wang Jingjing

Abstract: Beginning in 2009, China's environmental protection organizations began communicating with 29 IT companies on heavy metal pollution by the industry. For nearly two years, several rounds of discussions have been held on whether client companies should be held responsible for pollution along the supply chain, whether consumers should pay attention to the pollution resulting from the production process, and what the impact of consumption behavior is. Currently, a new element has been added to the concept of a green supply chain in China; that is, attention is being given to the impact of the production process upon neighboring communities, the environment, and the general public. Consumer engagement helps to make the supply chain greener, and green consumption campaigns have since been quietly launched in China.

Keywords: supply chain, consumer, Apple Inc., pollution

I. *Why Apple Inc. Out of All the IT Brands?*

China has become the world's IT factory, producing about half of the world's PCs, mobile phones, and digital cameras. However, China is also the victim of heavy metal pollution. Starting in 2009, heavy metal pollution incidents were reported in Liuyang (浏阳), Fengxiang (凤翔), Wugang (武岗), Shanghang (上杭), Jiyuan (济源), Yancheng (济源), and Qingyuan (济源). There were 12 heavy metal and metalloid pollution cases reported to the Ministry of Environmental Protection in 2009; having caused excessive blood lead content in 4,035 people, excessive blood cadmium content in 182 people, and 32 mass disturbances. Lead, cadmium, and other heavy metals are toxic and difficult to degrade in the environment. The accumulation of these heavy metals may lead to public hazards.

Since 2009, the Institute of Public and Environmental Affairs, Friends of Nature, and the Green Beagle Environment Institute have been leading investigations into heavy metal pollution by the IT industry and have produced four reports. This research shows serious rule violations and pollution by famous IT brands.

Based on this research, the supply relationship between IT manufacturers and IT brands has been sorted out. Thirty Chinese environmental protection NGOs jointly sent open letters to 29 IT brands concerning manufacturers' violations of environmental laws and regulations, in hopes that these IT brands would strengthen the management of supply chains. Up to now, 28 of the 29 IT enterprises have responded to the open letters and some of them such as Siemens, Alcatel, BT, and Sharp have been very responsive and cooperative. For example, HP has got its suppliers to organize the first third-party audit in the IT industry and Siemens has established its own search system in an effort to improve the management of its suppliers.

Among these 29 IT brands, Apple Inc. has been the most disappointing. In the face of accusations toward the management of its supply chain, Apple insists that it will solve its problems by itself. In addition, Apple has denied or avoided all complaints from its customers. At the end of 2010, Apple denied that Jianlian Tech (联建科技) (an enterprise accused of occupational pollution) was one of its suppliers. Instead, Apple claimed that it was the relevant environmental protection organizations that had to prove Apple's relationship with Jianlian. It was not until February 15, 2011 that Apple, in its report on the management of suppliers, admitted for first time that Jianlian, whose 137 workers were poisoned by n-hexane, was one of Apple's suppliers. However, the report claimed, contrary to the truth, that all these workers had been cured, and once again dodged environmental protection organizations' inquiries.

On August 31, 2011, five environmental protection organizations published a special issue entitled *Report on Heavy Metal Pollution of the IT Industry* (IT 行业重金属污染调研报告). In it, *The Other Side of Apple: Rampant Pollution* (苹果的另一面2—污染在黑幕下蔓延) pointed to the environmental and security problems of Apple's suppliers and the flawed management of Apple's supply chain. In mid-September, after dodging environmental protection organizations for 19 months, Apple Inc. had its first communication with environmental protection organizations, followed by more rounds of talks over the next two years. On the morning of November 15, 2011, the Institute of Public and Environmental Affairs, Friends of Nature, Universal Technology Inc., Green Stone, and Green Beagle Environment Institute sat down with Apple in Beijing. During their talks, Apple stated that it had taken some measures to solve the problems, commissioned a third-party auditor, and convened with a dozen of its suppliers. However, Apple refused to provide detailed information about the rectification of rule-violating suppliers on grounds of

Table 19.1: Ranking of supplier management effectiveness of IT brands.

IT Brands	Reply or Not	Background knowledge	Record of supplier's rule-violation		Strengthen supplier management with the use of public information		Supplier reform and information disclosure		Extensive environment management along the supply chain	
			Primary	In-depth	Considering searching system	Determined on search system	Reform with explanations	Regular information disclosure	Until key material supplier	First-level suppliers' search of second-level suppliers
Siemens	√	√	√	√	√	√	√	×	×	×
Vodafone	√	√	√	√	√	√	√	×	×	√
Philips	√	√	√	√	√	√	√	×	√	×
Nokia	√	√	√	√	√	√	√	×	×	×
ALU	√	√	√	√	√	√	√	×	×	×
British Telecom	√	√	√	√	√	√	×	×	×	√
HP	√	√	√	√	√	×	√	×	×	×
Samsung	√	√	√	√	√	×	√	×	×	×
Sanyo	√	√	√	√	√	×	√	×	×	×
Sony	√	√	√	√	√	√	×	×	×	×
Toshiba	√	√	√	√	√	×	×	×	×	×
Panasonic	√	√	√	√	√	×	×	×	×	×
Sharp	√	√	√	×	√	×	×	×	×	×
Lenovo	√	√	√	×	√	×	×	×	×	×
Intel	√	√	√	×	√	×	×	×	×	×
Seiko Epson	√	×	√	√	√	×	×	×	×	×
Motorola	√	√	√	√	√	×	×	×	×	×

Table 19.1 (*cont.*)

IT Brands	Reply or Not	Background knowledge	Record of supplier's rule-violation		Strengthen supplier management with the use of public information		Supplier reform and information disclosure		Extensive environment management along the supply chain	
			Primary	In-depth	Considering searching system	Determined on search system	Reform with explanations	Regular information disclosure	Until key material supplier	First-level suppliers' search of second-level suppliers
Hitachi	√	√	√	√	×	×	×	×	×	×
Canon	√	×	√	√	×	×	×	×	×	×
Cisco	√	×	√	√	×	×	×	×	×	×
Dell	√	√	√	√	×	×	×	×	×	×
Apple	√	√	√	√	√	×	×	×	×	×
Haier	√	√	√	×	×	×	×	×	×	×
TCL	√	√	√	×	×	×	×	×	×	×
Foxconn	√	√	√	×	×	×	×	×	×	×
Singtel	√	√	√	×	×	×	×	×	×	×
BYD	√	√	√	×	√	×	×	×	×	×
IBM	√	×	√	×	×	×	×	×	×	×
LG	√	×	√	×	×	×	×	×	×	×
Ericsson	√	×	√	×	×	×	×	×	×	×
Blackberry	√	×	√	×	×	×	×	×	×	×

protecting business secrets. These environmental protection organizations disagreed, emphasizing that environmental information disclosure was the key step and that information about rectification would not put any business secrets at risk. These one-month talks failed to produce any meaningful results and Apple, again, became silent.

II. *Should Apple Inc. Be Held Responsible for Supply Chain Pollution?*

Toxigenic incidents and environmental pollution by suppliers evidenced the loopholes in Apple's supply chain management. Yet, many voices, both in and out of China, argued that since these incidents did not happen in Apple's own factories, Apple is not to be blamed.

However, this argument does not hold ground, because this case not only highlights the complicated outsourcing relationship and original equipment manufacturer (OEM) supply chain as a result of globalization but also poses the question as to who should be held accountable for resulting environmental and social problems. Though suppliers and government departments have both failed in some ways, there are still good reasons to hold Apple accountable:

First of all, it has become a social consensus that an IT brand shoulders all the environmental and social responsibilities for its products. However, in the globalization age, Apple no longer has its own factories, but depends on OEM for even the smallest parts. But subcontracting does not mean the end of pollution and hazards commonly found in the production process of the IT industry. Dust-free plants, because of the concentration of harmful chemicals, have the most serious occupational hazards.

In the past nine months, environmental protection organizations, on the basis of research, attempted to call the 29 IT brands' attention to this problem. Most brands responded favorably, except Apple. Recently, though Apple, under public pressure, has come to admit the Lianjian toxigenic incident, and insists that suppliers should bear all the responsibility.

Apple's logic is thus: since Apple does not run any factory of its own, it is totally green, having nothing to do with pollution and occupational hazard in the production process. In other words, by subcontracting, Apple has cleared itself of any environmental and social responsibility and therefore is not guilty in these relevant incidents. If this logic is accepted, all major brands will follow suit, the result of which is the transfer of pollution across the globe.

Secondly, suppliers choose not to respect environmental protection rules and workers' health, so as to minimize costs and maximize benefits. Public materials show that Apple, as a high-end corporate buyer in the industry chain, has an overwhelming advantage in profit sharing.

According to analysis of the iPhone 4's supply chain profit sharing, for each 600 USD iPhone 4, Foxconn and other Chinese assemblers only get 6.54 USD while Apple reaps a profit of 360 USD. As common sense goes, the greater right, the greater responsibility. However, obtaining the biggest share of the profits, Apple refuses to take any responsibility. Is this reasonable?

Thirdly, Apple is not aware of the dangers of transforming its social responsibilities. In order to convince consumers that Apple is a green brand, it has made specific and high-profile promises concerning supply chain management; that is, "Apple promises to ensure safe working conditions in its supply chain, to respect its workers and to guarantee environment-friendly production."

However, a study shows that Apple has broken its promises in each of the three respects. Usually, if an enterprise fails to keep its promises, it feels guilty, endeavors to improve, and keeps the public informed of recent developments. Disappointingly, Apple has shown no regret about the pollution and toxigenic incidents but constantly passes the buck to others, putting its promises behind itself.

Fourthly, few people understand the difference between subcontracting and purchasing. Abundant evidence shows that Apple is actually deeply involved in the management of its supply chain, from materials to dust-free standards. Therefore, Apple is obliged to disclose and explain how its intervention affects pollution.

Taking advantage of globalization, Apple and other brands have attained global production and purchasing, relocating high-polluting and emission production processes to China and other developing countries. However, many developing countries are constantly plagued by localism, lack of law enforcement, and poor environmental awareness with enterprises turning a blind eye to environmental laws. In the past, many foreign brands, with the excuse of "not knowing who is polluting the environment," chose their suppliers based on price, without consideration of their environmental performance. However, with the development of environmental information disclosure in China, environmental law violation information has become readily available. Many international brands have also taken measures in a joint effort to curb the expansion of pollution in global production and purchasing.

Against such a backdrop, Apple chooses to stand alone. In the face of concrete charges against its suppliers, Apple refuses to take action on grounds that: "Our long-term policy is to not disclose our suppliers." Turning a blind eye to the violation records of many of its suppliers, Apple, by choice, continues to do business with them.

Apple has to make one of two choices: 1) continue to take advantage of the loopholes in environmental management in developing countries, and continue to partner with polluting enterprises to maximize profits at the expense of the environment and the public; or 2) cooperate with relevant interested parties and establish a more transparent and participatory supply chain social responsibility management system, so as to prevent pollution and toxigenicity in its supply chain and contribute to China's pollution reduction campaign.

Until now, Apple prefers the former, making itself an obstacle to China's pollution reduction campaign. More disappointingly, Apple chooses not to make its choice public, but to put on a "green mask" by claiming to enforce the so-called highest supply chain environmental and social responsibility standards. This kind of behavior on Apple's part not only harms the environment and communities, but also misguides its consumers worldwide in an unfair competition. Apple must be held responsible.

III. *How Should Consumers Take Action?*

Apple's final decision to cooperate is owed to the efforts of various interested parties, in which consumers have played a very positive role.

Since the 1990's, consumerism with wide public participation has developed in Europe and America. In addition to price and quality, consumers are more and more concerned about whether commodities are produced in a "clean" way; that is, whether the production process endangers neighboring communities, the environment, and the health of workers. For example, in June 1996, *Life* magazine carried photos of child laborers working in a Nike factory in Pakistan, triggering a boycott against Nike products, which eventually pressured Nike to prohibit its suppliers from using child labor. Nike's favorable response restored customer confidence.[1]

This present Apple case reflects the development of consumerism in China. More and more consumers are becoming concerned with the

[1] Zheng Guanghuai, "Social Forces Promote Transfers in Labor-Management Relations," *Dandu Daily*, January 16, 2011.

"greenness" of the production process. Since the second report on June 5, 2010, some consumers have expressed their concern to Apple and other brands:[2]

1) As of June 5, 2010, in response to the call of the South China Nature Society, Green Stone, China Green Student Forum, Blue Dalian, and the Institute of Community and Bio-Diversity Protection of Lanzhou University, over 300 Chinese consumers have expressed their concern with heavy metal emission in written form to Apple, IBM, Canon, and LG.
2) Since June 29, 2010, over 1,000 Apple consumers worldwide have responded to Pacific Environment's (an American environmental protection organization) initiative and written to Apple to inquire about its supply chain environmental management.
3) Beginning on June 30, 2010, America's Business and Human Rights Resource Center, with reference to the study reports of Chinese environmental protection organizations, have requested that Apple, Ericsson, and other IT brands provide information and shared the results with over 10,000 subscribers.
4) On September 9, 2011, the Internet campaign called "Green My Cell Phone" (随手拍给绿我手机) was launched by Chinese netizens championed by Green Stone, Dalian Environmental Protection Volunteers' Association, Green Kunming, Green Angel of Chongqing University of Science and Technology, and the China Green Student Forum. Since then hundreds of netizens have posted photos of their cell phones with green ribbons to their micro-blogs to call on IT brands to improve their supply chain management.

On top of demonstrations and organized initiatives, consumer participation is more importantly realized in daily consumer behavior. Consumers may refuse to buy products from polluting companies or express their concern to enterprises.

However, even for those Chinese consumers who are aware of the harms to workers, the environment, and communities, most of them take these as legal issues and in turn as government responsibility. Few people have realized that by purchasing products made in this fashion, they are helping

[2] On April 23, 2010, ICETT translated the open letters into Japanese and mailed them to the headquarters of these eight enterprises: Sony, Canon, Panasonic, Seiko Epson, Hitachi, Toshiba, Sanyo, and Sharp.

Figure 19.1: Consumers expressing their opinions at lectures organized by local environmental organizations.

the villain commit evil. In this sense, consumer participation and action is still in its primary stage.

To encourage Chinese consumers to shift their attention from the end product to the production process, environmental protection organizations, the media, and other social forces will have to, on the one hand, improve environmental education and promotion so as to enhance consumers' sense of responsibility. On the other hand, provide necessary information to help consumers learn about the production process. Only by exposing unfair labor practices and environmental and social hazards can consumers' sense of responsibility transform into green action, so as to ultimately promote behavior changes in enterprises.

PART FIVE

POLICY AND GOVERNANCE

GARBAGE PROBLEMS AND SOLUTIONS IN RURAL CHINA

Lü Zhongmei

Abstract: By analyzing the data collected in a field study in Hubei Province, this paper identifies the garbage problems and challenges in rural China and reveals the lack of a legal framework for solving these problems.

Keywords: rural garbage, consumer waste, garbage pollution, legal framework

Garbage problems in rural China under the "development of a new socialist countryside" were a hot issue in the media in 2010. In April, the Ministry of Environmental Protection issued *Technical Policies for the Prevention of Rural Domestic Pollution* (农村生活污染防治技术政策) (hereinafter referred to as the *Policies*), aimed at "implementing the *Environmental Protection Law* (环境保护法) and other laws and regulations, promoting development of a new socialist countryside, protecting and improving the rural environment, preventing rural domestic pollution, and advancing quality of life and health in the countryside." The *Policies* are tailor-made for rural environmental protection, which is gratifying, but at the same time they reflect the deficiencies that laws and regulations have in solving problems of rural environmental protection. More than thirty laws and hundreds of policies, regulations, and standards on environmental protection have been promulgated since the beginning of such policies in 1973 and the implementation of the *Environmental Protection Law* (*Trial*) in 1979. Yet, over 80% of these are made for cities and few for the countryside. Further observation reveals that among the scant legislation, almost all are lacking in feasibility. The lack of laws and regulations contributes to the current garbage problems in rural China and ensuing problems such as water pollution, air pollution, soil pollution, and agricultural product contamination.

I. *The Problem of Garbage in Rural China*

A phenomenon has emerged in the "development of the new socialist countryside" in recent years throughout China. While people have acquired new houses, roads, telecommunications, and cars, garbage has occupied every alley, street, and road in villages and has extended its territory to the highways, aqueducts, and woods outside the village. The garbage gives off a terrible stench and is a breeding ground for mosquitoes in summer, and when there is wind and rain, waste plastics and garbage bags are blown around, even to the top of the trees. It is unpleasant to the eye and harmful to the environment and human health.

The advancement of farmers' wealth and changes in lifestyle have brought about variation in agricultural production. Stalks are no longer a major fuel source but a waste, and are burned after summer and autumn harvests for lack of sufficient collection and disposal measures. The smoke from burning often results in highway and airport closures, resulting in huge economic losses. Moreover, fertilizers, pesticides, and the use of film seedling technology in low-temperature areas have been promoted to advance crop yields. The film that is left over settles into the soil and becomes an adverse influence on its quality. Waste from fertilizers and pesticides are tossed away at random in the fields or beside rivers and become a direct cause of water pollution.

The problem of rural garbage has aroused widespread concern. Many media outlets have conducted in-depth investigations into this problem and found that garbage disposal has not been part of the design for new countryside development in many areas. There is a lack of long-term mechanisms for rural garbage disposal, insufficient funding, out-dated disposal measures, and ineffective prohibition on the burning of stalks.

These might be the causes of garbage pollution in rural China, but we may continue asking: Who should be responsible for making garbage disposal part of the development design, the long-term mechanism, the funding, and the disposal measures? All in all, this is about the lack of laws and regulations on prevention of garbage pollution in the countryside, as well as the lack of sufficient systems and mechanisms. If China wants to settle its debt on rural garbage, it has to find its "debtor" first.

II. *The Causes of Garbage Problems in Rural China*

To better understand the situation of environmental protection in rural China, especially that of the garbage problem, several research teams from the Hubei University of Economics conducted a comprehensive study in the cities of Wuhan (武汉), Jinzhou (荆州), Yichang (宜昌), Xiangfan (襄樊), Shiyan (十堰), and Xiaogan (孝感). The teams visited provincial departments of environmental protection, construction, water resources, agriculture, and local administrative departments, and administered large-scale questionnaires. One hundred university students conducted interviews in the rural areas of Hubei Province (townships, towns, and below). The students, all of whom grew up in the rural reaches of Hubei Province, had sound knowledge of the local environment, interpersonal relationships, and the discourse systems of the countryside. One thousand questionnaires were handed out and 901 were returned, among which 803 were usable. Further, 68 interviews were collected from 62 interviewees.

Based on the study, the teams found two major causes for the garbage problems in rural Hubei. These are 1) domestic and agricultural pollution that is produced in rural areas and 2) pollution that is transferred from township enterprises and urban areas to rural ones.

A. *Domestic Pollution and Pollution Caused by the Rural Production Style*

The current life and production styles in the countryside are a major source of its garbage problem.

1. *Household Garbage Pollution*

With the advancement of quality of life in the countryside, household garbage has grown in variety. Household garbage increases in proportion to the wealth of the farmers: the richer they are, the more waste they produce. However, the collection, transportation, and treatment facilities for household garbage lag behind, while the safe treatment rate remains very low. A considerable part of household garbage is simply thrown and piled up in aqueducts, on roadsides, or in fields, creating "garbage villages." Such garbage not only gives off a stench and breeds mosquitoes, its leachate is also a direct cause for surface and ground water pollution. According to the survey, the safe treatment rate of household garbage in the rural areas of Hubei at the town, township, and farm levels is 32.03%, 21.70% and 13.37% respectively.

There exists no statistical data on garbage collection, transportation, and treatment rates of any of the villages in Hubei Province. But we know only 5,438 villages collect household garbage, 21.41% of the total number, and only 1,845 villages treat household garbage, a mere 7.26%.

2. *Agricultural Production Pollution*

a. *Pesticide Pollution.* A large quantity of chemical pesticides is used for prevention and control of farmland pests. Compared with 1990, the quantity of pesticides used in the province has increased seven times, and is still growing.[1] The water and air pollution potential caused by the spray of pesticides has caused wide concern, but the many unused pesticides and waste pesticide containers are also a serious pollutant risk not to be overlooked. Over the past 30 years, nearly 1,510 tons of waste pesticides have accumulated in 55 cities and counties in Hubei, among which 450 tons are extremely or highly toxic or with high residue, not to mention the 0.8 tons of waste pesticide containers (packages or bottles) left by farmers every year.[2] Not treated properly, these waste pesticides and pesticide containers will cause serious pollution to water and soil.

b. *Animal Waste Pollution.* Hubei Province has a thriving animal husbandry industry. However, as the animal waste from most husbandry and poultry farms is not properly treated and utilized, excrement and urine is discharged freely. Animal waste and wastewater from husbandry and poultry contain large amounts of nitrogen, phosphorus, suspended solids, and pathogens. Such highly concentrated pollutants are especially harmful to water. When discharged into rivers and lakes, they cause not only eutrophication but also the gradual death of aquatic creatures sensitive to organic pollution, posing a great threat to aquaculture. Moreover, animal waste does not only pollute surface water; its toxic and harmful pollutants will infiltrate into and pollute ground water.

c. *Stalk Burning Pollution.* With the improvement in quality of life, gas, coal, and electricity have replaced stalks as the major household fuel in rural areas. Moreover, with the mechanization of agriculture, rice and wheat harvests are now done by reapers. It has become more costly and time-consuming to either return the stalk to the fields or drag it home, adding a potential danger for fire. Most farmers therefore consider burning stalk the simplest and most thorough way of disposal, which fertilizes the fields at the same time. In recent years stalk burning has caused quite a few environmental problems, wasted resources, and economic losses: it turns easily into wildfire; the smoke from burning hangs in the air, posing a threat to people's health as well as traffic; it reduces the original humus in the soil, resulting in soil compaction, fertility decline, an increased

---

[1] Hu Jiusehng, Li Zhaohua, Xin Xiaoyan, Kang Qun, and Wang Jinzhou, "Study on Causes and Control Measures of Water Pollution in the Rural Areas of Hubei Province," *Journal of China Agricultural Resources and Regional Planning*, no. 1 (2009): 67.

[2] Wang Xiaodong, "The Pain of Waste Pesticide Disposal," *China Environment*, April 17, 2009.

degree of soil desertification, and weakened farmland water retention; it also wastes resources.[3]

B. *Transfer of Township Enterprise Pollution and Urban Pollution*

The countryside is not the only source of the rural garbage problem; township industries and cities have all played their role as well.

1. *Township Enterprise Pollution*

One characteristic of China's industrialization is the promotion of rural industrialization through township enterprise development. The fast-developing township enterprises in Hubei Province have increasingly contributed to economic growth and played a major role in the transfer of surplus labor in the countryside. The problem is that the township enterprise in Hubei is clearly industry-oriented in its structure, discharging untreated industrial solid waste pollutants and causing significant damage to the environment. Among the most seriously polluting industries are papermaking, printing and dyeing, electroplating, chemicals, metallurgy, mining, machine manufacturing, and building materials. Many township enterprises in Hubei are managed as workshops, with neither the awareness nor the capacity to deal with pollution. In the meantime, some out-dated and heavily polluting industrial projects and manufacturing equipment are relocating to the countryside in the name of township enterprises. This comes as urban industries are being upgraded in an effort to cleanse urban environments.

2. *Transfer of Urban Pollution*

In addition to the transfer of polluting industries, large quantities of pollutants have arrived in the countryside through other means. The suburbs have also become the dumpsite for urban solid wastes and heavy-polluting industries such as chemical, metallurgy, pharmacy, papermaking, and electroplating.

---

[3] Wang Li, Li Xuemin, and Xu Yan, "Study on Economic Losses Caused by Stalk Burning in the Chinese Mainland," *Journal of Arid Land Resources and Environment*, no. 2 (2008). They estimate that the direct losses of biological resources caused by stalk burning in the Chinese Mainland in 2004 was 11.34 billion RMB, and losses caused by air pollution was 19.64 billion RMB.

Surveys show that the total amount of collected and transported household garbage in cities of Hubei Province is 6.81 million tons. The amount of treated household garbage is 6.63 million tons, among which 3.61 million tons are harmless treated and buried in sanitary landfills in the nearby countryside. The safe treatment capacity is 0 in 20 cities. The total amount of collected and transported household garbage amongst Hubei's counties is 1.98 million tons. The amount of treated household garbage is 1.449 million tons, of which 0.19 million is safe treated. The household garbage treatment rate is 73.04%, while the safe treatment rate is only 9.95%.

The current status of the garbage problem in rural areas of Hubei reflects on the entire country. Our study shows that the increasing garbage pollution dramatically affects agricultural production, threatens the health of rural residents, and has an adverse effect on the development of the new socialist countryside.

III. *The Dilemma of Garbage Problems in Rural China*

All of the above suggests that the garbage problem in rural China is more than a rural problem. The problem has aroused wide concern largely because one of the objectives of a new socialist countryside is "a clean and tidy village," but the problem goes far beyond farmers' littering or a lack of garbage treatment facilities.

From an objective perspective, the garbage problem in rural China is closely related to the long tradition of life and production styles, characteristic of non-point source pollution in rural areas. Non-point source pollution refers to distribution and emission of pollutants from diffuse sources. Such pollution is dispersed, hidden, random, and difficult to monitor or measure; closely related to farmers' daily life and agricultural production, it pollutes extensively and is difficult to control. Non-point source pollution control has become a worldwide problem, with the countryside being its largest contributor and most difficult source to control.

From a subjective perspective, the garbage problem in rural China is an extension of urban pollution into the countryside, a consequence of China's unsustainable development of high-input, high-consumption, and high-pollution. The garbage problem in rural China accompanies its rapid economic growth. In the 1980s, with the soaring development of township enterprises, industrial pollution grew like a weed from the city to the countryside; some city enterprises transferred highly-polluting operations, technologies, and products into township enterprises with almost no pollution control capacity, causing water and soil contamination in certain rural areas, some of which remain unresolved today. In the 1990s, the increasing speed of urbanization and urban expansion drove the cities to find a "way out" for industrial and household garbage in the countryside. The garbage took up farmlands and formed "garbage hills" that experts dubbed as "malignant tumors" on the land. In recent years, while narrow-mindedly pursuing high yields in agriculture, overuse of pesticides and fertilizers in farming as well as overfeeding and overdosing medicines in animal breeding have become increasingly serious, producing large amounts of pollutants. Burning, the only way to treat garbage in some areas, only creates more pollution. What's more, some village leaders still consider economic development the sole goal of the new socialist countryside development, building roads and houses while ignoring the environment. The farmers have prospered at the expense of the rural environment.

It can now be concluded that the garbage problem in rural China is a complex problem of various interests involving the city and the countryside, industry and agriculture, life and production. Laws, regulations, and mechanisms should be established to stipulate the nature, agent, and responsibility of rural garbage prevention and control, specifying a monitoring system and jurisdiction, and defining consequences for violations so that all parties concerned can be held within a legal framework. However, faced with mountains of garbage in the countryside, we cannot find one applicable rule or effective mechanism. The essential problem behind the piling garbage is the lack of norms for conduct, lack of control for consequences, and lack of responsible monitoring due to insufficient legislation.

It is fair to say that environmental protection has been given high priority since the reform and opening up period. Legislation has developed quickly, and various laws and regulations have been promulgated since the implementation of the *Environmental Protection Law* (*Trial*) in 1979, playing a significant role in controlling environment pollution in China. But, regrettably, most of these laws and regulations target industrial pollution control and improving city environments, which, though in principle is applicable to the countryside, cannot be sustained in rural areas for lack of institutional guarantee. It is said that the legislators considered rigorous regulations on rural and agricultural pollution control disproportionate to the underdeveloped state of rural China, thus "sparing" the countryside. In reality this only contributes to the negligence of rural environmental protection and an unfavorable sense of urban superiority. Even worse, it exacerbates the rate and extent of rural environmental pollution as it creates a "legal vacuum" of rural pollution control, leaving all parties and their interests in disorder.

Take as an example the *Law on the Prevention and Control of Environmental Pollution by Solid Waste* (固体废物污染环境防治法), specifically targeted at garbage pollution. The Standing Committee of the Sixth National People's Congress (NPC) examined and adopted the law on October 30, 1995, and it went into effect on April 1, 1996. Although it is stipulated in the General Principles that "This law applies to the prevention of environmental pollution caused by solid waste within the boundaries of the People's Republic of China," only the "Prevention of Environmental Pollution Caused by Industrial Solid Waste" and the "Prevention of Environmental Pollution Caused by Urban Residential Refuse" are specified in its sections, with no mention of the prevention of solid waste pollution in the countryside. It can only be concluded that solid waste in the countryside is not included in the prevention and control system for pollution caused by solid waste.

In May and July 2003, an inspection team appointed by the NPC Standing Committee conducted a comprehensive investigation into the enforcement of the *Law on the Prevention and Control of Environmental Pollution by Solid Waste*. The results showed that the law has played an active role in promoting prevention and control of solid waste pollution in China, but also revealed that many new problems have emerged with the growing industrialization, urbanization, and quality of life, one of which is the increasing solid waste pollution in the countryside. The report suggests that the law be amended so that solid waste in the countryside is included as part of the urban-rural solid waste management system, which then made the legislation list of the Tenth NPC Standing Committee. However, opinion was divided on the prevention and control requirements for rural solid waste pollution during its drafting. Most people thought it fit that the country should adopt stricter measures to manage pollution problems caused by agricultural waste and rural garbage, while some maintained it was not the time to include rural solid waste into the country's solid waste management system. The amendment turned out to be a compromise of the two opinions: for areas with management capacity, specific requirements were made; for those without the capacity, only guidelines were given or local authorities could stipulate as they see appropriate.[4]

---

[4] Sun Youhai, "Specify Responsibilities, Demand Recycling, Take into Consideration the City and the Country: Comment on Amendment to Law on the Prevention and Control of Environmental Pollution by Solid Waste," *China Environment*, January 6, 2005.

On December 29, 2004, the 13th Conference of the Tenth NPC Standing Committee passed the amended *Law on the Prevention and Control of Environmental Pollution by Solid Waste*, effective from April 1, 2005. The amended law includes in its jurisdiction prevention and control of solid waste pollution in the countryside, with Article 49 stipulating: "Specific measures for prevention and control of environmental pollution by rural house refuse shall be prescribed by local regulations." Article 20 reads: "An entity engaged in raising livestock and poultry in a large scale shall, in conformity with relevant State regulations, collect, store, utilize, or treat the excrement and urine discharge by the livestock and poultry, in order to prevent environmental pollution." It also stipulates that "open-air burning of stalks in densely-populated areas, in the neighboring areas of airports, on the peripheries of the main lines of communications, and in the areas delimited by local people's governments is prohibited." Section 3 of Chapter 3 is titled "Prevention and Control of Environmental Pollution by Household Waste," no longer differentiating between urban and rural household waste. Other articles, such as 38, 44, and 45, can all be applied to rural household garbage.

Objectively speaking, the amended law is a big step forward, not only including prevention and control of rural garbage into the legislation, but also making provisions on rural household garbage, livestock and poultry excrement and urine, and open-air burning of stalks. However, compared with the prevention and control mechanism for pollution by urban household garbage and industrial solid waste, rural garbage pollution is clearly lagging behind: regulation is given over to local authorities, making the articles applicable to both the country and the city under Section 3 of Chapter 3 mostly void. So far, the author has only managed to locate three local regulations on the prevention and control of environmental pollution by solid waste that make uniform provisions concerning rural and urban household garbage disposal: in Jiangsu, Zhejiang, and Guangdong.[5] One of these is Jiangsu Province's *Regulations on Prevention and Control of Environmental Pollution by Solid Waste* (江苏省固体废物污染环境防治条例), passed by the 11th conference of the 11th Jiangsu People's Congress Standing Committee on September 23, 2009, and effective since January 1, 2010. The third chapter, "The Prevention and Control of Environmental Pollution by Rural and Urban Household Waste" has uniform provisions

[5] The author has painstakingly searched a number of websites for local regulations on prevention and control of environmental pollution by solid waste. Besides the three provinces, only the city of Nanjing has similar regulations.

for rural and urban household garbage disposal. Article 23 of the chapter clearly stipulates:

> Governments of cities with districts and counties should establish and improve mechanisms where rural household waste is cleaned by village groups, collected by villages, transported by townships (towns) and disposed of collectively by counties (cities, districts). Governments should also provide stipends and financial support for the cleaning, collection, transportation, and disposal of rural household waste... Governments of townships (towns) should strengthen the organization and implementation of the cleaning, collection and transportation of rural household garbage.

It may seem reasonable to suggest that it is not yet the time to provide uniform legislation for rural and urban areas concerning mechanisms for prevention and control of rural garbage pollution. We can, however, draw another conclusion: the "urban-rural binary" model in environmental protection legislation has long left rural garbage pollution unattended by institutional management, resulting in "no capacity, no norm for conduct, no agent, and no basis" for the prevention and control of environmental pollution by garbage in rural China.[6]

IV. *Solutions to the Garbage Problem in Rural China*

If we take inadequate legislation as the essential cause for the current garbage problems in China, it follows naturally that the problem can only be solved through the establishment of appropriate systems, mechanisms, laws, and regulations.

Both central and local governments are now taking various measures to control garbage pollution in rural China. Some useful explorations have been made, such as establishing collective disposal systems for rural garbage, promoting comprehensive stalk utilization technology, and adopting safe treatment measures. The Ministry of Environmental Protection issued the *Technical Policies for the Prevention of Rural Domestic Pollution* in April 2010, prescribing technology guidelines and supporting technical requirements such as "encouraging household garbage classification," "urban-rural uniform models where applicable," "adopting economical, proper, and safe disposal technology where a uniform model is

---

[6] Chen Wensheng and Wang Wenqiang, "Household Garbage is Seriously Polluting the Environment in the Countryside," *Journal on China Countryside Construction*, no. 15 (2009).

not applicable," "repeated cycling of non-organic waste," and "comprehensive use of organic waste."

These measures and policies certainly have a positive influence on establishing and improving prevention and control mechanisms for environmental protection from rural garbage. Nonetheless, we have to realize the complexity of the garbage problem in China: it is rural non-point source pollution characterized by diversified pollution agents, dispersed pollution behavior, and hidden pollution sources; it is also a transfer of urban pollution into the rural area characterized by the accumulation of pollution behavior, collective pollution agents, and indefinite pollution consequences. The problem is further complicated by its close connection with thousands of years of tradition in rural lifestyle and production, and its intertwined relationship with rural development, education, and infrastructure problems caused by a long-term "urban-rural binary" development model. The problem can hardly be solved by some pilot projects or technical policies. It has to be solved at its very root. It is hereby suggested that the *Law on the Prevention and Control of Environmental Pollution by Solid Waste* be further amended to break the "urban-rural binary" model, re-evaluate the nature, status, and function of the prevention and control of environmental pollution by rural solid waste, and establish a uniform mechanism for both the urban and rural areas. Supporting regulations should also be made to further improve the working mechanisms, policies, and measures in prevention and control of environmental pollution by rural solid waste. It is of utmost importance to clarify the following points:

1. Solving the garbage problem in rural China is not just a problem of livelihood, but also one that concerns urban-rural integration and economic, social, and scientific development. Rural garbage disposal should be a public service provided by the government, and the government is responsible for its planning, funding, management measures, and monitoring. It is therefore necessary to stipulate in law the government's responsibilities as an agent.
2. To solve the garbage problem in rural China means to balance relations between the countryside and the city, industry and agriculture, lifestyle and production. We should follow integrated plans for urban-rural environmental protection and give equal priority to industrial and agricultural pollutions. In dealing with rural pollution, we should take as our starting point to improve the rural environmental quality, evaluate in connection the building of a clean home, clean water, clean

countryside, and clean energy, and strengthen prevention and control of rural environmental pollution to promote the comprehensive utilization of agricultural resources and improve farmers' living and working environments. It is therefore necessary to stipulate in law the principles and specific measures for environmental and developmental policies.

3. Solving the garbage problem in rural China requires that the government shift from environmental management to environmental governance, adopting effective measures to boost farmers' engagement. On the one hand, more education and training should be provided to farmers to help them improve their agricultural skills and form good habits; on the other hand, measures should be taken to encourage farmers to take part in production reform and management activities. It is therefore necessary to establish, in law, public participation and interest stimulation mechanisms to change the mode of government management. For example, to solve the problem of stalk burning, governments on different levels have established enterprises for comprehensive utilization of stalks. However, the companies are not willing to go from door to door to collect stalks due to the high cost, and farmers are not willing to deliver them because of the huge inconvenience and no profit, so they still "burn it all." The companies are having a tough time with a lack of raw materials, and the government has yet to invest more resources in patrolling, monitoring, and managing stalk burning. The current measure is a management with only input and no, or even negative output; the farmers will not adopt it, and the government cannot have an eye on everybody. However, if only the government can set aside some funding as a stipend for the farmers so that they can profit from delivering stalks to the companies, or at least not at their own expense, the farmers will have no incentive to burn the stalks, the companies will have their raw materials for profit-making production, and the government can save monitoring costs. All parties will have benefited, and the problem of stalk burning is solved.

Of course, no problem is ever that simple. Based on a full investigation of the environmental problems in rural China, further analyses and discussions should be carried out on proposed mechanisms, measures, and plans to find a proper approach to a legal framework.

CHINA'S ENVIRONMENTAL LAW FAILS TO EFFECTIVELY ENSURE FAIRNESS

Feng Jia and Bo Xiaobo

Abstract: Coercive as a state power may be in the enforcement of laws, it is the degree of attention to fairness and the capacity of fairly distributing interests that really makes a law effective. A main reason for the ineffective enforcement of environmental laws in China is that the legislation fails to solve the conflict of interests and neglects justice and equality in environmental protection. Under such circumstances, no laws or regulations can be strict or coercive enough to make a real difference.

Keywords: environmental law, equality, justice, coercion, effect of law enforcement

I. *Review of Rule of Law in Environmental Protection in 2010*

China's rule of law in environmental protection went through a period of frustration in 2010. In retrospect, the situation has not changed much, and the overall picture still remains discouraging.

In legislation, the enactment of the *Tort Liability Law* (侵权责任法) was a highlight of 2010. Chapter Eight of the law is a special chapter that regulates the core issues relating to liabilities for environmental pollution. In contrast to the sporadic stipulations scattered in previous laws on liabilities for environmental pollution, using a special chapter in a fundamental law underpinning the state's civil law framework manifests the urgency and importance of controlling environmental pollution by civil means. From this point of view, provisions in the *Tort Liability Law* are an impetus to environmental protection. Besides the General Provisions of the law, it also provides the allocation of tort liabilities among several individuals and compensation for psychological damage, which may help protect victims of environmental pollution as well as the environment. However, the target and responsibility of the *Tort Liability Law* is not protecting the environment. Protecting the environment is by no means the original intention of the law in investigating and affixing polluter liability. The law still sticks to the scope of traditional laws of protecting personal or property rights. Only if the environment has been polluted and caused damages to individuals or property can the law be a weapon for the public to protect the environment. However, if the environment has been polluted but there are no obvious damages caused to individuals or property, no liability can be investigated and affixed to polluters. In addition, except for complex technical litigation issues, the *Tort Liability Law* actually has not broken through the existing laws and regulations. It merely collects and re-organizes provisions on liability for environmental pollution in laws such as the *Law on Prevention of Environmental Pollution Caused by Solid Waste* (固体废物污染环境防治法) and the *Law on Prevention and Control of Water Pollution* (水污染防治法), and extends the scope of the application of relevant provisions. In conclusion, the *Tort Liability Law* does not empower the public with any rights to institute environmental public interest litigation against polluters. Public interest litigation, which is a powerful weapon that allows public participation in environmental protection, is still not grounded well in definite legal basis. Therefore, the law is beneficial to environmental protection with limited intensity.

The *Law on Island Protection* (海岛保护法) is another important law for protecting the ecological environment. It was officially enacted on March 1, 2010. The law regards the ecological environment of islands as an important protected object, and stipulates a number of provisions that prohibit or limit development and exploitation of island resources. The law also stipulates that uninhabited islands belong to the state. Development and utilization of the resources of uninhabited islands shall follow the plan on protection and utilization of uninhabited island resources. This marks the end of a historically disordered, excessive, and free development of uninhabited islands. Whether the implementation of the *Law on Island Protection* can protect the ecological system of islands depends on its enforcement. If the enforcement is not strict, the law is best only on paper. It cannot produce any real effects.

On June 29, 2010, the Supreme People's Court issued *Several Opinions on Providing Judicial Guarantee and Service for Accelerating the Transformation of the Economic Development Mode* (关于为加快经济发展方式转变提供司法保障和服务的若干意见) (hereafter referred to as the *Opinions*). The *Opinions* are pivotal to environmental protection, because they show that judicial means will strengthen environmental protection as a new force of great importance. The *Opinions* explicitly stipulate, for the first time, the qualifications of all levels of environmental protection departments to bring public interest to civil litigation as plaintiffs. Meanwhile, the *Opinions* also stipulate that courts in jurisdictions where environmental cases frequently occur may establish environmental tribunals, which conduct specialized trials for environmental cases. This is the first time that the Supreme People's Court (SPC) defined environmental public interest litigation explicitly. It is also the first time that the SPC called for setting up environment tribunals. Both efforts have made special contributions to the enforcement of the environmental judicial function.

Despite steady progress on environmental legislation, the rule of law on environmental protection is far from optimistic. The pollution accident by the Zijin Mining Group (紫金矿业) in Fujian Province and the oil spill off Dalian's coast by the China National Petroleum Corporation (CNPC) cast a disgraceful stigma on China's environmental protection in 2010. The tragedy is reflected fully not only by the scope and degree of the pollution, but also by polluters' scorn toward the right to know environmental information and the right of action by the public. After the pollution accidents occurred, the Zijin Mining Group and CNPC kept silent as to the extent of damage caused by the accidents. Visits and petitions from victims were persistent, but all were stopped or blocked successfully by different forces. Zijin Mining Group and CNPC's responses to the incidents show contempt for environmental rule of law by enterprises and local governments that have interests linked with them.

II. *Reflection on the Huge Gap between Law and Reality*

Under such circumstances, one may ask why the enforcement of environmental legislation is far from satisfactory, despite the huge number of environmental laws and improving environmental legal system. It is known to all who study China's environmental law that the obvious characteristic is the "over focusing on administrative laws, but neglecting civil laws." Most of the content of China's environmental laws, regulations, and decrees are about relevant administrative bodies exercising coercive administrative power to protect the environment. There is scarcely any content about social subjects pertinent to safeguarding economic interests and environmental rights. In the existing environmental laws and regulations, the protagonist is always the state administrative body. Action on behalf of this eternal protagonist includes approval, prohibition, permission, and punishment, which reflects the mandatory role of the state. In such laws, regulations, and decrees, the public plays an insignificant role and their environmental and economic interests can hardly be fully protected.

What is environmental protection? When environmental problems occurred for the first time in human civilization, most people regarded environmental protection as purely a technical issue. It was thought people could rely totally on pollution control techniques and measures to radically solve environmental problems. Later on, this viewpoint was proven absurd and unrealistic. Therefore, the *Declaration of the United Nations Conference on the Human Environment* points out that environmental protection is not only a technical issue, but also a comprehensive issue involving economic, political, as well as other facets of society. As a result, environmental governance should be conducted in a comprehensive way, adopting all possible political, economic, legal, educational, and administrative measures. After decades of environmental governance, we find that environmental protection is also an issue of justice and equality, because all actions taken by mankind will inevitably influence the allocation of interests, which will affect judgments if equality has been achieved. Because of the connection between environmental protection and justice and equality, it is necessary and feasible to protect the environment by legal means. Only law can settle the conflict of interests in environmental protection and pave the way for implementing technical and administrative measures in environmental protection.

III. *Three Kinds of Fairness Issues China's Environmental Law Fails to Resolve*

Environmental protection is essentially an issue of interest distribution, and the law is necessary to guarantee fair distribution of interests. Whether and to what extent environmental laws are effective in environmental protection depends on whether and to what extent environmental laws pay attention to interest distribution in environmental protection. Any environmental protection measures will inevitability produce many conflicts of interests. Such conflicts of interest are ultimately the conflict between economic development and environmental protection. If such conflicts and contradictions cannot be solved fairly, they may reduce the public and enterprises' enthusiasm in participating in environmental protection, and finally pose a negative impact on the implementation of environmental protection measures. Generally speaking, there are three kinds of conflicts of interest in environmental protection that China's environmental laws fail to provide satisfactory solutions to.

A. *Polluters and Offenders of the Natural Environment Do Not Have to Pay*

It is a requirement of the principle of justice and equality in law that enterprises that have obtained economic gains by polluting and damaging the environment "foot the bill" for environmental losses and damage. Only by this means is it possible for enterprises to undertake their obligation to protect the environment while obtaining economic gains. The reciprocity of rights and obligations is an important criterion for justice and equality. Meanwhile, requiring that enterprises pay for environmental losses caused by production proves an effective way to promote their active participation in protecting the environment and saving natural resources. The compensation for environmental losses increases production costs. The best way for enterprises to lower their cost and reduce or be exempt from environmental protection fees is to conserve natural resources and control the discharge of pollutants. In theory, the fee charge system in environmental protection can motivate enterprises to actively take environmental protection measures and prompt the automatic implementation of environmental laws and regulations.

In order to control the discharge of pollutants by enterprises, China's existing environmental laws mainly adopt a few kinds of compulsory legal measures, such as requiring that an enterprise's environmental protection facilities be designed, constructed, and put into use simultaneously with their main construction project. Another example is requiring that enterprises that have committed excessive pollutant discharge or caused serious pollution accidents take compulsory measures such as redressing pollution within a certain time limit or paying fines. It goes without saying that these compulsory legal measures play a positive role in regulating the production of enterprises and protecting the environment. However, the challenge we are faced with is the low enforcement rate of the aforementioned compulsory legal measures. It is not uncommon that enterprises do not consciously operate environmental protection facilities and continue to discharge pollutants into the environment without any treatment. As to the law itself, the main reason is that China's environmental laws pay excessive attention to all kinds of compulsory legal measures but neglect the full application of environmental fee systems, and cannot mobilize enterprises' enthusiasm to protect the environment consciously. When coercive state powers appear to be weak, enterprises regulated by them will immediately cease taking environmental protection measures.

There are various forms of environmental fee systems, such as collecting waste fees, imposing fines, and compensation for damages. However, these types of fees face challenges under China's environmental laws.

1. *Collecting Waste Fees*

In theory, requiring polluting enterprises to pay waste fees may motivate them to take pollution control measures, but one prerequisite is that the fees collected must outweigh the cost for polluting enterprises to run environmental protection facilities. This point seems complicated, but the reason is very simple, because it is costly to operate environmental protection facilities. If waste fees are lower than the cost to run environmental protection facilities, businesses would rather pay the fees and fines than actively and consciously control pollution. That is exactly the case with the sewage fee collection system prescribed by China's environmental law. The collected sewage fee is lower than the cost for enterprises to operate environmental protection facilities, and thus it does not encourage enterprises' enthusiasm for environmental protection.

2. *Imposing Fines*

One principle that must be adhered to when imposing fines on polluting enterprises is that there should be no way for polluters to obtain benefits by illegally disposing pollutants. Otherwise, polluters would rather incur fines than abide by environmental laws and regulations. Unfortunately, there remain such flaws in China's environmental law. For example, the ceiling amount of fines prescribed in the *Law on Environmental Impact Assessments* (环境影响评价法) is as low as 200,000 RMB. However polluting enterprises that ignore the environmental impact appraisal system and start operation before approval can obtain economic profits well above 200,000 RMB. In China, it is a universally recognized fact that the cost of breaking environmental laws is low, whereas the cost of abiding by environmental laws is high. In foreign countries, penalties for pollution are usually calculated on a daily basis or by product amount, which means the daily pollution or each piece produced by the polluting enterprise is treated as one independent action in violation of the law. When calculating fines, every independent action in violation of the law shall be calculated individually, and the total amount of fines shall be the total fine prescribed in law multiplied by the number of each independent action in violation of the law. The fines calculated in this manner are, without fail, astonishingly high. This method of penalizing enterprises dramatically increases the cost and risk of violating environmental laws, and it prevents enterprises from blatantly breaking environmental laws and regulations. However, there has been difficulty for China's environmental law to endorse such penalty measures.

3. *Compensation for Losses*

If an enterprise has engaged in environmental pollution or damage in the production process, and further impaired the safety of individuals and property, it shall undertake civil tort liability. However, there are big defects in China's environmental judicial system. Court autonomy in trials cannot be ensured. When the victims of pollution bring a suit to the courts, all sorts of forces emerge to inappropriately disturb or interfere with the trial proceedings. As a result, many cases of compensation for pollution damages initiated by the public, especially cases of class action, commonly become "files hidden in the court's drawer," and can hardly be expected to see the light of day. That is why the attitude of China's polluting enterprises in compensating for environmental losses and damages are in sharp contrast to that of British Petroleum. China's polluting enterprises are bold enough to show contempt for the public's rights to know environmental information and their rights of action. It is thus not difficult to understand why the SPC specifically emphasized judicial functions in environmental protection in the *Several Opinions on Providing Judicial Guarantee and Service for Accelerating the Transformation of Economic Development Mode* issued in 2010. However, only one opinion issued by the SPC is remotely close to being sufficient to address the problem. The key lays in ruling out the inappropriate interference to court autonomy in trials, as well as establishing and perfecting a mechanism that can ensure the autonomy of the courts during trial.

In conclusion, while China's polluting enterprises obtain economic gains by damaging the environment, they have not undertaken their obligations to compensate for environmental loss and damage. While enterprises lack motivation in environmental protection, the innocent public suffers from ecological deterioration. It is absolute unfairness in the implementation of laws. It is a critical issue that needs more attention in the process of improving China's environmental laws.

B. *Environmental Protectors Receive No Economic Return*

People devoted to ecological protection or people who sacrifice on behalf of ecological protection have forgone personal economic benefit and opportunity for development of the public interest. It is rational for them to receive appropriate economic returns, so that a balance of costs and revenues can be reached, and the reciprocity of rights and obligations can be fulfilled. However, in China, people contributing to environmental protection hardly receive any economic benefit. It is obvious that rights and obligations are not reciprocal, which goes against the principle of justice and equality. A vivid example for this is China's policy of "removing hillsides to facilitate afforestation" (封山育林政策).

The policy of "removing hillsides to facilitate afforestation" is an important measure that reflects the carrying capacity of the ecological system. The policy can sustain forest resources by ensuring the self-renewal and self-recovery of forests. However, compulsory systems or measures should be implemented with a pre-condition of fair distribution of interest; otherwise it can never truly be implemented. The implementation of the "removing hillsides to facilitate afforestation" policy has had a huge impact on local farmers' livelihoods. Their original lifestyle, which was dependent on local mountain resources, has been changed. Local farmers are prohibited from hunting or gathering firewood in the mountains, nor can they exercise their right to use the forestland or their ownership of the forest. Local farmers have lost opportunities to build up family wealth on forestland. However, the laws and regulations do not stipulate full, timely, and adequate compensation for them. The "removing hillsides to facilitate afforestation" policy's lack of any effective compensation mechanism has resulted in farmers losing their land and their foundation for survival and development, resulting in their severe dissatisfaction with and even resistance to this policy. The implementation of the policy relies on voluntary compliance by farmers. Without the understanding and support from the public, how can implementation be sustained if it is purely based on coercive state power?

C. *Environmental Beneficiaries Need Not Pay*

Should environmental beneficiaries pay for enjoying a good natural environment, such as clean air and pure water? For a long time, many considered such resources as a 'gift' from nature and people need not to pay for access to such resources. However, they have failed to realize that costs will occur to keep the air clean and water pure. One typical example is watershed management.

Maqu County in Gansu Province's Gannan Tibetan Autonomous Prefecture (甘肃省甘南藏族自治州玛曲县)—in the upper reaches of the Yellow River—is known as the "kidney of the Yellow River." The Maqu Wetlands are an important water source for the Yellow River. According to the figures from the Department of Water Resources, the water supply from the wetlands accounts for 58.7% of the total runoff of the source area on the Yellow River, and 1/6 of the total runoff of the Yellow River basin. It is a genuine "plateau water tower." However, to meet the demands of local economic development in recent years, the wetland is experiencing disaster—it suffers from severe ecological deterioration. According to experts' estimates, if the wetland degradation maintains at the present rate, the Gannan Grassland where Maqu Wetland is located will become the fourth largest source of sand storms in China in less than ten years. The consequences of ecological deterioration are obvious. It is not only the threat of sand storms, but also the water supply capacity of Maqu Wetland will deteriorate dramatically intensifying the water shortage crisis in the middle and lower reaches of the Yellow River.

The middle and lower reaches of the Yangtze and Yellow rivers are densely populated and advanced in economic development, whereas the upper reaches are lagging economically. To increase GDP, if the upper reaches make efforts to continue deforestation and industrial development, it is inevitable that the environment will be seriously polluted and damaged. Meanwhile, along the entire river basin, the scope of environmental damage and pollution will break the administrative divisions and threaten the environmental quality in the middle and lower reaches of the river. In fact, the 1998 Yangtze River Flood in the basin was caused by the long-term and large-scale deforestation in the upper reaches of the river. The flood's upper reaches caused floods along the entire Yangtze River basin and resulted in heavy losses.

To ensure sustainable social and economic development in the middle and lower reaches of the Yellow and Yangtze Rivers, and ensure clean, abundant, and safe water resources in the lower reaches, it is necessary to adopt strict and timely environmental measures in the upper reaches, even to limit the social and economic development of the upper reaches to a certain degree. However, it is quite unfair for the upper reaches. Why is it necessary to limit the development opportunities of the upper reaches to ensure the economic interests of the lower reaches? Why is it necessary for the upper reaches to adopt environmental protection measures and bear all relevant costs to ensure a high-quality environment for the lower reaches? It is obvious that the interest distribution is not balanced between the upper and the lower reaches. The lower reaches enjoy both economic and environmental returns. In contrast, the upper reaches must limit or abandon development opportunities and bear all the responsibility for environmental governance, paying double the price. A scientific way to resolve this conflict of interests is that the lower reaches benefiting from the environment should compensate costs incurred or lost by the upper reaches. The funds from compensation shall be enough for the upper reaches to protect the environment and for the people in the upper reaches to acquire wealth without cutting forests or developing industries. This is the river basin ecological benefit compensation system advocated by academics. It is an important environmental legal system adhering to the principles of justice and equality. Its theory is to promote parties to adopt ecological and environmental protection measures consciously through the fair settlement of conflicts of interest in river basin protection. However, China's national laws have not endorsed the river basin ecological benefit compensation system. The disparity in interest distribution between the upper and lower reaches is still obvious. It results in unsustainable ecological and environmental protection measures in the upper reaches.

Despite the vast number of China's environmental laws, they fail to resolve the issue of justice and equality in environmental protection as mentioned above. This has made it extremely challenging to implement technical and administrative measures to environmental protection. One important reason for the lack of effective protection of China's ecology and environment lies in the failure to reflect the values of ecological functions. There are no economic sanctions imposed on individuals damaging environments, no reasonable economic benefit for individuals protecting the environment, or compensation from the environmental beneficiaries for those protecting the environment on their behalf. This, undoubtedly, is a major flaw in China's current environmental rule of law. Before this root problem can be eradicated there are many compulsory environmental laws and implementation measures of which none will be effective enough.

Now the focus on legislation should no longer be a matter of increasing the number of environmental laws and regulations, but on tackling the issues of justice and equality in environmental protection.

WHY DIDN'T THE MEP SUE THE ZIJIN MINING GROUP?

Qie Jianrong

Abstract: After the Fujian Zijin Mining Group's toxic pollution accident, Professor Wang Jin, a well-known environmental jurist at Peking University, published his expert views through the media. He urged the Ministry of Environmental Protection of China to bring a lawsuit against Zijin. Later, the Green Volunteer League of Chongqing applied to the MEP for administrative reconsideration, recommending prosecution. On December 13, 2010, the MEP issued an official response to the GVLC refusing to accept the application because the GVLC had no vital interest in Zijin. The way the MEP responded and its decision not to sue Zijin provide food for thought.

Keywords: Zijin Mining Group, Ministry of Environment Protection, pollution, NGOs

I. *Fujian Provincial Department of Environmental Protection Punishes Zijin Mining Group with Sky-High Fines*

On July 3, 2010, an acid waste spill occurred at the Zijin Mining Group's (紫金矿业) copper mine in Fujian Province. As much as 9,100 cubic meters of waste flowed into the Ting River (汀江) through a drainage culvert. Parts of the Ting River were polluted and at least 1.89 million kilograms of fish died. After the pollution accident, residents along the Ting River had to buy "mountain spring water" at a price of 2 RMB per bottle instead of using tap water.[1]

Even though the accident took place on July 3, the public was not notified of it until July 12. The Zijin Mining Group thus concealed the pollution for nine days.

The Zijin Mining Group, a listed company, not only had a pollution incident that shocked the nation, but also concealed the accident from the public. Compelled by heavy media condemnation and pressure, the Fujian Provincial Department of Environmental Protection issued a fine of over 9.5 million RMB to Zijin on September 26, 2010.

---

[1] Zeng Ming, "When They No Longer Believe in Their River," *Xiaoxiang Morning*, July 22, 2010.

This was the first time such a large fine had been issued to a polluting company since the founding of China. Some in the media exclaimed that the issuance of the "sky-high" fine was a "milestone in China's environmental law enforcement."

However, the "sky-high" fine did not win praise, but questions from environmental law jurists. According to Prof. Wang Jin (汪劲), a supervisor of PhD candidates at Peking University and a well-known environmental jurist, the penalty imposed on Zijin by the Fujian Provincial Department of Environmental Protection was merely for political purposes. It was not fully based on legal means and was limited to a penalty of an administrative nature.[2]

As Prof. Wang Jin stated, this was not new. Even in the Tuo River (沱江) pollution incident in Sichuan Province and the Songhua River (松花江) pollution incident in Heilongjiang Province, the state also adopted administrative sanctions rather than legal means to handle these serious water pollution incidents. He severely criticized the administrative departments' way of handling such cases. Prof. Wang said that when dealing with these major water pollution cases, measures adopted by the state administrative departments are astonishingly consistent: "There is nothing more than the following steps. At first, the polluting enterprise conceals the accident with an excuse of 'maintaining stability.' And then they try to reduce the aftermath to a minor scale. The third step is that government departments initiate an investigation, and finally they bargain with the enterprise on compensation." After going through these steps, government departments exert an administrative penalty against the person in charge and finally against the enterprises.[3]

In Prof. Wang's view, relevant departments of the state have devised a formula to deal with significant water pollution cases. Basically, the administrative penalty is the final action. China has thus never adopted legal means to deal with significant water pollution incidents.

II. *Law Experts Suggest that the MEP Bring Charges against Zijin Mining Group*

Although the Fujian Provincial Department of Environmental Protection imposed the administrative penalty of a fine of 9.5 million RMB, this did

---

[2] Qie Jianrong, "Law Experts: MEP Should Sue Zijin Mining," *Legal Daily*, October 11, 2011.

[3] *Ibid.*

not necessarily mark the end of punishment for the Zijin Mining Group. Prof. Wang told a journalist that following the exhausting of administrative punishments, legal means could still be employed.

The Supreme People's Court (SPC) issued the *Opinions on Providing Judicial Guarantee and Service for Accelerating the Transfer of Economic Development Modes* (关于为加快经济发展方式转变提供司法保障和服务的若干意见) (hereafter referred to as the *Opinions*) in June 2010. The *Opinions* clearly states that "the people's courts at all levels shall accept environmental tort cases brought by environmental protection departments on behalf of the state." The issuance of the *Opinions* marked the elimination of judicial obstacles limiting environmental protection departments from initiating environmental tort litigation on behalf of the state.

As for the plaintiff, Prof. Wang Jin held the opinion that if the environmental protection departments, especially the Ministry of Environmental Protection (MEP) or local governments, would act as plaintiff in suing Zijin, they were very likely to win the case. One of reasons was that both the MEP and local governments had participated in the investigation on the Zijin pollution accident after it occurred. They had collected a lot of first-hand evidence. For example, the administrative penalty decision issued by Fujian's Department of Environmental Protection to Zijin on October 8 cited as many as 17 pieces of evidence.

On October 11, 2010, a report in the *Legal Daily* called "The MEP Should Sue Zijin Mining" published suggestions for the MEP on bringing a lawsuit against Zijin by Prof. Wang Jin together with Prof. Wang Canfa (王灿发), the well-known environmental jurist and PhD supervisor at China's University of Political Science and Law. The suggestion from the two professors put the MEP in a very embarrassing situation and the report struck a chord in society. Dozens of the most popular websites, including People.com and Xinhua.net, re-published the report. Surprisingly, though, despite the strong reactions raised on the internet, the MEP did not take a position.

III. *Suggestion to Sue Zijin Raised again by the Chongqing Green Volunteer Union*

If Prof. Wang Jin's suggestion to the MEP to sue Zijin Mining could be regarded as the first time that the MEP was put into an embarrassing situation, the second time would be the document, "Suggestion to the MEP on bringing ecological damage compensation litigation against Zijin Mining Group," which was sent to the MEP on October 29, 2010.

In this "Suggestion," the Chongqing Green Volunteer Union (CGVU) (重庆绿联) pointed out that the enactment of the SPC's *Opinions* had cleared the obstacles for the MEP being qualified as a plaintiff in environmental public interest litigation. Furthermore, the CGVU cited the case of the Xinyi (信宜市) municipal government in Guangdong Province versus Xinyi Zijin Mining in the Yinyan Tin Mine's (银岩锡矿) dam-break accident. In this case, the claim for compensating economic losses was as much as 19.5 million RMB, and the case was accepted by the Xinyi municipal court.

In the 3500-word "Suggestion," the CGVU explicitly advised the MEP to: respond to the *Opinions* issued by the SPC; cherish the opportunity to be qualified as plaintiff in environmental public interest litigation; pilot and popularize environmental public interest litigation in the environmental administration system throughout China; strengthen guidance for investigation and building up of the system; and take actions in the handling and supervision of major and significant cases.

In order to bring litigation against the Zijin Mining Group, the CGVU also advised the MEP to start preparing evidence collection and evaluation work. In addition, the MEP should initiate a lawsuit against Zijin on behalf of the state as soon as possible; require Zijin to compensate the environmental losses caused by years of illegal discharge of sewage, bear the reasonable cost for controlling the pollution, and restore the ecological system and expenses for relevant emergent actions; and apply to the court to retrieve Zijin's illegal gains generated from discharging sewage illegally.

The CGVU also suggested that the MEP seize the opportunity of this case to establish an All-China Fund for Environmental Public Interest Litigation and that partial compensation be allocated as a start-up. When the conditions are mature, a foundation could be established to raise funds and provide support for environmental public interest litigation and to receive and properly use the compensation from environmental public interest litigation for expenses of ecological recovery and other public interest purposes.

IV. *The MEP Dismisses the Administrative Reconsideration from the CGVU*

On December 28, 2010, Prof. Wang Jin was informed that the CGVU's application to the MEP might be dismissed. According to well-informed sources, several days earlier, officials from the MEP and Fujian Provincial Department of Environmental Protection had called on Wu Dengming (吴登明), the president of CGVU, to ask him to withdraw the application. This request from the MEP and Fujian Provincial Department of Environmental Protection was rejected by the CGVU.

On December 29, Prof. Xia Jun (夏军), a lawyer, said that the director of the Administrative Reconsideration Division of the Department of Regulations and Laws of the MEP had called on him to discuss the CGVU's application for administrative reconsideration and its "Suggestion" to the MEP. Prof. Xia Jun responded to the officials that it was expected that the MEP could handle the case according to law. Moreover, he also conveyed to the director that CGVU would not withdraw the application and expected the MEP to provide legal documents regardless of the result.

On the last day of 2010, CGVU received a notification from the MEP, in which they claimed:

> According to Article 28 of the Regulation for Implementing the Law on Administrative Reconsideration, it is one of the prerequisites for accepting an administrative reconsideration application that the applicant must have a relation of vital interest with specific administrative acts. In this case, the administrative penalty imposed on the Zijin Mountain Copper Mine by the Fujian Provincial Department of Environmental Protection has no real impact on your institution's rights or interests. There is no relation between your institution and the administrative penalty imposed on the Zijin Mountain Copper Mine by the Fujian Provincial Department of Environmental Protection. Therefore, the administrative reconsideration application by your institution does not comply with the conditions for accepting the administrative reconsideration prescribed by law.

As a result, all the efforts suggesting the MEP sue the Zijin Mining Group, regardless of whether they came from experts or environmental civil society organizations, were in vain.

V. *The U.S. Department of Justice Can Sue BP, Why Can't the MEP Sue Zijin Mining?*

A. *The U.S. Department of Justice Sues BP Exploration and Production*

On December 15, 2010, the Obama administration sued BP Exploration and Production Inc. and other companies to ask that they be held liable for the largest offshore oil spill in U.S. history. The lawsuit alleged that environmental laws and several safety and operating regulations were violated. Alabama Attorney General Luther Strange said at a press conference on December 15 that "the case begins with the civil litigation which is not yet the final step. The lawsuit is also under civil and criminal investigations to ensure that American taxpayers are not forced to undertake the cost of reconstruction in the Gulf region."

According to Caixin.com, in republished news originally from Reuters, the prosecution considered that nine defendants holding ownership of the oil wells, including BP, Transocean Holdings, MOEX Offshore (the American branch of Mitsui & Co), Anadarko Petroleum Corp (APC.N), and QBE Underwriting/Lloyd's Syndicate 1036, caused damage to the U.S. in the Gulf of Mexico oil spill.

Reuters reported that the U.S. Department of Justice suit charges that these companies violated safety management regulations before the oil spill accident. They also failed to maintain continuous surveillance and failed to maintain equipment and material that were available and necessary to ensure the safety and protection of personnel, equipment, natural resources, and the environment. The U.S. government also claimed that this was a serious violation of the *Clean Water Act* and the *Oil Pollution Act*. The accusation did not put forward a specific amount for compensation, which would be decided based on the judge's discretion.

BP told the media that it was estimated that 40 billion USD would be compensated for the Gulf of Mexico oil spill, including fines and the cost of oil clean up. BP sold a huge amount of assets and business to raise 30 billion USD. It even considered selling its Canadian natural gas liquids business.[4]

B. *The MEP Dares Not Sue Zijin Mining, the Supreme People Court's "Opinions" Becomes the Buck Passed*

If the MEP did not sue Zijin Mining Group before the enactment of the SPC's *Opinions* in June 2010, it would have seemed forgivable, as there was no legal basis to back up the environmental protection administrative organs acting as plaintiffs in environmental public interest litigations. The lack of legal basis became an excuse for environmental protection administrative organs to shirk their responsibility in initiating litigation against serious polluters.

---

[4] Xu Wenqing, "The U.S. Sued BP and Other Parties Liable for the Oil Spill Accident," http://international.caixin.com/2010-12-16/100208036.html, December 16, 2010.

In Prof. Wang Jin's opinion, this was "passing the buck." In the past, environmental protection administrative organs always claimed that they could not initiate litigation against polluters on behalf of the state, because their qualification as plaintiffs lacked a legal foundation. For this reason, environmental protection administrative organs constantly urged the state's legislative organs to amend relevant procedure laws or judicial departments to issue judicial interpretations, and to qualify them as plaintiffs to initiate environmental public interest litigation on behalf of the state. Before the enactment of the SPC's *Opinions*, environmental protection administrative organs could shift the blame onto the court for not initiating environmental public interest litigation. However, once the court could accept environmental public interest litigation, the public shifted the target to the environmental protection administrative organs. It became a critical issue whether or not the environmental protection administrative organs dared to initiate a lawsuit.

An investigative report entitled "Why Didn't the MEP Take the Initiative to Sue Zijin Mining" published in the *Legal Daily* on November 14, 2010 was conducted. But faced with plenty of advice from experts as the MEP was, it continued its silence.

C. *Reasons the MEP Did Not Sue Zijin Mining*

As for the reasons why the MEP did not sue Zijin Mining, experts concluded that although the judicial obstacles had been wiped away, environmental protection administrative organs were still under pressure, mainly from the Party and governmental departments at all levels. In addition, the judicial organs were under the same pressure, because the Party and government departments had the final say on court decisions of accepting environmental public interest litigation.

Prof. Wang Jin stated that China's rule of law has a basic organizational principle. That is, all important issues are decided by the Party committees. If the environmental protection administrative organ is to file a lawsuit in the court, it must also seek permission from the leaders of Party committees or governments at the same level. Similarly, if the court is to make a decision on accepting a case, it must also seek guidance from the leaders of Party committees or governments at the same level. Therefore, as the Fujian Provincial Department of Environmental Protection dared not sue the Zijin Mining Group, the MEP also did not want to bring a lawsuit against Zijin.

According to Prof. Wang, besides the pressure from Party committees and governments, there are also technical barriers such as internal discussion among environmental protection administrative organs to clarify the plaintiff qualification. The MEP may entrust the subordinated Supervision Center to initiate the lawsuit on their behalf. However, besides environmental protection administrative organs at the ministerial level, there are also environmental protection administrative organs at the provincial, municipal, district, and county levels. Are all of them qualified to initiate the lawsuit? Therefore, before suing Zijin Mining, it is necessary for the environmental protection administrative organs at various levels to divide the powers of initiating lawsuits among themselves based on a clear division of the scope of the jurisdiction. For example, the MEP is responsible for prosecuting inter-basin, cross-provincial, or significant pollution cases. Provincial environmental protection departments are responsible for prosecuting pollution cases within a province. It is necessary for the MEP to establish regulations to guard against confusion among different environmental organs when initiating lawsuits.

Another technical barrier limiting the MEP from suing Zijin is the scope of compensation and the method of calculating the losses. If the MEP were to sue Zijin, redefining the scope of pollution compensation, direct economic losses and indirect economic losses, as well as cause and effect would be necessary. "The losses considered as indirect economic losses to the public may actually be direct economic losses to the state. After all, the environment and resources belong to the state," said Prof. Wang. According to him, if the MEP were to sue Zijin Mining, it should invite experts, scholars, and officials to define the scope of compensation, and direct and indirect losses.

Difficulty also lies in calculating the losses to the state. For example, it is very difficult to calculate ecological losses. For one loss, the fee could be 100 million, 200 million, or even 1 billion RMB. It is necessary for the MEP to study this issue and formulate a relevant regulation.

The third technical barrier is evidence, which is actually not that significant. When a major water pollution accident occurs, governments will exercise state power to order monitoring agencies affiliated with state organs to collect evidence. Therefore, in most cases, evidence disclosed by state organs always has greater credibility and admissibility than common evidence. Prof. Wang Jin did not consider evidence to be a big problem if the MEP were to sue Zijin Mining. The challenge lies in preventing local monitoring agencies from refusing to provide evidence or providing fake evidence.

VI. *The CGVU Prepares to Apply to the State Council to Urge the MEP to Adjudicate According to the Law*

The suggestion that the CGVU put forward was not accepted by the MEP. According to Xia Jun, the MEP dismissal was not beyond their expectations. However, Xia thought that the CGVU should be prepared to apply to the Legal Affairs Office of the State Council to request the MEP to withdraw the dismissal, accept the case, and adjudicate in compliance with the law, according to Article 20 of the *Law on Administrative Reconsideration*. At the same time he said: "We will continue to communicate with the MEP for the proposal of suing Zijin Mining. The possibility of applying for another administrative reconsideration has not been ruled out."

The MEP needs a judicial precedent to bring forth environmental public interest litigation. Prof. Wang Jin said after the enactment of the SPC's *Opinions*, the Zijin Mining Group's acid spill, and Dalian's oil spill, the environmental protection administrative organs, especially the MEP, should initiate a lawsuit as soon as possible. "The judicial organ shall create a precedent, even in a case in which the defendant should compensate 100 million RMB but the court only orders 10 million RMB." In Prof. Wang's opinion, the creation of the precedent will have a series of consequent effects. There will be a second and third case only if there is a first case.[5]

Prof. Wang Jin thought that the first case could be an exemplary one for study. People could assess and consider it from economic, technical, legal, and other perspectives. "Only in this way can we know whether the compensation amount is high or low." Such a case would be an important means to initiate state compensation for major pollution cases to control the occurrence of major environmental pollution accidents. However, whether or not to initiate a lawsuit, and whether or not polluting enterprises compensate, are issues of different natures. Even if the compensation amount is very limited, and controversy of the case is very severe, it is important to take action.

In conclusion, if the U.S. Department of Justice can sue BP, why can't China's MEP sue Zijin Mining? The MEP must respond to this question.

---

[5] *Ibid.*

ENVIRONMENTAL INFORMATION DISCLOSURE THREE YEARS AFTER IMPLEMENTATION*

Hu Jing and Shi Tie

Abstract: The Chinese government is more proactive in disclosing environmental information than before, but to different extents in terms of region and information type. Companies remain in the early stage of disclosing environmental information. Environmental non-governmental organizations (NGOs) have been working with environmental authorities to provide training, but fail to receive large numbers of applications from companies. Some organizations' gentle approaches to promoting environmental information disclosure among companies are rather impressive. In several typical cases, few applicants could access the environmental information. Nonetheless, the latest judicial interpretation shows encouragement.

Keywords: official environmental information, corporate environmental information, information disclosure, environmental NGO's

It has been more than three years since the *Governmental Information Disclosure Rules* (政府信息公开条例) (the *Rules*) and the *Environmental Information Disclosure Measures (Trial)* (环境信息公开办法(试行)) (the *Measures*) were both officially implemented on May 1, 2008. How is environmental information disclosure going in China? This article presents a review and outlook for this new concept in China.

I. *Proactive Disclosure of Environmental Information by Governments*

A. *Environmental Authorities at All Levels Become More Proactive in Information Disclosure*

Both the Ministry of Environmental Protection (MEP) and most environmental authorities in China's provinces, autonomous regions, and municipalities directly under the central government have created an

* We quoted several Pollution Information Transparency Index (PITI) assessment results and would like to express our gratitude to this index's developers here.

Information Disclosure section at an easy-to-see location on their websites. Moreover, they publish annual reports on official environmental information disclosure.

Co-developed by the Institute of Public & Environmental Affairs (IPE), the Natural Resources Defense Council (NRDC), and others, the Pollution Information Transparency Index (PITI) is one of the most important indices for assessing environmental information transparency in China. They co-published the 2008 and 2009–2010 assessment results. The average score of 113 Chinese cities assessed for the 2009–2010 period was 36, or about five points higher than in 2008.[1]

B. *Interregional Differences in the Extent of Information Disclosure*

China's coastal regions surpassed other regions, as the average scores for Shanghai and cities in Fujian, Jiangsu, Zhejiang, and Guangdong Provinces were all among the highest, according to the 2008 and 2009–2010 PITI assessment results. In contrast, some central and western regions saw their average scores drop, as Jilin, Jiangxi, Guizhou, Gansu Provinces, and the Inner Mongolia Autonomous Region had the lowest scores.[2] In 2010, the Policy Research Center for Environment and Economy and the MEP launched a project called Research on Public Engagement in Water Pollution Prevention & Control and Information Disclosure Regime & Demonstration (水污染防治公众参与和信息公开制度及示范研究课题) (hereinafter the Information Disclosure Project), in which it assessed the websites of the environmental protection bureaus of 112 cities across China using data as of April 16, 2010. Generally speaking, economically developed regions were better than less developed ones in terms of the extent of information disclosure. To verify the correlations between environmental information disclosure and economic factors, the project team conducted correlation analysis using regional GDPs, industrial value-added outputs, and per-capita GDPs as well as the cities' scores in environmental information disclosure. The results showed that the industrial value-added outputs and the cities' overall scores were the most correlated factors while the industrial value-added outputs, regional GDPs, and per-capita GDPs were the least correlated with corporate environmental information

[1] IPE and NRDC, *Changes in Environmental Information Disclosure: Results of 2009–2010 Promotion Information Transparency Index (PITI) Assessment on 113 Chinese Cities and Comparative Analysis*, 15.

[2] *Ibid.*, 3.

disclosure. It suggests that, pertaining to corporate environmental information, local governments did not disclose more information as the economy grew because of constraints such as economic development and the need to deal with corporate relationships. On the other hand, the weaker correlation between the per-capita GDP and environmental information disclosure suggests that there is much room to increase public awareness in this area.[3] There is great similarity between the results of the two assessments.

C. *Corporate Environmental Noncompliance Information Has the Lowest Transparency among All Types of Information*

With support from the Center for Legal Assistance to Pollution Victims (CLAPV) and Article 19, seven environmental NGOs including the Home of Gansu Green Volunteers, the Greener Beijing Institute (Greener Beijing), the Youth Environment Association of Chongqing (YEA), Hebei Green Friend Association, the Resources and Environmental Protection Institute of Zhaotong, the Environmental Protection Association of Xinxiang, and the Friends of Nature (FON) in Shanghai conducted environmental information transparency tests amongst local environmental authorities and large companies from July through October 2010. Among the scores for environmental information that should be disclosed on their own initiative, the environmental authorities received the lowest scores for publishing lists of environmentally non-compliant companies. The assessment and analysis results from the Information Disclosure Project also show that the main problem with current environmental information disclosure includes publishing the lists of companies that cause pollution incidents, of heavily polluting companies, and of environmental management.[4] The PITI assessment results also show that, as the most important indicator, releasing local companies' daily non-compliance and accident records remains a weakness in terms of information transparency. Only 45 Chinese cities had scores higher than the lowest limit (5.6 points) for this indicator. There was no information on local companies' daily non-compliance records in sections such as the Administrative Penalty Bulletin and Environmental Law Enforcement Bulletin on the websites of the

---

[3] Zhou Jun, et al., "Assessment on Environmental Information Disclosure by the Chinese Government and Policy Recommendations," *Environmental Protection* (Issue 13, 2011), 34.

[4] Zhou Jun et al., "Assessment on environmental information disclosure by the Chinese government and policy recommendations," *Environmental Protection* (Issue 13, 2011), 13.

environmental authorities in cities such as Pingdingshan (平顶山) and Jinzhou (锦州), among which Jinzhou had a score of zero because no information released by any municipal or provincial authorities was available from the website of the local environmental authority.[5]

D. *Significant Improvement in Disclosing Information on Environmental Non-Compliance by Companies*

While the overall disclosure of information on environmental non-compliance by companies is insufficient, the amount of such information released by governments presents a trend of rapid growth. Ma Jun (马军), IPE Director, said that only 2,500 articles of information on environmental non-compliance by companies were released, nationwide, by governments in 2006. But as of November 4, 2011, there were already 90,168 articles available from the Pollution Map published on the IPE website and all of them had been released by governments. In addition, the MEP published the *Report on the Excessive Emission of Main Pollutants by Major State-monitored Companies and Wastewater Treatment Plants in 2009* (关于2009年国家重点监控企业及污水处理厂主要污染物全年排放超标情况的通报) and the *Report on the Environmental Impact Assessment on China Petrochemical Corporation for the Purpose of Listing* (关于对中国石油化工股份有限公司上市环保核查情况的公示) in March and October 2010, respectively. In September 2011, the MEP published the *List of Non-compliant Major State-monitored Wastewater Treatment Plants Identified through Supervisory Monitoring in the First Half of 2011* (2011年上半年国家重点监控污水处理厂监督性监测结果超标名单). All these suggest that the MEP is becoming better in terms of disclosing information on environmental non-compliance by companies.

II. *Corporate Environmental Information Disclosure*

The fact that corporate environmental information disclosure remains in the early stages in China is connected to insufficient legislation. Neither the *Rules* nor the *Measures* have any general requirements on mandatory environmental information disclosure by companies. They do not require

5 IPE and NRDC, *Changes in Environmental Information Disclosure: Results of 2009–2010 Promotion Information Transparency Index (PITI) Assessment on 113 Chinese Cities and Comparative Analysis*, 23–24.

that polluting companies be obliged to disclose corporate environmental information, much less specify penalties in the event of companies' failure to fulfill their duty of disclosure nor incentives for them to provide relevant information on their own initiative. According to the *Measures*, only companies that cause heavy pollution by emitting pollutants in excess of relevant local/national standards or discharge quotas set by local governments must disclose relevant environmental information to the public. Despite this requirement, few companies that are subject to mandatory disclosure make their environmental information available to the public.

According to a Greenpeace report, *The Silent Majority—Survey on Corporate Pollutant Information Disclosure* ("沉默"的大多数—企业污染物信息公开状况调查), 25 China-based facilities of 18 Fortune 500 companies in addition to China's Top 100 public companies involved in the survey were identified by environmental authorities on their websites in 2009, for discharging excessive pollutants into water. According to the *Measures*, they are required to disclose pollution information to the public. None of the 25 facilities disclosed pollution information by the deadline specified in the *Measures*.[6] Published by nine environmental NGO's such as Green Watershed, the *Environmental Records of China's Banking Industry (2010)* (中国银行业环境记录 (2010)) presents environmental policy execution by 14 listed Chinese banks from 2009 through 2010. These banks improved slightly in terms of environmental information disclosure in the reported period than before, but disclosed very limited information. According to the report:

> The main forms of environmental information disclosure remain corporate social responsibility (CSR) reports and annual reports published by the banks to differing extents of disclosed contents and details. Most banks tend to disclose information on loans to energy-conservation and environmental projects and internal efforts to save energy and protect the environment, while disclosing very limited information on loans to projects that come with high financial benefits as well as environmental and social risks, or simply keeping silent.[7]

---

[6] Greenpeace, "The Silent Majority—Survey on Corporate Pollutant Information Disclosure," 9.

[7] Chen Yuanyuan, "Why Is It So Hard to Access Corporate Information?" *China Environment News*, May 18, 2011.

Given the performance of large companies and banks in this respect, the situation among other businesses is much less optimistic.

III. *Environmental NGOs and Citizens' Efforts to Promote Environmental Information Disclosure*

A. *Interactions between Environmental NGOs and Environmental Authorities*

Several Chinese cities, including Jiaxing (嘉兴), Beijing, Foshan (佛山), Zhongshan (中山), Yantai (烟台), Baoding (保定), and Yinchuan (银川), conducted direct communication about environmental information disclosure with environmental NGOs by such means as meetings and forums in the process of PITI assessment.[8] Such communication has led to greater information disclosure in some regions. Additionally, applicants communicate continually with local environmental authorities, and some cities have thus begun releasing regulatory information on pollution sources on a regular basis. Jiaxing, for example, releases information on polluting companies and penalties for non-compliant companies monthly. Some large companies also appear on the list of non-compliance. Moreover, the city publishes specific data about local sewage outfalls.[9]

B. *Organizing Environmental Information Disclosure Training for Environmental NGOs*

The CLAPV and Article 19 co-hosted two training sessions on environmental information disclosure in May 2010. Additionally, they co-hosted the Environmental Information Disclosure Test Result Analysis and Discussion Conference (环境信息公开测试结果分析研讨会) in November 2010 in order to present, summarize, and analyze the Environmental Information Disclosure Test Project that they conducted. The attendants discussed the state of environmental information disclosure in China. The CLAPV and the Xiamen Green Cross Association co-hosted two training sessions on environmental information disclosure in April and June 2011,

[8] IPE and NRDC, *Changes in Environmental Information Disclosure: Results of 2009–2010 Promotion Information Transparency Index (PITI) Assessment on 113 Chinese Cities and Comparative Analysis*, 23–24.

[9] Wang Shuo, "On the Third Anniversary of Environmental Information Disclosure: It Remains Necessary to Overcome Challenges," *Renmin Zhengxie Bao*, May 5, 2011.

where representatives of environmental NGO's and lawyers dedicated to environmental public interest litigation from more than ten municipalities and provinces in China were in attendance. The training was focused on the legal mechanism for environmental information disclosure, effective ways of public engagement in environmental protection, and discussion on cases of environmental information disclosure, as well as other issues.

C. *Environmental NGO's and Citizens Applying to Access Official Environmental Information*

Both the MEP and most environmental authorities in China's provinces, autonomous regions, and municipalities directly under the central government have published their respective annual reports on official environmental information disclosure. Below is a table that summarizes the number of applications for accessing official environmental information over the past three years on the basis of reports on official environmental disclosure published by provincial environmental authorities.

It can thus be seen that the total number of applications for accessing official environmental information is rapidly increasing every year. The trend is shared by the data available from the MEP, which received 72 and 226 such applications in 2009 and 2010 respectively, with a growth rate of 205%. It suggests that the public is becoming increasingly more aware of applying for access to such information, along with a growing knowledge of the environmental right-to-know. Nonetheless, there are obvious differences among regions. In 2010, for example, economically developed regions such as Beijing, Tianjin, Shanghai, as well as Jiangsu and Zhejiang Provinces received 110 applications combined, or 56.1% of the total in China. In contrast, the environmental authority of Yunnan Province received no applications for three consecutive years.

On May 20, 2010, *Southern Weekend* sent applications to the environmental authorities of 29 municipalities directly under the central government and capital cities of provinces and autonomous regions in order to access the lists of businesses in their respective jurisdictions that received administrative penalties for environmental non-compliance that year. There were big differences in attitudes among those authorities. Twelve of them issued proper replies; three bluntly refused; another raised improper requirements; and the remaining thirteen kept silent after *Southern Weekend* reporters applied twice. The reasons for refusal included the lack of such information (Xining, Qinghai Province), the possibility of affecting

Table 23.1: Number of applications for accessing environmental information received by provincial environmental authorities, 2008–2010.

	2010	2009	2008*
Beijing	14	27	17
Tianjin	10	4	1
Hebei	1	–	–
Shanxi	5	3	–
Inner Mongolia	–	–	–
Liaoning	4	1	1
Jilin	0	0	–
Heilongjiang	1	–	–
Shanghai	45	–	–
Jiangsu	33	18	6
Zhejiang	10	4	2
Anhui	13	16	20
Jiangxi	7	0	0
Shandong	5	2	0
Henan	8	–	–
Hubei	5	4	2
Hunan	7	5	1
Guangdong	–	–	–
Guangxi	6	12	4
Hainan	0	–	–
Chongqing	11	5	1
Sichuan	3	–	–
Guizhou	1	–	–
Yunnan	0	0	0
Tibet	–	–	–
Shaanxi	1	2	–
Gansu	2	2	–
Ningxia	1	1	2
Qinghai	0	5	0
Xinjiang	–	–	–
Total	196	115	64
Subtotal for provinces with complete three-year statistics	129	108	64

* Some provinces had whole-year statistics for 2008, while others had statistics only after the *Measures* became effective after May 1.

national/public/economic security and social stability (Tianjin), and threats to businesses' goodwill, trade secrets, and personal privacy of the self-employed (Guiyang, Guizhou Province).[10]

As for the Environmental Information Disclosure Test, environmental NGOs did receive most of the required information by submitting applications, partly because the applicants had good relationships with local environmental authorities. It also suggests that those authorities have certain knowledge of the *Rules*, the *Measures*, and the public's right-to-know; they realize their obligation to provide applicants with information on environmental protection; and they have created certain mechanisms. As an example, environmental authorities have all set up organizations responsible for information disclosure in order to respond to the public's applications. Nonetheless, most of them gave no replies within the required period of time. The applicants had to urge them many times before finally receiving the desired information. The YEA applied twice and saw rather different results in the test in which it was to submit an application to the local environmental authority for access to information. The Chongqing Municipal Environmental Protection Bureau responded actively to the YEA's first application for releasing the list of businesses that emitted excessive pollutants in 2009 as well as information on the allocation and implementation of the pollutant discharge quotas in the year, as it informed the YEA that such information had been published on its website. In contrast, it gave no reply to the YEA's second application, which asked it to disclose information on the amount of various pollutants such as industrial solid wastes discharged across Chongqing in 2009. The reason for no reply was presumed to be that the information asked for by the first application was explicitly specified in the *Measures* but that the information asked for by the second application was not, with exception to the amount of discharged solid waste. Regardless, the unspecified information is not naturally within the scope of information that cannot be disclosed.

The amount of discharged solid waste is the same type of environmental information as the amount of any other discharged waste, so it is obviously unreasonable to treat them differently. Greener Beijing applied for

---

[10] Yuan Duanduan and Xu Nan, "Why It Is So Hard to Disclose Environmental Information: The Story of 29 Applications for Information Transparency," *Southern Weekend*, June 23, 2010.

information fundamentally identical to what was applied for by the YEA, but it was told that the desired information did not belong to what the environmental authority should release on its own initiative and could be released only when it was related to the applicant's production, living, or research. Greener Beijing was not eligible for submitting such an application. Applicants' eligibility tends to be used by administrations as the reason for refusing to provide official information. However, according to Article 5 of the *Measures*: "Citizens, corporations, and other organizations may submit applications to the environmental authorities for accessing official environmental information." It can thus be seen that the *Measures* specifies no limitations on applicants' eligibility. Moreover, Greener Beijing is a Beijing-based environmental NGO and should have no eligibility problem at all.

The IPE and NDRC's application for disclosing regulatory information on pollution sources received better results than previously. The assessment team submitted applications to 113 cities and 49 environmental authorities gave replies. Thirty-two of them provided the desired data and represented 28% of the cities to which the applications were submitted, or four percentage points higher than in 2008.[11]

D. *Environmental NGO's Remind Businesses to Disclose Environmental Information*

Neither the *Rules* nor the *Measures* have any provisions about environmental NGO's submitting applications directly to businesses for access to environmental information. Nonetheless, some Chinese environmental NGO's prompted businesses to disclose environmental information through gentle approaches such as communication and reminders.

The IPE collates information on environmental non-compliance by businesses published on websites of environmental authorities at all levels and by the media, thereby creating a database of such information. FON, Green Camel Bell (GCB), the South China Nature Society (SCNS), the Green Longjiang (GLJ), the Green Volunteer League of Chongqing, and Green Zhejiang have used the published official data stored in this database to remind local environmental authorities that there are environmental problems with some listed companies. The authorities have

---

[11] IPE and NRDC, *Changes in Environmental Information Disclosure: Results of 2009–2010 Promotion Information Transparency Index (PITI) Assessment on 113 Chinese Cities and Comparative Analysis*, 24.

responded to most of those problems and suggested that relevant companies communicate with environmental NGO's.[12] The Golden Dragon Precise Copper Tube Group (金龙精密铜管集团股份有限公司), Zhejiang Beingmate Technology Industry & Trade Company (浙江贝因美科工贸股份有限公司), and Tangshan Sanyou Group's Xingda Chemical Fiber Company (唐山三友集团化纤有限公司), for example, have communicated with environmental NGOs, the latter having since provided an environmental NGO with information showing that it had made improvements since its non-compliance record in 2007.

According to the *Measures*, companies on the list of businesses that cause heavy pollution by emitting pollutants in excess of local/national standards or discharge quotas set by local governments are obliged to disclose relevant environmental information to the public. Chinese environmental NGO's represented by the IPE and the FON have opted to remind some of the above-mentioned companies by sending them "Reminders of Disclosing Corporate Environmental Information." Many Chinese environmental NGOs take such approaches to persuade companies subject to mandatory disclosure of environmental information to make relevant environmental information available to the public. On October 25, 2010, for example, nineteen environmental organizations such as the IPE, Green Beagle Environment Institute (Green Beagle), the SCNS, and FON co-signed a reminder and faxed it to the Shenzhen Zhongjin Lingnan Nonfemet Company (中金岭南有色金属股份有限公司) (NONFEMET) and its Shaoguan Smelting Plant (韶关冶炼厂), reminding them that they should disclose emission data, including the amounts of discharged toxic heavy metals such as cadmium, thallium, and lead as well as the extents to which they exceeded standards and quotas, by November 21, 2010. But the company has not disclosed the information. There are also good examples of communication and reminders. Pan-Asia PET Resin (Guangzhou) Company (广州泛亚聚酯有限公司) ("Pan-Asia Guangzhou"), for example, was punished by the Guangzhou Municipal Environmental Protection Bureau for having discharged excessive water pollutants. It was obliged to disclose environmental information under the *Measures*. On September 29, 2011, Pan-Asia Guangzhou called the IPE to talk about its environmental regulatory records in 2009 and 2010. It then provided the IPE with the 2010 and 2011 wastewater monitoring reports and the 2010

---

12 Chen Yuanyuan, "Why Is It So Hard to Access Corporate Information?" *China Environment News*, May 18, 2011.

and 2011 flue gas monitoring reports. The IPE published the reports in the "Information on Follow-up Regulation by Environmental Authorities" section under the Corporate Environmental Regulatory Records.[13]

IV. *Cases of Applying for Administrative Review and Administration Litigation*

In 2010, the MEP received 25 applications for administrative review of official information disclosure, including 24 applications for official information disclosure by the MEP (23 of which were from the same applicant) and one application for official information disclosure by a local environmental authority. It received no litigation or appeal about official information disclosure that year. In 2009, it received nine applications for administrative review of official information disclosure, including six applications about official information disclosure by the MEP submitted by the same applicant (the MEP approved all of them) and three applications for official information disclosure by local environmental authorities. It received no litigation or appeals about official information disclosure that year.

There were only a few cases of administrative review, administrative litigation, and appeal in 2009 and 2010, according to annual reports on official information disclosure published by provincial environmental authorities. A lot of these authorities saw no cases at all for several consecutive years.

In addition to the cases in which Huang Jianxin (黄建新) filed a lawsuit against the Suzhou Municipal Environmental Protection Bureau for non-disclosure, Sun Nong (孙农) filed a lawsuit against the Zhuhai (珠海) Municipal Environmental Protection Bureau for non-disclosure, Xu Taisheng (许太生) filed a lawsuit against the Shanghai Municipal Environmental Protection Bureau for non-disclosure, and Yan Yiming (严义明) submitted an application to the Anhui Provincial Department of Environmental Protection for environmental information disclosure, there have been several typical cases of environmental information disclosure over the past couple of years:

---

[13] Source: The IPE Bulletin section on the IPE's website.

A. *Green Beagle Applies for Administrative Review of Information Disclosure*[14]

With FON's support, Green Beagle submitted an application to the environmental authority of Hai'an County (海安县), Jiangsu Province in February 2011 for disclosing three articles of information including the environmental impact assessment report on the county's domestic garbage-fueled power plant project. Green Beagle asked for a written reply. It received an email reply from the environmental authority saying that this project received an approval document (SHS (2010) No.223) from the Jiangsu Provincial Department of Environmental Protection on September 15, 2010. This reply had nothing to do with the applicant's request. Later on, Green Beagle communicated with the authority through phone calls and face-to-face talks, but the county's environmental protection bureau gave no official reply. Green Beagle submitted an application to the Nantong (南通市) Municipal Environmental Bureau in late April for administrative review, requesting it to revoke the reply given by Hai'an County's environmental protection bureau and to order the latter to disclose the three requested articles of information. On July 8, 2011, the Nantong Municipal Environmental Protection Bureau issued a letter of decision on administrative review to support the applicant's request. On July 22, 2011, Hai'an County's environmental authority gave a reply saying: The Green Beagle should ask the environmental assessor and the project implementer for the environmental impact report in accordance to a relevant reply from the MEP; there was no report on the disposal of emitted gas, fly ash, slag, and water percolating through garbage from the county's garbage-fueled power plant in the period from June 2006 to October 2009; and it believed that, as for public access to monitoring data and reports, it was necessary to submit a written request for instructions to its upper-level environmental authority since Article 7 of the *Administrative Measures for Environmental Monitoring* (环境监测管理办法) specifies that "no individual or organization may disclose environmental monitoring information unless it is released under law," and that it would give a reply after its upper-level environmental authority gave it instructions.

[14] This case information is from the Friends of Nature.

B. *Lu Jinsong Files Administrative Litigation against the Environmental Protection Bureau of Qidong, Jiangsu Province*[15]

With the FON's support, Lu Jinsong (陆锦松), a citizen in Qidong (启东市), submitted an application to the Qidong Municipal Environmental Protection Bureau on July 6, 2011 for disclosure of environmental information including the flue gas emission monitoring data as well as the data and report on the disposal of water percolating through garbage, slag, and fly ash from the Qidong Domestic Garbage-fueled Power Plant Project (Phase 1) (启东生活垃圾焚烧发电一期项目) in the period from 2009 to May 2011. On July 26, 2011, the municipal environmental authority gave a reply that both the *Environmental Information Disclosure Measures* and the *Administrative Measures for Environmental Monitoring* specify that no individual or organization may disclose environmental monitoring information unless it is released under law. It also said that, as to whether to disclose the information requested by the application or not, it would send a request for instructions to the upper-level environmental authority and would obey the enforcement interpretation made by the competent national environmental authority. On September 1, 2011, Lu Jinsong filed administrative litigation against Qidong Municipal Environmental Protection Bureau over official information disclosure. He requested that: The defendant's reply dated July 26, 2011 to the plaintiff's application for information disclosure be revoked; the defendant be ordered to provide the flue gas emission monitoring data as well as the data and report on the disposal of water percolating through garbage, slag, and fly ash from the Qidong Domestic Garbage-fueled Power Plant Project (Phase 1) in the period from 2009 to May 2011 after this project passed environmental acceptance procedures in 2009. The court has not heard this case so far.

C. *Mao Da Submits an Application to the MEP for Disclosure of the Results of the Nationwide Persistent Organic Pollutant Investigation Project*[16]

On February 27, 2011, Mao Da (毛达), a then-doctoral student at Beijing Normal University, submitted an application to the MEP for disclosure of the results of the Nationwide Persistent Organic Pollutant Investigation Project (全国持久性有机污染物调查) that the MEP conducted in the

15 This case information is from Friends of Nature.

16 Feng Yongfeng, "Being Open, Transparent, and Independent—the Value of Disclosing Environmental Risk Information," *Green Leaves* (Issue 4), 2011.

period from 2006 to 2008. On March 10, 2012, the MEP rejected Mao Da's request. The MEP claimed that, according to the *Rules* and the *Opinions of the General Office of the State Council on Governmental Information Disclosure on a per-Application Basis* (国务院办公厅关于做好政府信息依申请公开工作的意见) (the *Opinions*), the project results belonged to the MEP's internal information and should not be made public. The MEP refused to disclose the requested information in accordance with Article 2 of the *Opinions*, which specifies that "both the internal management information that an administrative authority creates or receives in daily operations and the process information under discussion, study, or review are generally outside the scope of governmental information that the *Rules* specifies should be made public."

D. *Yang Zi Applies for Disclosure of Flue Gas Measurement Data from a Medical Waste Incineration Facility*

In January 2010, Beijing citizen Yang Zi (杨子) asked the Beijing Municipal Environmental Protection Bureau a simple question—did it have the flue gas emission measurement data on a particular medical waste incineration facility? She requested the bureau to disclose them to her if they were available. Nonetheless, the bureau did not answer her question in its written reply, as it said that it was still doing measurements. An unsatisfied Yang Zi filed administration litigation with the People's Court in Haidian District, Beijing. Yet, the judge ruled that Yang Zi was ineligible to submit such an application because she did not live within 800 meters of the incineration facility. Yang Zi then appealed to the Beijing Municipal Intermediate Court, but again, the latter made the same ruling for the same reason.[17] Both the bureau and the courts gave the following reason: the *Technical Specifications on Centralized Medical Waste Disposal (Trial)* (医疗废物集中处置技术规范(试行)) specify that a medical waste incineration facility must be at least 800 meters away from residential areas. Yang Zi lived 2.5 kilometers away from the facility. Whether the bureau responded to her application or not, therefore, had no effects on her rights, and she was ineligible to submit such an application or file an administrative litigation.[18]

---

[17] Wang Shuo, "On the Third Anniversary of Environmental Information Disclosure: It Remains Necessary to Overcome Challenges," *Renmin Zhengxie Bao*, May 5, 2011.

[18] Mao Da, "Why Is It So Hard to Disclose Environmental Information?" *Southern Weekend*, June 30, 2010.

E. *FON Submits an Application to the Ministry of Agriculture for Information Disclosure*[19]

In January 2011, the MEP published the "No. 1 Bulletin" on its website, saying that "the application for making adjustments to the Upper Yangtze Rare and Endemic Fish National Nature Reserve (长江上游珍稀特有鱼类国家级自然保护区) has been approved by the National Nature Reserve Judging Panel." FON submitted an application to the Ministry of Agriculture (MOA) to disclose the comprehensive investigation report on making the adjustments to the nature reserve. Even so, the MOA as the approver decided not to disclose the requested information with the reason that it belonged to process information under discussion, study, or review as specified by the *Rules* and the *Opinions* (GBF [2010] No. 5). FON, as the applicant, did not accept the MOA reply and submitted an application to the latter on March 24, 2011 for administrative review, asking it to revoke the said decision. Moreover, it also applied for legality review of the "process information" provision specified in the *Opinions*. On May 13, 2011, the MOA issued a letter of decision on administrative review, confirming the fact that it would not disclose the information requested by the applicant. It claimed that "before they were approved by the State Council, the proposed adjustments to this nature reserve as well as relevant materials were always under review and study rather than finalized," thereby supporting the interpretation that the requested information belonged to "process information" and maintaining the decision "not to disclose the information." As for FON's application for legality review of the *Opinions*, the MOA said in the letter of decision that it did not comply with Article 7 of the *Administrative Review Law*, so it did not respond to this application. FON then submitted an application to the State Council for administrative ruling, which has not been addressed as of this writing.

As seen in the above-mentioned cases, very few applicants could easily receive requested environmental information. The reasons for non-disclosure primarily included: the applicant had no legal relationship with the issue; the requested governmental information did not exist; the issue was outside the approver's disclosure jurisdiction; the requested governmental information should not be made public; and the requested governmental information belonged to "internal management information" or "process information."

---

[19] This case information is from the Friends of Nature.

The Supreme People's Court promulgated the *Regulations on Issues in Hearing Administrative Cases of Governmental Information Disclosure* (关于审理政府信息公开行政案件若干问题的规定) on July 29, 2011. These regulations became effective on August 13, 2011. The issuance of this judicial interpretation in itself suggested that the judicial authorities were paying much attention to cases of governmental information disclosure. Moreover, there are two highlights:

Firstly, it deleted the provision that scenarios where the disclosed information may threaten national, public, and economic security and social stability, and where the requested governmental information is still under discussion, study, or review shall be within the non-disclosure scope identified by the court. Instead, it specifies that governmental information that shall not be disclosed is information that involves national secrets, trade secrets, or personal privacy.

Secondly, this judicial interpretation has stringent requirements on administrative authorities' burden of proof. It specifies that, if an administrative authority refuses to provide the plaintiff with governmental information, then it should provide evidence for the reasons for refusal and the performance of its legal obligation to keep the plaintiff informed and explain relevant reasons. It also specifies that, if the administrative authority claims that the requested governmental information does not exist, the plaintiff may apply for the court to demand evidence, provided that the plaintiff is able to supply evidence that the requested information was developed or kept by the defendant. This will definitely reduce the applicant's burden of proof, since it is in an obviously disadvantageous position in terms of information availability.

Unfortunately, this judicial interpretation still holds to the provision that the applicant should request governmental information for his or her own special purposes such as production, living, and research, which is one of the *Rules*' provisions that is most criticized. This clearly adds to the difficulty of applying for information disclosure. When this judicial interpretation was first developed, an official from the Administrative Court of the Supreme People's Court say that they would no longer require this.[20] Nevertheless, the officially issued judicial interpretation on governmental information disclosure still specifies that, if the defendant refuses to

---

[20] Ye Doudou, "The Judicial Interpretation on Governmental Information Disclosure: There Are Detailed Provisions on the Scope of Disclosable Information and Proof of Duty," *Caixin*, http://china.caixin.com/2011-08-15/100290933.html (accessed November 10, 2011).

provide the requested information since it has nothing to do with the applicant's special purposes, then the court may require that the plaintiff explain such special purposes.

V. *Predictions for Environmental Information Disclosure*

China began implementing the environmental information disclosure regime more than three years ago. It has made advancements in multiple areas despite many things that are barely satisfactory. China's environmental authorities are paying more attention to releasing official environmental information and beginning to work with environmental NGOs. On the other hand, both environmental NGOs and citizens are increasingly proactive in submitting information disclosure applications. We have reason to believe that a larger quantity and more variety of environmental information will be made public and that the public will access such information more easily.

The author believes that environmental information disclosure in China will show the following trends in the next few years:

First, environmental authorities will shift their focus to the content of disclosed information from the form of information disclosure. Today, environmental authorities, which are among China's best authorities in terms of governmental information disclosure, are making more efforts to build disclosure platforms and to solve the issue of whether to disclose something or not. As the public has higher requirements for environmental information disclosure, performance assessment of environmental authorities will be more about whether the official environmental information is accurate, complete, timely, and easy to understand.

Second, environmental NGOs will shift their focus to usage of environmental information away from information disclosure. Nowadays, when Chinese environmental NGO's apply for environmental information disclosure, they tend to do it for the purpose of disclosure. They will probably shift their efforts toward how to use such information, which will definitely raise requirements on their professionalism in environmental protection. Environmental information disclosure should ultimately be converted into effective environmental protection and improvement, which is its real value. Ma Jun believes that environmental NGOs should not only be an important force that promotes environmental information disclosure, but they should also help develop better use of such information.

Third, there will likely be major breakthroughs in mandatory environmental information disclosure among companies. So far, mandatory corporate information disclosure remains a weakness in environmental information disclosure, since the current law requires that only businesses that have discharged pollutants in excess of standards or quotas and that have been identified by environmental authorities are subject to mandatory environmental information disclosure. This has led to very limited scope in terms of entities subject to mandatory environmental information disclosure and the information disclosed. In contrast, both the U.S. Toxics Release Inventory and the pollutant release and transfer registry (PRTR) specify and expand the scope of businesses subject to mandatory pollutant release information disclosure, while specifying a list of pollutants subject to information disclosure. They emphasize the disclosure of toxic and harmful pollutant release and transfer information, in particular. In December 2008, Shanghai and Tianjin were identified as the cities for a pilot PRTR project. It is foreseeable that this regime will be turned into a national law if the pilot project turns out to be effective, thereby greatly pushing the disclosure of corporate environmental information.

THE STRUGGLING ENVIRONMENTAL COURTS AND ENVIRONMENTAL PUBLIC-INTEREST LITIGATION

Qie Jianrong

Abstract: On October 19, 2011, the Intermediate People's Court of Qujing, Yunnan Province accepted a public-interest lawsuit filed by Friends of Nature and other non-governmental green organizations on the chromium slag pollution incident in Qujing. This lawsuit was considered a groundbreaking case of environmental public-interest litigation in China. Although environmental courts have been set up in nine provinces and two municipalities, it remains difficult for environmental public-interest lawsuits to be accepted, leaving most of these courts in disuse. To solve this problem, legislation needs to be facilitated in an active and appropriate way. The first task is to ensure the legal qualification of environmental organizations as subjects of action.

Keywords: environmental public-interest lawsuit, environmental court, environmental organization

I. *Two Landmark Environmental Public-Interest Lawsuits in 2011*

A. *Breaking the Ice for Environmental Public-Interest Litigation in China*

On September 20, 2011, grassroots environmental organizations including Friends of Nature (FON) and the Green Volunteer League of Chongqing filed a public-interest lawsuit at the Intermediate People's Court of Qujing (曲靖), Yunnan, on the chromium slag pollution incident in Qujing, asking for 10 million RMB in compensation for environmental damages.

The two organizations filed the following claims against the defendants: 1) the defendants should be ordered to immediately cease the infringement, which was the act of stockpiling chromium slag illegally that caused environmental damages; 2) the defendants should be ordered to eliminate the hazard immediately by taking concrete and effective measures to fully remove the pollution caused by the slag that had already been dumped and stockpiled; 3) the measures taken by the defendants to clean up the pollution should be legally appraised by a third-party organization, publicized, and supervised by the plaintiffs and a third party; 4) the defendants should compensate (at a provisional amount of 10 million RMB) for the environmental losses resulting from the chromium slag, with the final damages to be based on forensic appraisal reports; 5) the compensation should be paid into a public welfare fund account specially set up by a third party for ecological restoration and be used for curbing environmental damages caused by the defendants and restoring the environment under the joint supervision of green organizations including the plaintiffs, the court, and a third party; 6) the defendants should cover the plaintiffs' expenses within reason resulting from the litigation and its execution, including expenses from travel, investigation and evidence collection, appraisal, and hiring of experts (at a provisional amount of 50,000 RMB), with the final payment to be based on actual expenses; 7) the defendants should pay for all litigation expenses stemming from the suit; and 8) the defendants should assume joint liability for all claims of the plaintiffs. In their allegations, the two environmental organizations named the Yunnan Luliang Chemical Industry Corporation (云南省陆良化工实业有限公司) and the Yunnan Luliang Peace Technology Company (云南省陆良和平科技有限公司) as defendants and the environmental protection bureau of Qujing as the third party.

The lawsuit immediately aroused broad public attention for having plaintiffs that were grassroots environmental organizations without any government background and for the amount of damages reaching 10 million RMB—the biggest ever compensation claim in the history of China's environmental public-interest litigation.

While China's civil law procedure demands a reply from the court within seven days of the filing, the Intermediate People's Court of Qujing remained silent on whether to accept the case within the allotted timeframe. It was undoubtedly a test for the court to accept this environmental public-interest action brought on by grassroots green organizations.

Nearly a month later, FON officials told the press on the afternoon of October 19 that the court had formally accepted the lawsuit.

This being the first environmental public-interest lawsuit filed by grassroots environmental organizations in China, the case drew praise from environmental jurists. Wang Canfa (王灿发), professor of environmental law at the China University of Political Science and Law, believed the case was a historical breakthrough in China's environmental public-interest litigation and a good beginning for non-stakeholders to bring public-interest actions. The practice of accepting and hearing the case would render profound influence on the on-going revision of the *Civil Procedure Law* (民事诉讼法) and the *Administrative Procedure Law* (行政诉讼法), he said.

FON was pleased by the court decision to accept the case, saying it affirmed and accepted the identity of non-governmental organizations (NGO's) as plaintiffs in public-interest litigation and was a landmark in such litigation filed by grassroots NGO's.

B. *All-China Environment Federation (ACEF) Sues the Environmental Protection Bureau of Xiuwen County, Guizhou Province*

Known as China's largest environmental organization, the ACEF has been actively engaged in promoting environmental public-interest litigation for years.

As an environmental organization, the ACEF filed China's first environmental public-interest lawsuit at the Environmental Court of Wuxi (无锡), Jiangsu Province, on July 8, 2009. It filed more suits later on, including an environmental administration case against the land and resources bureau of Qingzhen (清镇市), Guizhou Province, at the environmental court of Qingzhen. In recent years, the ACEF has also sued the Jiangsu Jiangyin Port Container Company (江阴港集装箱有限公司), the land and resources bureau of Qingzhen, and the Dingpa Paper Factory (定扒造纸厂) of Guiyang, Guizhou Province and won all the cases.

Although these cases were public-interest lawsuits, their public value was not fully recognized. Senior environmental legislation experts believed one of the reasons behind the sluggish environmental public-interest litigation in China was the lack of such lawsuits in a real sense. The ACEF was often questioned over its identity because it was directly under the authority of the Ministry of Environmental Protection.

However, the environmental court of Qingzhen accepted the ACEF's indictment against the Xiuwen (修文) Environmental Protection Bureau for refusing to publicize environmental information in December 2011. The decision changed public opinions that once saw a lack of public value in environmental public-interest lawsuits filed by the ACEF.

On October 17, 2011, the ACEF brought a lawsuit against the Guizhou Hao1duo Dairy Company (贵州好一多乳业股份有限公司), a major dairy firm in Guizhou Province, demanding compensation for environmental pollution at an amount five times that of the pollutant discharge fees the company is obligated to pay in order to fund environmental restoration. Unlike previous lawsuits, this case filed by the ACEF was more substantive by demanding compensation for the purpose of environmental restoration.

During the litigation process, the ACEF submitted a notarized request through express mail service to the Xiuwen Environmental Protection Bureau on October 28, 2011, for the disclosure of environmental information related to the dairy company. However, the bureau failed to publicize such information or reply to the request within the legal time limit for government information disclosure.

On November 24, 2011, the ACEF sent a letter to the environmental protection bureau of Guiyang, suggesting it urge Xiuwen's environmental protection bureau to publicize the information regarding the case, but did not receive any replies from either bureau by December 9, 2011, which was 15 days past the legal time limit for government information disclosure. Therefore, the ACEF believed the Xiuwen Environmental Protection Bureau was in obvious violation of laws and regulations such as the *Regulation for the Disclosure of Government Information* (政府信息公开条例) and the *Regulation on the Disclosure of Environmental Information* (环境信息公开办法).

On December 12, 2011, the ACEF filed a public interest lawsuit regarding environmental information disclosure to the environmental court of Qingzhen in Guizhou Province, demanding the Xiuwen Environmental Protection Bureau reply to the ACEF's request for government information disclosure within three days and provide relative information in line with the ACEF's request within 10 days. The court examined and accepted the case on the same day.

On January 11, 2012, the environmental court of Qingzhen pronounced in court its support for all the ACEF's demands after hearing the case, marking a victory for China's first public interest lawsuit regarding environmental information disclosure.

The reason that the significance of this case was fully recognized by the public was that the ACEF, which was under the authority of the Ministry of Environmental Protection, actually sued its fellow institutions, which were local environmental protection bureaus in this case. In addition, the amount of compensation it demanded was set at five times the pollutant discharge fees the defendant should pay. Ma Yong (马勇), head of the inspection and litigation department of the ACEF's environmental law center, told the author that the specific amount of the claim was not yet determined and would be calculated based on the actual payment of pollutant discharge fees by Guizhou Hao1duo Dairy.

II. *The Establishment of Environmental Courts in China and the Handling of Environmental Public-Interest Litigation*

A. *The Beginning of Litigations Filed by Social Organizations in 2009*

Can social organizations become the subject of environmental public-interest litigation? China's *Environmental Protection Law* (环境保护法) provides no answers to this question and there were no such cases until 2009.

On July 6, 2009, the Intermediate People's Court of Wuxi, Jiangsu Province accepted the ACEF's indictment against the Jiangsu Jiangyin Port Container Company for environmental pollution. It was the first environmental public-interest lawsuit filed by a social organization in China and achieved a breakthrough in making social organizations qualified subjects in such lawsuits.

In May 2009, the ACEF received complaints from more than 80 families dwelling on Junshan North Road (君山北路) in Jiangyin, Jiangsu Province, accusing the Jiangsu Jiangyin Port Container Company of polluting the air and water, causing noise pollution when loading and unloading, and littering iron ore powder. The families said the company's operations seriously affected the air quality, the quality of drinking water from the Yangtze River, and the living environment of residents in nearby regions.

After nearly two months of investigation, the ACEF sued the private company and asked the court to order an immediate cessation of the infringements. They asked the company to make sure the air quality near the port is up to standard and stop operations that infringed on the interests of residents near the port; immediately dispose of rinse water that was used to wash iron ore powder, and eliminate hazards to drinking water source and intake areas; restore the sewers near Huangtian Port (黄田港) and Jiangyin Port and safely dispose of the iron ore powder silt; and pay the litigation fees including those for appraisal and hiring of a lawyer.

After this case, the ACEF then brought lawsuits against the land and resources bureau of Qingzhen, Guizhou Province, and the Dingpa Paper Factory of Guiyang, in succession.

According to Cai Shouqiu (蔡守秋), an environmental law professor at Wuhan University, there was no more than five environmental public-interest lawsuits in China before the ACEF sued the Jiangsu Jiangyin Port Container Company. The five precedents included a case on pollution of the Shiliugang River (石榴岗河) in Guangzhou, one on sea pollution by the Tasman Sea oil tanker, one filed by local authorities against the

Guizhou Tianfeng Chemical Industrial Company (贵州天峰化工有限公司) on environmental pollution, one on green land of the Qingjiayuan Neighborhood (清嘉园小区) in the Haidian District (海淀区) of Beijing, and another filed by a Peking University professor on behalf of polluted fish in the Songhua River against a PetroChina benzene production plant in Jilin Province. The subjects of those litigations included local procurators, water area management institutions, and individuals, but none of those cases were filed by social organizations.

B. *Idleness of Environmental Courts*

China's earliest environmental court, the Environmental Protection Court under the People's Court of Qiaokou District (硚口区) in Wuhan, Hubei Province, was established in 1989, followed by a large number of environmental courts in other areas.

April of 2002 saw the establishment of the Environmental Protection Court under the People's Court of Dongling District (东陵区) in Shenyang, Liaoning Province, and the Environmental Protection Circuit Court under the People's Court of Shahekou District (沙河口区) in Dalian, Liaoning Province. The environmental protection courts of the cities of Jinzhou (晋州市) and Shijiazhuang (石家庄市) in Hebei Province were also established in 2002.

In 2006, the Environmental Protection Circuit Court of Chiping County in Liaocheng (聊城市茌平县), Shandong Province was established. The Environmental Protection Court under the People's Court of Tiexi District (铁西区) in Shenyang, Liaoning Province was set up in June of that year and located in the Tiexi Environmental Protection Bureau. The court was mainly responsible for accepting law enforcement cases regarding environmental protection and providing consulting services on environmental laws and regulations. The court was in the charge of the administrative division of the district court, which takes measures such as prior intervention, mediation in advance and rectification orders before launching judicial procedures when mediation is refused.

In November 2007, the Environmental Protection Trial Division under the Intermediate People's Court of Guiyang and the Environmental Protection Court under the People's Court of Qingzhen in Guizhou Province were established. They were responsible for hearing civil, administrative and criminal cases of first and second instance and relative enforcement cases such as lawsuits on pollution-caused infringements, damage compensation, and environmental issues of public interest regarding water resource protection at the local Hongfeng Lake, Baihua Lake, and Aha Reservoir (两湖一库) as well as water, soil, and forest protection in regions under their jurisdiction.

In February 2008, the Environmental Protection Circuit Court of Jianye District (建邺区) in Nanjing, Jiangsu Province was set up in the Jianye Environmental Protection Bureau. It conducted on-site hearings and enforcement of environmental administrative cases within its jurisdiction two days a week, carried out publicity and consulting activities regarding environmental laws and regulations, and took part in resolving and coordinating environmental administrative disputes. In May, the Environmental Protection Trial Division under the Intermediate People's Court of Wuxi, Jiangsu Province and an environmental protection collegiate panel within the court's jurisdiction were established. Civil, administrative, and criminal cases regarding environmental protection were heard collectively at these courts, which were responsible for handling various criminal, civil, and administrative lawsuits of first and second instances involving environmental protection within their jurisdiction and were in charge of relative enforcement. It was also their duty to strengthen ties with local governments and environmental protection authorities and bring forward judicial advice. The trial division of the intermediate court was also responsible for supervising the trials at environmental protection collegiate panels under grass-roots courts and offering them professional guidance. In August, the Environmental Protection Circuit Court under the People's Court of Xinbei District in Changzhou (常州市新北区), Jiangsu Province was established in the environmental monitoring branch of Changzhou. The district court appointed officers to work on the spot on a regular basis every week and they were mainly responsible for accepting legal consulting on environmental protection, conducting pre-litigation mediation of environmental administrative disputes, as well as hearing administrative cases involving environmental law enforcement and damage compensation cases regarding environmental pollution. By the end of 2008, environmental protection trial divisions were established, in succession, under the intermediate people's courts of Kunming and Yuxi (玉溪) in Yunnan Province. Such divisions were also established under the county courts of Chengjiang (澄江县) and Tonghai (通海县) in Yuxi.

There are now environmental courts under three intermediate courts and six grass-roots courts in Yunnan Province. As Tian Chengyou (田成有), the deputy chief justice of the Higher People's Court of Yunnan Province, puts it, Yunnan has the most environmental courts among all provinces in China.

According to calculations by the ACEF, environmental courts, trial divisions, collegiate panels, and circuit courts have been set up so far in 11 provinces and municipalities and in 37 cities and counties, among which 10 cities and counties have environmental and resource courts.

However, despite their names, those courts rarely accept lawsuits regarding environmental pollution. An official with the Higher People's Court of Yunnan Province said that from more than two years of practice in Yunnan, the total number of environmental cases was too small and that even fewer cases involved pollution as most were related to natural resources such as forests, land, and mineral resources. Moreover, most of those courts face difficulties in finding enough work to do.

In Kunming, Yunnan Province, special environmental institutions have been established in public security bureaus, procurators, and courts, making the city's environmental judicial institutions the most specialized and extensive in the country. However, these institutions have been troubled by the shortage of lawsuits, which has sparked controversy among the public.

Tian says people have pinned great hopes on environmental courts in Yunnan since they were established but reality is far from meeting their expectations, with environmental courts facing embarrassingly few cases and waiting for work. He notes that the trials at environmental courts have drawn criticism for having no substantial differences from traditional courts and not being professional enough.

The problem not only occurs in environmental courts, but also in the environmental and resource division under the People's Procuratorate of Kunming, which was established on December 8, 2008 and faces the same dilemma of inadequate cases. The environmental protection branch of the Public Security Bureau of Kunming is the earliest and only such branch in China, but its biggest problem is also that there are few cases to deal with.

At present, having no cases to hear is a universal phenomenon in the nation's environmental courts. As one of the earliest provinces to institute environmental courts, Liaoning has several environmental tribunals but not even one environmental public-interest lawsuit has been heard in 15 years since its establishment.[1]

---

[1] Fan Chunsheng, "Zero Environmental Public-Interest Lawsuits in 15 Years Reveals Legislation Flaws," *Xinhua Net*, May 15, 2011, http://news.xinhuanet.com/2011-05/15/c_121418066.htm.

III. *Advance Legislation on Environmental Public-Interest Litigation in an Active and Steady Manner*

A. *Major Reason behind Sluggish Progress in Environmental Public-Interest Litigation Is Lagging Legislation*

According to an official with the Higher People's Court of Yunnan Province, the problem of few environmental lawsuits troubles not only Yunnan but also almost all other places with environmental courts. The root cause behind this is the slow progress in environmental legislation in China, the official says.

Although the environmental and judicial circles have been appealing for environmental public-interest litigation for many years, progress has been slow. The major reason for this is the lagging environmental legislation, which has never managed to achieve a breakthrough in defining the legitimate subjects of such litigation. Currently, the basis for Chinese citizens or social organizations to become the subjects of environmental public-interest lawsuits are merely a few policies and regulations, such as circulars issued by the State Council on environmental protection, but no laws in this country provide specific clauses that clearly stipulate that citizens or social organizations can bring such lawsuits as subjects. The debate has never ceased on who can be the subject of such lawsuits, making the legislation process difficult.

Another reason is that even though some local environmental courts receive environmental public-interest indictments, they usually cannot decide on their own whether they can accept the cases and have to repeatedly ask for the opinions of superiors and even need the tacit consent of superior courts.

For example, when Friends of Nature sued the Yunnan Luliang Chemical Industry Corporation and the Yunnan Luliang Peace Technology Company for chromium slag pollution but did not receive a court reply long past the legal timeframe, a Friends of Nature official in charge of the case telephoned the Intermediate People's Court of Qujing. The court said they had submitted the indictment materials to the Higher People's Court of Yunnan Province and had to wait for replies from the higher court.

B. *First Make Environmental Organizations Legitimate Subjects of Public-Interest Lawsuits*

There have been repeated public calls in China for accelerating legislation on environmental public-interest litigation to make social organizations and citizens legitimate subjects of such litigation. However, little progress has been made. The academic community has been debating over who can be the subjects and cannot reach consensus, which greatly affects the decision making of legislators.

The schools of thought supporting a greater role of plaintiffs advocate that environmental groups, procuratorates, environmental administrative and law enforcement departments, as well as citizens can all become subjects of environmental public-interest litigation. However, the government and some scholars believe it wouldn't be realistic under current circumstances in China. Yi Yuanyuan (伊媛媛), a doctor of environmental law at Wuhan University, argues that citizens are not suitable subjects for such litigation at present. According to her, environmental public-interest is characterized by integrity and universality and the decentralization of individual interests makes it difficult to form uniform appeals. Meanwhile, the limited individual resources, inadequate law consciousness, and low awareness of rights will affect the actual outcome of citizens participating in environmental public-interest lawsuits and may lead to abuse of litigation, heavy burdens for courts, waste of public resources, and disorder in such litigation.[2]

As part of an amendment to the *Environmental Protection Law*, the legislation regarding environmental public-interest litigation has been raised high on the agenda. The author suggests two steps: social organizations should be made legal subjects of such litigation first, or the law should clearly stipulate that social organizations can bring such lawsuits as subjects, and the law can be applied to certain qualified and capable social organizations in each province on a trial basis. Based on the practices, restrictions can be lifted gradually on other potential subjects.

The lack of cases in environmental courts has posed a sharp contrast with the reality of severe environmental pollution and the large number of pollution incidents in various places throughout the country. We should realize that the key issue at present is not the abuse of litigation but the lack of legal protection and assistance for citizens whose environmental interests are hurt. Therefore, local environmental courts should accept environmental public-interest lawsuits in an active and bold manner.

[2] Xi Jianrong, "Slow Progress in Environmental Public-Interest Litigation with No Cases to Hear in Local Environmental Courts," *Legal Daily*, June 20, 2011.

PART SIX

CIVIL SOCIETY AND PUBLIC PARTICIPATION

NGOs PLAY A POSITIVE ROLE IN HANDLING SHANGHAI RICHINA LEATHER POLLUTION

Gu Xunzhong

Abstract: In 2009, Friends of Nature (FON), the Institute of Public and Environmental Affairs (IPE), and community residents joined together to handle environmental violations by the Shanghai Richina Leather Industrial Company. Through publicity of pollution information, supply chain tracing, and media involvement, the company eventually changed its ways and the conflict was resolved in a rational, peaceful, and mutually beneficial way.

Keywords: environmental pollution, environmental NGO, public participation, consensus and mutual benefit

I. *Richina Leather Had a Pollution Record for Six Consecutive Years*

The Shanghai Richina Leather Industrial Company (上海富国皮革工业股份有限公司) began as a joint venture in 1995 and became a foreign-invested joint-stock company in 2008. Through many years of development, it became a big brand. At the same time, it discharged a large amount of illegal pollution during its expansion. According to files kept by both the Shanghai Municipal and Baoshan District (宝山区) Environmental Protection Bureaus, Richina Leather had a pollution record for six years from 2004 to 2009. It left air pollution control facilities unused without authorization and was listed as an environmental law breaking company many times. Residents of Shanghai University's new campus, Beizhang Village (北章村), the neighborhood of Jufengyuan (聚丰园), and the Qilian New Village (祁连新村), were troubled by the odor over the years. Many residents developed respiratory ailments such as tracheitis, bronchitis, and chronic pneumonia, while some even died of lung cancer.

The pollution from Richina Leather is mainly produced during the process of tanning leather. For many years, residents from the neighborhood filed complaints to relevant departments, leaders, and the court. They successfully gained the attention of the Shanghai Municipal Government, which took measures to prevent the deterioration of the problem.

Unfortunately, the government's plans to improve and renovate the neighborhood were not made known to the public, preventing the public from seeing any real hope of improving the problem. The odor from Richina Leather continued to haunt the neighborhood. Some residents decided to turn to the law for protection, only to be rejected by the court with no specific reasons and no written responses. Therefore, on June 25, 2009 the residents of the Jufengyuan neighborhood wrote and signed a letter addressed to the Party Secretary of the Shanghai Municipal CPC Committee, the mayor, and the Ministry of Environmental Protection (MEP) complaining about the pollution that had bothered them for so many years. Some more radical residents even considered more violent measures.

On September 11, 2009, these residents wrote again to the MEP to express their anger about the air pollution from Richina Leather.

II. *Environmental NGOs in Action*

The first environmental NGO that paid attention to the pollution from Shanghai Richina Leather was the Institute of Public and Environmental Affairs (IPE) in Beijing. On the IPE's air pollution map of China, Shanghai Richina Leather was on the top of the list for six years. According to environmentalist Ma Jun (马军) and the IPE, the pollution from Shanghai Richina Leather was very serious. Therefore, when the Shanghai team of Friends of Nature (FON) asked the IPE for information about polluting companies in Shanghai, IPE informed FON's Beijing office and Shanghai team of Shanghai Richina Leather's ranking.

A. *Action 1: Access of the Site by the Shanghai Team*

The Beijing office and Shanghai team of FON were very concerned after receiving the message from the IPE. They held many discussions within the presence of lawyers to decide on a plan.

Volunteers from FON's Shanghai team took the responsibility to ask for information on Richina Leather's air pollution, permits, sanctions, and corrective actions from the district government, the Environmental Protection Bureau, and the Economic Committee of Baoshan District.

With some knowledge of the air pollution of Richina Leather, Director Yi Xiaowu (易晓武), the leader of FON's Shanghai team, and Liu Hongming (刘红明), a member of the FON Working Committee, conducted field research in the neighboring communities in late February and June

2009. Warmly welcomed by the residents, they learned about the residents' constant efforts against pollution and their attitudes toward the current situation. FON managed to win the trust of local residents, paving the way for future cooperation and laying down the foundation for the development of an operational strategy.

B. *Action 2: The Environmental NGOs Jointly Send an Environmental Information Inquiry Letter*

Based on the progress of the Shanghai team, FON's Beijing office and the IPE held a meeting on June 22, 2009 to formulate their next step.

Eighteen influential Chinese environmental NGOs, including FON and the IPE, jointly sent an environmental information inquiry letter to Richina Leather, asking the company to publicize data on the main pollutants by July 5, 2009. This move was made within the framework of the law, and offered Richina Leather a chance to take the initiative. However, Richina Leather did not respond to these NGOs.

C. *Action 3: Tracing the Supply Chain of Richina Leather*

As part of the operational plan, FON's Beijing office and the IPE traced Richina Leather's supply chain. This action was taken while the inquiry letter was being prepared.

Richina Leather is a major raw materials supplier to Timberland, an American top outdoor leisure brand for products such as clothes, shoes, raingear, and watches. Therefore, FON and the IPE wrote about Richina Leather's pollution problems to the Timberland headquarters in America and its CEO, Jeff Swartz, on July 13, 2009.

This letter proved effective. On July 16, the Timberland office in China called the IPE to ask for information and on the same day, Richina Leather also called IPE to discuss and explain the issue. This was Richina Leather's first official response to the Chinese environmental NGOs, although this response was obviously made under pressure from Timberland.

Nevertheless, Timberland did not reply the first moment they received the letter. Moreover, except for the phone call made by Timberland's China office, no response was made from the headquarters of this giant multinational corporation, ignoring the concerns of Chinese environmental NGOs.

As a result, the IPE checked through an American environmental protection organization on July 20, 2009 whether the headquarters of Timberland had received the email. One day later, IPE received an email from this organization, saying that the headquarters of Timberland had received the letter and would give an answer at a proper time. But the organization reminded the IPE that the answer was not about to come any time soon.

D. *Action 4: Media Exposure Finally Pressures Richina Leather to Apologize*

The involvement of the media was a critical part of the FON and IPE plan. FON's Beijing office and IPE disclosed information to the media about Richina Leather's pollution and the silent treatment from the company.

Some media outlets responded promptly. ChinaEconomy.cn reported the serious pollution by Shanghai Richina Leather twice, on July 31 and August 6, 2009. *East China Morning Post* published an article entitled "Timberland Linked to Polluting Factories," exposing to English readers around the world the serious pollution caused by Shanghai Richina Leather and many other international brands that have a business relationship with Richina.

Media coverage, particularly the article carried by the *East China Morning Post,* finally pressured Timberland and Richina Leather to respond. On August 8, the day after the publication of the article by the *East China Morning Post,* Arne May from the Timberland American headquarters emailed FON and the IPE. This led IPE to realize that even giant American companies like Timberland would concede under pressure from the media.

Since then, both Timberland and Richina Leather have made several moves. Timberland's China office called IPE to ask for details about Richina Leather's pollution on August 10, 2009 and explained that Timberland usually depended on BLC (British Leather Technology Center), an independent center that does product testing, audits, problem solving, etc.) to review its suppliers. Therefore, it did not pay attention to the operational process of the supplier companies. In BLC's reports, Richina Leather was ranked as a "silver prize" supplier in 2007 and 2009, certifying the qualification of Richina as a supplier. Timberland's China office had no knowledge of the pollution records and promised that the Timberland headquarters would recheck the review process of BLC.

On August 11, Richina Leather called IPE to discuss the pollution situation.

On August 12, Timberland's China office called IPE again, saying that they had already reduced their orders from Richina Leather and intended to end business with it in two months. Timberland also promised to contact other companies that had business with Richina Leather and exert pressure in a joint effort.

In early August, FON and the IPE continued exchanging emails with Timberland's American headquarters on how to deal with the pollution.

On August 19, a representative from Timberland's China office visited IPE for discussions. According to the Timberland representative, Richina Leather had been dealing with its own pollution problem half-heartedly and Timberland intended to end business with it. Moreover, the representative admitted the impropriety of having the review process supervised by a third-party and promised that Timberland would take the advice of FON and the IPE to take public opinions into its review process in the future.

On August 19, Richina Leather wrote to the chief editor of the *East China Morning Post*, expressing regret over their actions.

III. *The Involvement of Environmental NGOs Brings Positive Interaction*

It was under the pressure of environmental NGOs that Richina was forced to take action. During the process, positive interactions between residents, environmental NGOs, and polluters enabled residents to seek justice, companies to mend themselves, and the environment to improve.

A. *"Public Open Day" at Richina Leather*

Under pressure from various sources, the CEO of Richina Leather, Bao Bohua (鲍博华), finally decided to tackle the issue directly and agreed that a neutral party should be invited to review the situation after discussions with FON's Beijing office and the IPE. Eventually, on the "public open day" on September 23, 2009, Richina Leather opened its door to some residents and organizations.

While the change of attitude by Richina Leather was worth noting, environmental NGOs and residents faced other problems. After all, in China, both residents and environmental NGOs are still inexperienced in communicating face to face with companies that are polluting the environment. What if there were to be a conflict? How could we make sure different parties could exchange opinions in a rational, peaceful, and mutually beneficial way? How could we make sure the demands of residents were fully expressed? How could we achieve the goal of improving the company's actions and satisfying the residents' environmental demands? What role should environmental NGOs play in this process?

To solve these new problems, FON's Beijing office, the IPE, members from the FON Shanghai team, resident representatives, and lawyers exchanged emails actively beforehand. Approximately 77 emails were exchanged among these parties in addition to several video conferences via Skype. People eventually agreed on the following issues:

First, residents would play the main role in this round of face-to-face discussions so that they could fully express their opinions. For this purpose, FON's Beijing office and the IPE successfully revised the agenda set by Richina Leather to give residents a chance to speak their mind.

Second, environmental NGOs would, in line with their positions as environmental protection volunteer organizations, abide by the rules and policies of public environmental protection participation; follow the principles of openness, fairness, equity, and transparency of information; strive to achieve the goal of win-win for both the company and the community; and encourage the roles the community played in improving the environment. Broadly speaking, environmental NGOs represent public environmental interests, but to avoid potential conflicts, they would not formally represent the neighborhood on this occasion.

Third, since Richina Leather had not put forward a plan concerning a third-party reviewer, the Shanghai team held several preparatory meetings with residents, lawyers, and journalists to work on a proposal together. Eventually, eight resident representatives and two representatives from FON's Shanghai team visited Richina Leather on the public open day.

To people's surprise, the public open day was a success. Encouraged by the representatives from FON's Shanghai team, residents took the front seats and each of them got the chance to speak directly to the senior management about their concerns. Bao Bohua and other senior managers also introduced the measures already taken and those on the way. Bao promised that in the future the residents could call him personally if they smelled the odor again. Members of FON's Shanghai team suggested a memorandum, a proposal endorsed by Richina Leather. The success of the public open day greatly boosted the confidence of the residents and environmental NGOs, encouraging them to work with Richina Leather to solve the pollution problem.

B. *Richina Leather's Open Inspection and Residents' Complaints*

As an effort to honor its promise on the public open day, Richina Leather invited resident representatives and FON's Shanghai team to inspect the air quality of the factory area monthly. However, this invitation was rejected by resident representatives on the grounds that the air quality monitoring was not going to be performed by a third-party organization as advised by FON's Beijing office and the IPE, but by an inspection agency subsequently commissioned by Richina Leather. Residents believed that this kind of review might cover up the pollution for the company. This showed that the residents still did not trust Richina Leather.

The "community inspection" incident which happened later more typically showed the lack of mutual trust.

On October 13, 2009, the familiar odor from Richina Leather once again lingered over the neighborhood of Jufengyuan. Some residents called Bao Bohua directly to complain about it, but Bao did not answer the phone at first. He later invited the residents to inspect the factory. The residents were worried about their personal safety, as it was nighttime. They rejected Bao's offer twice. Bao thought the residents were not sincere about this issue, while the residents complained Bao was just playing tricks with them. After Bao went into the factory, the odor gradually cleared away.

This incident showed that the long-term distrust and conflict between the residents and Richina Leather did not disappear after one successful conversation. Therefore, FON's Beijing office and the IPE recommended a neutral third-party review and public disclosure of information, the only effective way to build mutual trust.

C. *Third-Party Review*

On February 1, 2010, Bao Bohua, the president and CEO of Richina Leather, visited FON's Beijing office and discussed the pollution issue with FON's Director Li Bo (李波) and environmentalist Ma Jun. Ma Jun gave four pieces of advice. Richina Leather should: 1) explain its former misbehavior and measures for improvement; 2) publicize the government environmental inspection reports and other subsequent supervising reports; 3) disclose emissions data regularly and the third-party review report; and 4) prove its renovation measures to the public. Bao Bohua agreed to have the third-party review and invite environmental NGOs and resident representatives to supervise the review process.

During the discussion, FON's Beijing office and the IPE learned that Richina Leather had closed the tanning workshops that produced the odor in Shanghai and were preparing to relocate the new factory to the Qinghe District of Fuxin (阜新市清河区), Liaoning Province. FON and IPE accordingly suggested that Richina Leather should take strict environmental protection measures at the new factory and disclose the emissions data to the public from the beginning.

The third-party reviewer entered Richina Leather on April 19 and 20, 2010. By then, there were 34 (originally 18) environmental NGOs actively involved.

As recommended by FON's Beijing office and the IPE, and authorized by Richina Leather, the international engineering and design consulting company AECOM was commissioned as the third-party reviewer. Moreover, this time residents of the neighboring community played a role as supervisors. In addition to an on-the-spot inspection and document examination, residents were allowed to conduct targeted and random interviews with environmental NGO representatives, resident supervisors, and residents in the neighboring community as interviewees.

During interviews with the residents, the inspectors learned about a pipe discharging polluted water outside the factory. Staff from the Environmental Protection Bureau of Baoshan District examined the pipe. However, there was no evidence to show where this pipe came from and therefore it was difficult to decipher the source of the polluted water. AECOM required Richina Leather to prove that the factory was not the source and invited NGOs and residents to supervise this process. Richina Leather inspected this pipe again on July 15, 2010, with representatives from AECOM, the Shanghai teams of IPE and FON, and residents. However, with the pipe blocked and buried, this joint inspection did not reach any conclusions.

During the review process, ShanghaiEnvironment.org.cn published the annual enterprise "Green List" and "Black List," as certified by Shanghai's Environmental Protection Bureau. Richina Leather was listed as one of only two on the "Black List."

By now, the third-party review report was finished. Since Richina Leather failed to present sufficient information to prove that the problems that had put it on the list were resolved, FON's Beijing office and IPE advised that they should keep Richina Leather's pollution records from 2004 to 2009 on the China air pollution map and decide whether to remove the records after Richina Leather had worked with departments to solve these problems. In the meantime, these two environmental protection organizations expressed regret over the fact that Richina Leather had been put on the "Black List" again for failing to provide the necessary information during the review.

IV. *Pollution Problems Remain Unresolved*

At this point, the pollution problem of Richina Leather has not been resolved.

Richina Leather is still working on environmental treatment to get itself removed from the "Black List." During the second public open day on October 15, 2010, Richina Leather held meetings with residents and environmental NGOs, and invited them to participate in the environmental treatment process. Environmental NGOs proposed three measures: renovate the company; strengthen environmental control at the new factory in Liaoning and disclose environmental protection information promptly; and strengthen the management of the supply chain, identify environmental protection regulation violations promptly, and push reforms.

While residents and environmental protection NGOs are still discussing these issues with Richina Leather, the conflicts have been replaced by conversation and interaction in a rational, peaceful, and open way, giving residents a channel to file complaints. Now, many fewer complaints have been filed against Richina Leather.

On October 30, 2009, Richina Leather's tanning factory at Fuxin laid its foundation. For the residents of Baoshan District in Shanghai, a main source of pollution may have left. But, for residents of Liaoning it is not clear if the factory will bring severe pollution or not, and for environmental NGOs such as FON and IPE, there is still a long way to go in learning how to deal directly with corporate polluters.

TINY PARTICLE, BIG ACTION: PUBLIC PARTICIPATION IN ENVIRONMENTAL MANAGEMENT AS SEEN IN THE $PM_{2.5}$ CASE

Wang Tao

Abstract: The smog in 2011 caused public concern and the term $PM_{2.5}$ has since become known to the public. The public, after learning of the adverse health effects of $PM_{2.5}$, started to participate in and contribute to the environmental management process. This paper traces the process of public attention on $PM_{2.5}$ in 2011, and describes the actions and contributions of the media, government, experts, NGOs, and individuals during the three stages of the half-year event. It also introduces the history of China's monitoring and research of $PM_{2.5}$ and analyzes the reasons why public participation in 2011 was able to accelerate the environmental management process. Finally, it makes some suggestions for all stakeholders to better encourage public participation in environmental management.

Keywords: $PM_{2.5}$, stakeholder, public participation

Particulate matter, or PM, is the term for particles found in the air. Particles less than 2.5 micrometers in diameter ($PM_{2.5}$) are referred to as "fine" particles. At the end of 2011, this atmospheric and environmental chemistry term made its way to one of the top ten buzz words in China.[1] In the same year, more and more individuals and non-governmental bodies started to monitor air quality and $PM_{2.5}$ drew extensive attention from the public, compelling environmental authorities to release their monitoring results. This development led to the nationwide monitoring of $PM_{2.5}$ and is thus a milestone of public participation in environmental management.

I. *How the Chinese Got Concerned about $PM_{2.5}$ in 2011*

Looking back on the Chinese public's concern over smoggy days and $PM_{2.5}$, we may discern three stages in public participation and environmental management.

---

[1] "Top Ten Chinese Buzz Words of 2011," http://book.sina.com.cn/news/c/2011-12-14/1726293679_2.shtml.

A. *Spread of $PM_{2.5}$-related Information*

February 21 was the first severely polluted day in Beijing in 2011. According to a *Global Times* report on February 23, the U.S. Embassy assessment found that the pollution level was "beyond index" (below the worst measureable rating). In 2008, the U.S. Embassy in Beijing set up a private air monitoring station and has broadcast real-time pollution readings on the Sina Weibo microblog since then. However, as its readings are often inconsistent with the official data, an Embassy spokesperson said in July 2009 that its readings only reflected the air quality at the Embassy compound and were not representative of the entire city. In reality, the major reason for the discrepancy is that $PM_{2.5}$ was monitored by the U.S. Embassy while it had never been incorporated in China's air quality standards. Since then, the public has become aware of $PM_{2.5}$.[2]

Dense fog shrouded Beijing for several days in a row in the early fall of 2011. On October 27, the *Southern Weekend* published a story entitled "I Gauge the Air Quality for my Motherland," introducing the citizen air quality monitoring campaign. The same newspaper also carried a chart comparing the daily air quality published by the Beijing Municipal Environmental Protection Bureau and that published by the U.S. Embassy. In the meantime, more and more people began following the U.S. Embassy readings and downloading the application on their smart phones. On October 31, a representative from the Beijing Municipal Environmental Protection Bureau gave an interview under public pressure, explaining the reasons behind the difference. However, the interview did not satisfy the public.

On November 1, the *Ambient Air Quality PM_{10} and $PM_{2.5}$ Weight Measurement Law* (HJ618–2011) (环境空气 PM_{10} 和 $PM_{2.5}$ 的测定重量法), which was drafted by the Ministry of Environmental Protection (MEP), came into effect. The law stipulates for the first time the way to measure $PM_{2.5}$. By then, there had been extensive media coverage and public attention on the subject. Two topics were highlighted in the Chinese media: (1) explaining the adverse health effects of $PM_{2.5}$ and reporting on related public action, and (2) describing the status quo of China's monitoring of $PM_{2.5}$ and appealing for the introduction of environmental standards and the release of data. At the same time, some public figures made their voices heard through microblogs in the hope that the government

[2] Li Ying and Zhang Wang, "U.S. Embassy Claims Beijing Air Worse than Hazardous, Chinese Experts Object," *Global Times*, February 23, 2011.

would take action. For instance, Pan Shiyi (潘石屹), a real estate tycoon, re-tweeted the air quality readings of the U.S. Embassy every day, and Zheng Yuanjie (郑渊洁), a famous writer, threw seven harsh questions on air quality at the Beijing Municipal Environmental Protection Bureau, drawing the attention of netizens, who actively spread the word. Experts on $PM_{2.5}$ were interviewed frequently and they discussed extensively the research development and adverse health effects of $PM_{2.5}$ both in and out of China from such perspectives as atmospheric and environmental chemistry and public health.

B. *Environmental Authorities Speed Up Their Efforts in Revising Ambient Air Quality Standards*

Such media coverage and public attention helped accelerate the revision of the *Ambient Air Quality Standards* (环境空气质量标准). On November 16, the MEP issued the *Official Letter on Soliciting Opinions for Two National Environmental Protection Standards—"Ambient Air Quality Standards" (2nd Draft for Soliciting Opinions) and "Technical Specifications for Daily Publication of Air Quality Index (AQI)" (3rd Draft for Soliciting Opinions)* (关于征求<环境空气质量标准> (二次征求意见稿) 和<环境空气质量指数 (AQI) 日报技术规定> (三次征求意见稿) 两项国家环境保护标准意见的函). $PM_{2.5}$ was added to the revised standards. At the same time, an MEP official, during a public interview, explained the background, guidelines, and conditions for revising the *Ambient Air Quality Standards*, WHO guidelines, meaning of relevant indicators, status quo of air quality monitoring throughout China, and why "the public's personal feelings are often inconsistent with API."[3]

However, the official added, "As nationwide monitoring of $PM_{2.5}$ requires extensive preparations and capacity building, such as equipment procurement and installation, data quality control, training of professionals, and funding, it is fairly difficult to do so throughout the country immediately."[4] The revised *Ambient Air Quality Standards* will become effective nationwide on January 1, 2016. Although the MEP encouraged certain localities to follow the new standards earlier, the late effective date still drew criticism from the public. The media, together with public figures, local environmental officials, experts on smog and diseases, energy experts, volunteers, experts on environmental laws, environmental monitoring

[3] MEP document, http://www.zhb.gov.cn/gkml/hbb/qt/201111/t20111116_220068.htm.
[4] MEP document, http://www.mep.gov.cn/gkml/hbb/qt/201111/t20111116_220067.htm.

equipment manufacturers, and NGOs, began to make more suggestions for the new standards and request local governments announce their timetables for implementing the new standards and release available monitoring data. Different local governments, however, adopted different attitudes. For instance, the Beijing Municipal Environmental Protection Bureau said it had monitored $PM_{2.5}$ and kept all the data, but refused to disclose it as the data were "reserved for scientific study only." Further, on December 2, it officially rejected netizens' requests to release $PM_{2.5}$ data.[5] By contrast, some other localities were more positive. For instance, the Shandong Provincial Government announced on November 22 that 17 cities in the province will start to monitor $PM_{2.5}$ from 2012.[6]

On December 5, the opinion solicitation period for *Ambient Air Quality Standards* (*2nd Draft for Soliciting Opinions*) officially ended. According to the MEP, experts and people from all walks of life as well as the media responded enthusiastically and put forward many suggestions. Over 1,500 opinions were received via email, fax, and mail. The public reached a consensus that $PM_{2.5}$ should be incorporated into the new standards. It was suggested that different localities should follow the standards on a gradual basis while some localities suffering from heavy pollution or those that are ready should implement the standards before January 1, 2016. [7]

C. *Public Participation: from Making Appeals to Taking Action*

By then, the government still had not released $PM_{2.5}$ data. However, the public's attitude underwent some changes. They began to think: "Whether you monitor it or not, $PM_{2.5}$ is here with us." Many people began to take action. For instance, Zheng Yuanjie published a daily poll on the air quality in Beijing via a microblog; the sales of masks and air filters boomed, and the citizen air quality monitoring campaign with the slogan of "I gauge air quality for my motherland" expanded to more regions. In the meantime, media coverage became more diversified to include stories on how to properly use masks, what air filter to buy, and how to reduce concentrations of $PM_{2.5}$.

---

[5] "Air Quality Monitoring Experts: $PM_{2.5}$ in Beijing Triples or Quadruples Every Year," *Economic 30 Minutes*, CCTV.

[6] "Shandong to Monitor and Publish $PM_{2.5}$ Data," *Qilu Evening News*, November 22, 2011.

[7] Wu Ting, "MEP Encourages Localities with Right Facilities to Monitor $PM_{2.5}$ Data by 2016," *Beijing Morning News*, December 7, 2011.

At the same time, the MEP put forward a more detailed implementation timetable for the monitoring of $PM_{2.5}$. On December 21, Zhou Shengxian (周生贤), Minister of the MEP, announced the timetable for the monitoring of $PM_{2.5}$ and ozone at the Seventh National Environmental Protection Conference (第七次全国环境保护工作大会). Specifically, by 2012, such key regions as the Beijing-Tianjin-Hebei area, the Yangtze Delta area, and the Pearl Delta area, as well as provincial capitals and municipalities directly under the central government, should start monitoring $PM_{2.5}$ and ozone; by 2013, 113 key environmental protection cities and environmental model cities should start the work; by 2015, all prefecture-level cities should do it; and by 2016, all localities in China should monitor and publish the data.

Thereafter, the regions that had monitored $PM_{2.5}$ began to disclose the data. By the end of December, the Shanghai Environmental Monitoring Center of the Shanghai Environmental Protection Bureau and the Nanjing Environmental Protection Bureau released the 2006–2010 $PM_{2.5}$ readings and the 2008–2011 readings respectively. Neither city met the standards set forth in the *Ambient Air Quality Standards* (*2nd Draft for Soliciting Opinions*).

According to the news release published on the MEP website, on December 30, Minister Zhou Shengxian chaired an MEP executive meeting at which the *Ambient Air Quality Standards* was reviewed and approved. By then, the public appeal for the disclosure and inclusion of $PM_{2.5}$ data in the new national standards was addressed.

II. *Public Participation Helps Promote the Inclusion of $PM_{2.5}$ in the New National Standards*

In fact, China was one of the earliest countries to monitor and study $PM_{2.5}$. According to experts in atmospheric and environmental chemistry, China imported equipment that monitors $PM_{2.5}$ from the U.S. in the early 1980s. Since then, Chinese researchers have been monitoring $PM_{2.5}$ in a variety of pollution crises from acid rain to the photo-chemical smog in Lanzhou. Realizing the danger and severity of $PM_{2.5}$, scholars urged the government to monitor and study it in the mid-1990s. By the end of the 1990s, Beijing witnessed more and more smog days in fall and winter, drawing the attention of the public and the government alike. Research institutes and universities began to study the relationship between $PM_{2.5}$ and visibility. Such research was deepened when China was bidding to host the Olympic Games. Since 2000, monitoring of $PM_{2.5}$ has been more extensive, and by now, there are over 40 monitoring sites in Beijing.

$PM_{2.5}$ is composed of a mixture of particles, including oxides of nitrogen (NOX), sulfide, and volatile organic compounds. It may lead to secondary pollution and area pollution. However, there is still a lot to be known about the particulate matter, such as the composition of different pollutants and mechanism of reaction. Therefore, scholars are trying to analyze the sources and changing patterns of pollutants as well as the epidemiological relationship between pollutants and health.

On the one hand, $PM_{2.5}$ data gathered at monitoring sites in urban and suburban areas of Beijing from 2005 to 2007 show that sources of pollution and meteorological conditions such as wind and precipitation are closely related to concentrations of $PM_{2.5}$. In any given year, for the urban area, concentration is highest in winter and lowest in summer; for the suburban area, concentration is highest is spring and summer, and lowest in winter. On any given day, for the urban area, the two peaks of $PM_{2.5}$ concentration are during 7:00–8:00 and 19:00–23:00, and concentration is the lowest at noon; for the suburban area, the peak appears during 17:00–23:00. Epidemiological studies have found that during the 2008 Olympic Games, the incidence rate of acute respiratory inflammation in children decreased as a result of the lowering of $PM_{2.5}$ concentration.[8]

Apart from the progress in research, ambient air quality monitoring with the special focus on such pollutants as $PM_{2.5}$ and ozone was gradually implemented by cities around the country. Thanks to some major international sports games, the monitoring in Beijing, Shanghai, and Guangzhou began earlier. According to employees of the China National Environmental Monitoring Center, since 2007, municipalities directly under the central government and Olympic Game Cities have set up monitoring sites for smog and ozone and have been equipped with facilities to monitor $PM_{2.5}$. In particular, Beijing, Shanghai, and Tianjin have been equipped with facilities to monitor PM_1. In June 2011, pilot monitoring in compliance with the *Measures for the Performance Evaluation of Urban Ambient Air Quality* (*for Trial Implementation*) (城市环境空气质量评价办法(试行)) were conducted in 26 cities in China. Apart from the three "conventional" indicators—SO_2, inhalable particles, and NO_2—the pilot cities were required to monitor as many indicators as possible, including $PM_{2.5}$.

---

[8] Lin Weiwei, et al. "Acute Respiratory Inflammation in Children and Black Carbon in Ambient Air before and during the 2008 Beijing Olympics," *Environmental Health Perspectives*, 2011, 119(10):1507–1512.

However, prior to the widespread public concern over $PM_{2.5}$, the monitoring data had been exclusively reserved for research purposes.

In reality, $PM_{2.5}$ is not a new topic, and media coverage on $PM_{2.5}$ and smog appeared much earlier. As the meteorological conditions in fall may easily result in smog days, the media covered the issue from time to time, and there were some well-written investigative stories.[9] However, none of the coverage led to a social impact comparable to the one in 2011. Why did 2011, especially the second half of the year, witness an extensive public participation campaign, which led to some tangible results? The major reason is that in the environmental management of $PM_{2.5}$, all stakeholders—the media, citizens, NGOs, scholars, and the government—interacted with one another in a positive manner.

A. *The Media*

The media is the major channel through which the public gets information. However, traditional media—newspapers, magazines, radio, and television—alone cannot achieve a sustained impact on an issue. As described earlier, news stories and investigative stories on $PM_{2.5}$ appeared much earlier. One major reason why the issue received sustained public attention in 2011 was that new media, especially microblogging, became much more popular. During the new media era, both individuals and organizations may voice their opinions via the Internet, and microblogging served as an ideal platform. The U.S. Embassy published $PM_{2.5}$ readings via its microblog, and followers got the information and forwarded it. In the meantime, individuals, especially public figures, called for action via microblogging, further expressing their opinions and spreading the information.

The traditional media made a nice summary of the case and served as an educator of environmental issues. For instance, the story "I Gauge Air Quality for my Motherland" published by the *Southern Weekend*, which introduced the citizen air quality monitoring campaign and quoted many microblog posts, received widespread attention. Later on, many newspapers, magazines, radio, and TV stations extensively discussed the adverse health effects of $PM_{2.5}$ and the development of China's monitoring and research on the matter. In the meantime, the new media and traditional

---

[9] Yao Yijiang, Qin Wang, Guo Liping, and Wang Xia, "Smog Days Kill 300,000 a Year, Experts Warn of the Dangers," *Southern Weekend*, April 2, 2008.

media interacted with each other in a positive manner—stories published by the traditional media were quickly forwarded via the new media, while discussions on the new media, especially microblogs, got reported by the traditional media. For instance, all of the public figures active in the microblogosphere were interviewed by the traditional media.

B. *NGOs*

Tree planting, bird watching, and garbage collection used to be the three traditional activities of environmental NGOs. However, their activities have become more diversified and professional. In May 2011, for instance, Green Beagle Environment Institute set up a monitoring center, recruited volunteers, and started to monitor $PM_{2.5}$ concentrations. In the same month, the Energy Foundation and Beijing Global Village jointly sponsored a Sustainable Energy Journalist Forum (中国可持续能源记者论坛) with the theme of "Healthy Travel—Fine Particle Pollution and Its Prevention," building a platform of dialogue and exchange for experts, NGOs and the media.[10] After the general public became concerned with $PM_{2.5}$, many NGOs invited experts to give public lectures and conduct seminars. In November, when people were mobilized to make suggestions for the *Ambient Air Quality Standards* (*2nd Draft for Soliciting Opinions*), the media, together with six environmental NGOs, made an aggressive and professional appeal. Later, 21 environmental NGOs joined hands again and appealed to the public to take part in environmental monitoring pursuant to the section on private environmental monitoring in the *Regulation on Environmental Monitoring* (*Draft for Soliciting Opinions*) (环境监测管理条例) promulgated in April 2009.

C. *Scholars*

Chinese experts on $PM_{2.5}$, from those of atmospheric and environmental chemistry to public health, gave numerous interviews and exchanged with the public through public lectures or journalist salons. Pan Xiaochuan (潘小川), a professor with the School of Public Health, Peking University, opened his microblog account on December 22 and answered questions from netizens online. These scholars helped spread the information and

[10] Beijing Global Village, http://www.gvbchina.org.cn/chuanbo/luntan/luntan_40/luntan_40_jianxun.html.

address the doubts of the general public on some professional matters. For instance, one netizen asked this question: "Now that the AQI in Beijing is already 'beyond index,' what else could we do to stay away from toxic air apart from wearing masks?" "I have to clarify the issue of 'beyond index' first," Pan answered. He went on:

> Both China and the WHO have standards for daily/annual average concentrations, but there has not been any hourly average standard. Therefore, it is unscientific to say the AQI is 'beyond index' based solely on the reading of a given hour. As to the health effects, the daily average level implies adverse effects will ensue if a person has been continuously exposed to a concentration higher than this level for 24 hours. It does not mean adverse effects will occur with the exposure for less than one hour.

D. *Government*

Faced with the challenge from the public and the media, the government's initial response was silence. Then, it publically explained the discrepancy between the official data and U.S. Embassy readings. At that time, it still refused to release the monitoring data already collected. However, as the public outcry grew, the government began to be more active and made some positive moves, such as accelerating the revision process, promulgating the *Ambient Air Quality Standards* (*2nd Draft for Soliciting Opinions*), and conducting press conferences to directly take questions of public concern. Later, some local governments took a proactive attitude in releasing monitoring data and the monitoring plan. At the end-of-year environmental protection conference, the MEP announced the implementation timetable of monitoring, while Shanghai and Nanjing environmental authorities released their respective monitoring data gathered over the past few years. In this sense, the public's request to access the data was fulfilled.

III. *Proposals on Better Public Participation in Environmental Management*

Public participation played a positive role in making the terminology of $PM_{2.5}$ known to the general public and pushing forward China's ambient air quality environmental management. Looking back on the entire process, the author intends to make the following proposals on how to facilitate public participation and improve China's environmental management.

Reporters should learn more about the subject and do more "participatory reporting." Some experts complained that due to the lack of professional knowledge, certain reporters asked silly or irrelevant questions that experts found difficult to answer. Other times, reporters failed to get the message of the experts, quoted them out of context, and misled the public. Without proper background knowledge, the general public could only passively receive the information from the media. This may lead to over-concern or even public panic. Therefore, when the media plays up such stories, they should be more scientifically conscious and properly guide public opinion. At the same time, reporters should do more "participatory reporting" by initiating and organizing stakeholders to have discussions, conduct opinion polls, and solve the problem through positive interaction of all parties concerned.

The government should respect the public's environmental right to know and optimize the procedures for public participation. Environmental protection is the obligation of every citizen, but the prerequisite is the environmental right to know. The public should have access to the information on the current environmental situation and the relationship between the environment and individual health and behavior. In the $PM_{2.5}$ case, the public, with the help of the media and experts, actively sought information on air quality and compelled the government to respond. However, the public still knows little about China's environmental status in general. Despite the government's annual *Report on the State of China's Environment* (环境质量状况公报), the information provided is too little for the public to take an active part in environmental management. In China today, the major means of public participation is environment impact assessment. Therefore, the government should take the initiative in updating the public on program progress and providing relevant information so as to help the public understand the state of the environment. Further, even though the public made many suggestions on the *Ambient Air Quality Standards* (*2nd Draft for Soliciting Opinions*), the document was officially circulated in the ministries of the central government and local environmental authorities. It is still not clear what procedures the general public should follow, and at what stage and how they could take part in the suggestion-making process.

NGOs should be more professional. These organizations are closer and more approachable by the public. They can organize different kinds of activities, and are more flexible. In the $PM_{2.5}$ case, the professionalism of NGOs was given full play and they offered a channel for public participation with limited resources. In the future, NGOs should work more closely with the government and experts in terms of advocacy, education, and research.

Scholars played a positive role in educating the public. If they could do so via the mass media on a regular basis, it would help the public take part in environmental management and the formulation of national policies.

The $PM_{2.5}$ case saw little participation from the private sector. Probably due to the diversified sources of pollutants and regional characteristics of $PM_{2.5}$, it is hard to identify which companies are more relevant and responsible. However, as a crucial stakeholder in environmental management, the private sector should take the initiative in disclosing environmental information.

To sum up, if all the stakeholders make their own efforts and interact positively with one another, public participation in environmental management will be advanced.

NUMBERS OF "GREEN CITIZENS" INCREASE AS COMMUNITIES PARTICIPATE IN SOCIAL AFFAIRS

Chen Yuanyuan

Abstract: The public in China has steadily become more involved in offering suggestions and participating in social management. It is a trend that is having profound impacts on the country's political, economic, and social affairs. In the field of environmental protection, a growing number of citizens are acting on their sense of social responsibility, seeking ways to protect their environment. Their primary goal is no longer solely to solve the region's environmental problems, but to focus more on both social and environmental interests. They are trying to collaborate with the government to find solutions to environmental problems through their own practice and exploration.

Keywords: environmental protection, green citizen, environmental rights

In recent years, more citizens are becoming "green citizens," participating in growing environmental movements. As this change continues, the Chinese government should increase efforts to promote communities' participation in environmental movements, thereby improving the government's role in environmental protection. By the same token, citizens should focus on enhancing their abilities in handling social affairs.

I. *"Green Citizens" with Better Environmental Awareness*

A. *Nanjing Citizens Fight to Save Chinese Parasol Trees*

In March 2011, citizens in Nanjing launched a movement to rescue local parasol trees. Mostly planted in the mid-20th century in the main urban districts of the city, the trees were scheduled to be chopped down or transplanted due to the construction of subway lines. This decision triggered strong opposition among local residents. Some 300,000 citizens participated in a movement to protect the trees. They tied green ribbons on the trunks, took photos with the trees, and posted information over various social networking services. Their calm projection of their opinions showcased the developing generation of green citizens in China.

Corresponding media reports brought the "parasol incident" to public attention. Chinese celebrities like Huang Jianxiang (黄健翔) and Meng

Fei (孟非) expressed their concern about the incident on their microblogs, supporting the demonstrations. In response to the public's outcry, the Nanjing government initially explained that the trees were to be transplanted, rather than cut down. A local government spokesman said he "understood people's misunderstanding and anger," but the problem still continued.

In opposition to the government's decision, local residents tied green ribbons onto the parasol trees alongside Zhongshan East Road (中山东路), initiating the Green Ribbon Campaign to protect the trees from being moved. The next day, March 15, Chinese Nationalist Party (Kuomintang) legislator Qiu Yi (邱毅) responded to the issue on his microblog, saying that if the Nanjing Municipal Government did not stop their plans to cut down the trees, he would submit a proposal to the Party's Central Standing Committee in support of their protection. The following day, Qiu submitted the proposal in the hopes of coordinating a solution with the Association for Relations across the Taiwan Straits and the Straits Exchange Foundation. It was then that the event reached a turning point. The State Council's Taiwan Affairs Office held a press conference on March 17, saying it believed the Nanjing Municipal Government would handle the matter properly.

On the afternoon of March 19, thousands of Nanjing citizens gathered in front of the Nanjing City Library for a peaceful protest. The crowd did not clash with the police at the scene. On March 20, Lu Bing (陆冰), deputy mayor of the city, said the transplanting of trees along Metro Line 3 had been completely stopped, adding that the government would be open to collecting public opinions to further optimize the subway construction program. On September 11, after 41 discussions and hearings, the Nanjing Urban Management Bureau announced the approved results for tree transplanting and cutting near the Nanjing Subway's City Hall Station, releasing a document to further protect the city's ancient and street side trees.

B. *Two "Small Potatoes" and the Controversy over Waste Incineration*

In the past few years, as citizens become more involved in environmental protection, similar incidents are being reported where public opinion has affected the government's decision-making.

Since 2009, waste incineration has become a major issue that has drawn large-scale public attention. In places like Jiangqiao (江桥) in Shanghai, Wujiang (吴江) in Jiangsu Province, and Panyu (番禺) in Guangzhou, waste incineration projects led by the government were resisted by the public. Disputes, conflicts, and even group violence were triggered. Facing the anti-incineration surge, the government felt pressure to change their methods of handling the issue.

In March 2010, various departments of the Beijing Municipal Government invited online opinion leader Huang Xiaoshan (黄小山) (online name “Lüshidaner” (驴屎蛋儿)) who opposed the construction of the Beijing Asuwei Incineration Plant (北京阿苏卫垃圾焚烧厂), to join a government delegation as a citizen representative to visit waste disposal projects in Japan and Macau. The visit highly impacted Huang and altered his position on the waste disposal problem. He changed his extreme promotion of the anti-incineration campaign and tried to open a dialogue with the government in a rational way.

In 2010, the Beijing Municipal Government promoted waste classification in almost 600 residential areas, and in 2011, the method was spread even further. However, the classification process had a number of difficulties. Some residents can easily sort their garbage, while for others, it is more difficult. Additionally, some garbage trucks transported classified bags of garbage and mixed them with others in landfills. This created a situation of pointing fingers, where the government blamed residents, and residents blamed the government.

In 2011, under the support of the Beijing Municipal Management Committee, Huang designed and created a waste classifying and processing station for the community he lives in. Named the “Green House” facility, it was established to re-classify residents’ garbage before being transported, collected, and processed based on the “Green House” classification. This was the second time a new method to classify garbage was explored. Huang said he spent a lot of money and effort building the “Green House,” in an effort to more effectively fulfill citizens’ responsibilities. “I don’t want to blame or complain about anyone. I only want to prove through my actions that the things an ordinary citizen like me can do, could be done better by the government.” Huang pushed the government, the experts, as well as the public, to contemplate the issue.

The development of anti-construction incidents against a waste incineration power plant in Guangzhou took similar paths. In 2009, the government proposed building a waste incineration power plant in Panyu, a district in the Guangzhou suburbs. The announcement triggered activist campaigns throughout nearby areas. On April 9, 2010, online activists “Basuo Fengyun” (巴索风云) and “Agassi” (阿加西) accepted an invitation from the local government to visit the waste incineration plant in

Macau. Activist "Basuo Fengyun" said he gained a better view from the experience, understanding how the government and the public could communicate positively with each other. He hoped that the government would listen and respond to the public's voices, and seek a solution for the situation together. He believes that the more people care about how the government formulates and implements policies, the more transparent policies will become.

With the efforts of all parties, the public and the government moved from confrontation to dialogue, from battle to compromise, and ultimately to collaboration, making the Panyu waste incineration incident a classic example of positive interaction between the government and the people. Since March 2010, residents in Panyu voluntarily initiated a "Green House" campaign in their own communities, publishing a classification guide and acquiring knowledge through real practice.

C. *"Beijing Blue Sky Diary" Written by Lu Weiwei and Fan Tao*

In 2010, two Beijing photography enthusiasts, Lu Weiwei (卢为薇) and Fan Tao (范涛), came up with a simple idea to count the number of "blue-sky" days in Beijing. After one year, they compiled a series of photos into a "Beijing Blue Sky Diary" (北京蓝天日记). Following this, they continued their path as environmentalists, cooperating with a non-governmental organization in passing their cameras to children of migrant workers and continuing their documentation of Beijing's air quality.

II. *The New Features of Green Citizens and Their Participation in Social Affairs*

The activists mentioned previously, including "Lüshidaner," "Basuo Fengyun," Lu Weiwei, and Fan Tao, as well as 300,000 citizens of Nanjing, have become the face of the growing generation of green citizens. These environmentally conscious residents continue their efforts to protect the environment voluntarily, and their numbers are expanding. The group is growing, covering a much larger range of citizens with varying social statuses, some with strong public influence. While some of the country's activists focus on issues like flight punctuality and the safety of high-speed trains, which concern their own interests, green citizens are more concerned with environmental problems that represent the public's interests as a whole.

One significant feature of this new generation of green citizens is their knowledge and capability to engage in social affairs. With a general grasp of basic legal knowledge, they can actively use laws and regulations to protect their interests in both substantive and procedural justice. They understand the government's decision-making process, and they are skilled at finding written work favorable to their efforts in documents and government provisions. They maximize the effect of interpersonal social circles to obtain the support from important or influential figures, to guide the development of their cause. They are adept at using the media to monitor the government's decision-making process. As a result, though they are few in numbers, they have become an important force to promote social change in China.

The other feature of their efforts is that they do not deal with their problems from a place of fierce resistance, and instead use more mild approaches. Public participation and environmental rights are changing from a relatively radical way of expression to more rational methods. When the public feels that their interests are violated, they tend to defend their rights with emotional investment. At first, residents around Asuwei protested by hanging banners and organizing marches, but forcing their demands would have put strain on the public and the government's relationship, sending it into a tense impasse. Instead residents learned to actively cooperate with the government, to express their demands with a moderate and rational attitude, and solve the problem together.

In the incident of the Chinese parasol trees in Nanjing, the public strictly abided by the law, and made their voice politely heard throughout the entire progress. They were initiating, responding, and making appointments respectively, largely reflecting their lawful awareness of their rights. Some people call this "soft power," which is not weak but in fact tougher, stronger, and more sustainable than some "hard power."

Besides presenting opposing views, green citizens also provide constructive comments. In 2011, Huang submitted his own design of the "Green House" to the Beijing municipal government and received affirmation from government departments. This method of garbage separation emphasized the process of waste disposal in prior stages (within communities) providing a reference for future policies of the government.

With good will and expectations from the public, the "Beijing Blue Sky Diary" produced by Lu Weiwei and Fan Tao created an atmosphere of dialogue between ordinary people and environmental officials. Environmental officials not only responded to their questions regarding air quality monitoring and current problems, but also expressed an encouraging and supportive attitude towards the behavior regarding environmental affairs.

The third feature is how the public takes advantage of the use of media wisely. Public participation usually works best with combined efforts in many areas, of which the media plays a key role. Media amplifies the demands of the public and communicates information. Green citizens use new media—namely the internet and microblogging platforms—both wisely and actively. Within the online world, the citizens gather to communicate and find an effective method to solve the problem. The "Nanjing parasol trees" event first drew public attention from online posts, responses, and forwarding, and later attracted media coverage through several tree-protection online discussion groups. Some traditional mainstream media also reported the incident and finally gained heavy attention from the government.

III. *Improving Citizens' Capabilities to Participate in Public Affairs and the Government's Ability to Manage Them*

The concept and theories of public participation were first introduced to China from the West in the 1990s, and have developed and become popular since then. On the one hand, public participation allows people to influence and participate in decisions and actions that could affect their lives. On the other hand, public institutions are able to listen to groups of citizens, to take their opinions into account, and finally reach a decision in an open and transparent way.[1]

Many citizens have a strong willingness to participate in environmental areas that relate to public health. The Chinese government has been encouraging people to participate in environmental protection, and has issued a set of rules to protect the public's right to information and supervision. The policy of environmental protection, issued in 1972, states "We rely on the masses, everyone should participate." This kind of support continued through the years. In 2011, the *Advice for Strengthening Key Environmental Protection Projects* (关于加强环境保护重点工作的意见) announced, through the State Council, stated that people should carry out the National Environmental Publicity and Education Action

[1] Cai Dingjian, *Issues and Prospects in Public Participation in China, Democracy and Science*, November 20, 2010.

Plan, foster volunteer teams for environmental protection, and guide and support non-governmental organizations. For nearly three decades, China has encouraged people's participation in the cause of environmental protection.

After decades of development, the environmental awareness of the public has increased, and the scope of public participation has expanded. The public has become an important force in promoting the development of China's environmental protection cause. Discovering how to gather more involvement in public affairs is an interactive process between the government and the public in which the two sides can learn from one another. In decades past, angered residents would show their feelings through ineffective means, as a result of having to endure ordeals if their environmental rights were violated. But over time it became clear that a peaceful solution served all parties, including the environment, best.

Over time, laws, regulations, and policies have been improved upon to better protect public participation. Some people changed their radical ways, used their creativity, took voluntary actions to try new methods, and also viewed the issue from a higher level. As a result, they won more government recognition and acceptance.

People began to realize that if the public is given a dominant position on environmental issues, they can better participate in activities of environmental protection, thus providing strong support for government's decision-making and management. As a result, the legitimacy and feasibility of the decisions are improved and the concept of a "people-oriented" management style is better realized.

The Xiamen (厦门) PX incident in 2007 is a typical case. Driven by public opinions, the Fujian Provincial Government and Xiamen Municipal Government finally changed their mind. On the other hand, the public learned new ways to carry out civil society movements. What began as participation in a specific event, turned into a change in policy or even a regulation or law. Influenced by the PX Incident in Xiamen, people took similar paths in many other environment-related events and have contributed to the government's decisions.

However, influenced by a traditional mindset, some government officials still ignore the strong willingness from the public to participate in public affairs and urban governance. They question their capacity and disregard their voices or opinions. They essentially underestimate the role of the community. There have been cases in which display cards for Environmental Impact Assessment were hung in conspicuous corners when launching projects that matter to the public. In some instances, displayed content is concealed, its display time shortened, or people are interviewed with no direct interest in the project. In the worst known cases, some projects begun with force, despite strong public opposition. "Black Box Operations" have become a norm in some areas, in efforts to escape the supervision of the society, the public, and the media by blocking information.

The next question is how to improve the quality of communication between the government and the public. A wise government should know how to utilize folk wisdom, stimulate the public's enthusiasm to participate in environmental protection activities, and provide support and guidance. The above-mentioned cases are very good examples for the public, but it will be a long-term learning process to improve the capacity and quality of public participation. The key to future success will be further exploration.

PART SEVEN

CHINA AND THE WORLD IN AN ENVIRONMENTAL PERSPECTIVE

TRANSGRESSING GLOBAL AND LOCAL: ENVIRONMENTAL NGOs AND CHINA'S OVERSEAS INVESTMENT

Lo Sze Ping and Yi Yimin

Abstract: As discussions on the social and environmental impacts of China's overseas investment have gotten increasingly heated in recent years, disparate views from inside and outside China contradict each other. This article centers on the relationship between the status quo of China's overseas investment and the "China Model." It examines China's economic development in the context of globalization and explores the neutral perspectives taken in reviewing overseas investment. The article proposes that non-governmental environmental organizations should transgress the concept of global and local, promote the idea of sustainable development beyond individual interests of nation-states, and encourage the green transformation of China, as well as the world.

Keywords: China's overseas investment, China Model, sustainable development, environmental NGOs, local/global

In recent years, both China's investments and loans overseas have grown rapidly, as target countries and industries have expanded continuously. A recent study released by the *Financial Times* shows that the China State Development Bank (国家开发银行) and China Export-Import Bank (中国进出口银行) issued at least 110 billion USD in loans to other developing countries between 2009 and 2010. This was more than the loan commitments of 100.3 billion USD made by the World Bank in the same period. China's loans to other developing countries, for the first time surpassing the World Bank, highlights the increasing influence of China in the global economy.

For the past thirty years, the economic growth of China has on the one hand brought remarkable economic achievements and produced a "China Model" which saved hundreds of millions of people from poverty. On the other hand, huge social and environmental costs are hidden behind this economic growth and prosperity. As China's economy heated up, the influence of the "China Model" also radiated overseas. At the 15th CPC Central Committee's Plenary Meeting in 1997, President Jiang Zemin proposed the "Going Global Strategy" (走出去战略) for the first time, formally opening the curtains for a large-scale investment from China to overseas.

For the next few years, Chinese corporations executed a number of mergers and acquisitions overseas, including the successful case of Lenovo's acquisition of IBM's PC business in March 2005, as well as the unsuccessful case of China's National Offshore Oil Corporation's attempted acquisition of Unocal Corporation. State-owned and private enterprises in China have been involved in overseas projects like assets acquisition, factory building and investing, and natural resource exploitation, with footprints throughout many developing countries in Asia, Africa, Latin America, and even in Europe and North America. Before China's opening up in 1979, China's capital invested overseas basically took the form of international assistance, with a tint of ideology to gain political support from developing countries in the international arena. After 2000, China's foreign direct investment (FDI) grew fast, reaching 26.51 billion USD by the end of 2007. This was up from 1 billion USD in 2000. In 2009, China's FDI reached a historical net value of 56.53 billion USD, sustaining a 1.1% annual increase during the global financial turmoil. It ranked first among developing countries and regions and fifth around the globe, accounting for 5.1% of the global traffic.[1] China became the world's second most potential foreign investor in 2010–2012, surpassing Germany, Great Britain, and France, according to a report published by the United Nations Conference on Trade and Development (OECD).[2] According to figures recently released by the Ministry of Commerce, non-financial FDI in 2010 totaled a record high of 59 billion USD, increasing 36.3% annually. China's non-financial FDI has maintained an upward slope for nine consecutive years, totaling 258.8 billion USD by the end of 2010.[3]

The Economist published a special report, consisting of over 10 articles on the cover story "China's Quest for Resources," in 2008. The articles discussed China's rapid economic development and its thirst for resources. The feature used the metaphor of "a ravenous dragon" to describe China, conveying the anxiety of mainstream Western media towards Chinese investment overseas.[4] Western media reports on the investments of

---

[1] "2009 Statistical Bulletin of China's Outward Foreign Direct Investment," Ministry of Commerce, National Bureau of Statistics, the State Administration of Foreign Exchange, 2010.

[2] Zhu Jingruo and Yu Ronghua, "UNCTAD Releases World Investment Prospects Survey," http://world.people.com.cn/GB/1029/42408/12660904.html, September 7, 2010.

[3] Wang Xi, "China's FDI in 2010 Reaches 59 billion USD," http://policy.caing.com/2011-01-18/100218387.html, January 18, 2011.

[4] "A survey of China's quest for resources," *The Economist*, March 13th 2008, http://www.economist.com/node/10795714.

the Chinese government and enterprises have been increasing significantly, emphasizing problems regarding those projects including human resources, environment, and social justice. Some even argue that China is exploiting Africa and elsewhere in a new colonialist pattern. Some developing countries also worry that cheap Chinese products will affect the local manufacturing industry.

It is noteworthy that a number of intellectuals and organizations from the Third World, especially some scholars from African NGOs, have suggested that China's overseas investment is suspected of being neo-colonialist. They said they did not see a fundamental difference between the way China invests overseas and how the earlier Western colonial states controlled natural resources in their own favor. They accused some investment projects of neglecting local interests or threatening the local ecological balance.[5]

In recent years, many of China's overseas investment projects indeed caused protests from local people, attracting attention from NGOs. For example, International Rivers has been concerned about 269 dam projects by Chinese companies and banks in 67 countries. Mekong Watch and Save the Mekong Coalition are concerned with the long-term ecological and social impacts of investments from China's Yunnan Province and Southeast Asian countries in the Mekong River Basin.

To be fair, China's foreign direct investment stock accounted for only 1.3% of the global stock by the end of 2009. Its cumulative scale of investment is not comparable to that of developed countries in the West.[6] China's overseas investments in recent years have caught international attention as anxiety over China's economic development grows. Some scholars have even pointed out that the Western media's criticism of "China in Africa" reflects a nationalism and moral superiority that some Western elites hold toward China's "strategic competition."[7] In fact, as the world's factory, China is facing an increasingly severe shortage of resources, including

---

[5] Manji, Firoze and Stephen Marks (ed.), *African Perspectives on China in Africa*, Pambazuka Press, 2007.
Naidu, Sanusha and Haley Herman, "Africa's New Development Partners: China and India—Challenging the Status Quo?" in Hakima Abbas and Yves Niyiragira (ed.), *Aid to Africa: Redeemer or Coloniser?* Pambazuka Press, 2009.
Bello, Walden, "China Eyes Africa: The New Imperialism?" *Multinational Monitor*, Vol 28. No. 1, 2007.

[6] "China Overseas Investment Development Report 2010," Ministry of Commerce, November 2010.

[7] Sha Boli and Yan Hairong, "China in Africa: Triple Misunderstanding," *South Wind*, April 3, 2010.

land, minerals, forestry products, as well as energy and power. China is thirsting for all resources and many Chinese companies are trying to gain control over the exploitation of resources through overseas investment, which inevitably plays a part in the international geopolitical game.

Facing growing international concern, President Hu Jintao delivered a high-profile speech during his visit to South Africa in 2007 addressing China's rapidly expanding influence in Africa. He said: "China did not, is not, and will not impose its own will and unequal practices on other countries, let alone do anything harmful to African countries and the African people."[8] Some government-affiliated institutions also began to respond to questions from overseas and defend the Chinese government.[9]

I. *China's Rise in the Context of Globalization*

In recent years, academia has been trying to seek an interpretation for China's economic success and its rise on the international stage. From the "Beijing Consensus"[10] to the "China Model,"[11] they emphasize China's distinctness. China's development model, in fact, is not unique. It is a typical government-led and growth-oriented development, and its essential development model is no different from that of the Four Asian Tigers (South Korea, Taiwan, Hong Kong, and Singapore)—all under the auspices of authoritarian regimes, and with economic development as the primary national goal.[12] They were all fighting for global preferences with enormous environmental costs, cheap labor, and government subsidies, in order to seize the market share of consumer goods in developed

Sha Boli and Yan Hairong, "China in Africa: The Plight of the Global System," *China's Economy*, 2010 Issue 1.

[8] "China's Hu Quashes 'Colonial Africa' Talk," *Financial Times* Chinese edition online, http://www.ftchinese.com/story/001009492, February 8, 2007.

[9] Liu Hongwu and Shen Beili, NGOs in Africa and Sino-Africa Relations, World Knowledge Publishing House, 2009.

[10] Huang Ping and Cui Zhiyuan, *China and Globalization: Washington Consensus or Beijing Consensus*, Social Sciences Documentation Publishing House, August 2005.

[11] Pan Wei, *Contemporary Chinese System: the Chinese Model of Economic, Political and Social Analysis*, Joint Publishing (Hong Kong), 2010.

Pan Wei, *Chinese Model: Interpretation of 60 Years of the People's Republic of China*, The Central Compilation and Translation Press, 2009.

[12] Sachs, Wolfgang (ed.), *The Development Dictionary: A Guide to Knowledge as Power*, London: Zed Book, 1992.

Xu Baoqiang and Wang Hui, *The Illusion of Development*, The Central Compilation and Translation Press, 2003.

countries.[13] China is positioned as the middle link in the global production chain. Its title of the "world's factory" demonstrates the importance of China's manufacturing exports to the world. If viewed from a worldwide perspective, the chain's upper links of product research and development, and the lower links of value-added marketing are still basically controlled by developed countries and multinational corporations.

Qin Hui (秦晖), a Tsinghua University professor, said in 2008 that China's development over the past few years, though exhibiting amazing competitiveness compared with both free-market or welfare states, is based on the "triple advantages" of low wages, low benefits, and low human rights. Moreover, factors other than the "triple lows" may include the "triple highs," including high pollution, energy consumption, and water consumption. China's prosperity and development are more or less based on the "triple lows" and "triple highs."

In fact, the rise of China should be attributed both to China itself and to the world. It has been fostered by economic globalization. To meet the ever-expanding need of the developed world, China needed to accelerate the exploitation of natural resources, including deforestation, excavation of mineral resources, petroleum refining, and the construction of dams and power plants. China's expansion of resources to the external world is rooted in over-consumption all over the globe. It is important for developing countries to follow the "China Model" of the "Beijing Consensus" if they want to compete with a rising China.

II. *Review China's Overseas Investment in a Comprehensive and Objective Way*

The heated discussion by the international community on the "China Model" largely shows that some developing countries hope to explore the path to development following China's example. However, the cost to environment and social equity in achieving this should not be forgotten. While the developing countries are learning China's experience via aid and investment, it is important for them to also learn the lessons from China's development process. That is because once a model of development is fixed, is difficult to change paths. China has also been promoting "a change in its development model," stating a clear will to discontinue

[13] Chen Zhiwu, *No such thing as Chinese Model*, Gusa Publishing (Taiwan), 2010.

the "pollution first, treatment later" model of the past.[14] Nevertheless, after this policy direction was proposed, its implementation has been far from sufficient.

After over thirty years of economic development, Chinese society has finally come to realize that the sole pursuit of economic growth is not sustainable. As an ancient Chinese saying goes, "do not impose on others what you yourself do not desire." So when China examines its overseas aid and investment projects in other countries, it should not only focus on promoting their economic growth. Instead, China should take a broader perspective to analyze the pros and cons of an investment project and its impact on local communities from economic, social, and environmental dimensions.

The following questions are often ignored. By answering these questions, we could analyze the impact of a specific overseas investment by China on a specific region in a more comprehensive and objective way.

A. *Who Is Influenced? And in What Way?*

When we hear the words "cooperation" or "win-win," we need to figure out who are the ones we are "cooperating" with or who are we creating this "win-win" situation with. We also need to figure out who benefits and who sacrifices during China's overseas aid and investment process. If we do not differentiate between different groups of people facing different consequences, we will not be able to see where the problem lies in a general discussion of a rise in economic indicators or employment rate.

All stakeholders in China's overseas aid and investment should be considered. This includes Chinese government, enterprises, financial institutions, workers, and society, as well as the recipient countries' government, enterprises, local communities, society, workers, citizens, media, and the environment.

Moreover, after the recipient country gets support via funding or training, an imbalance in different sectors of society may be exacerbated. Take China's assistance to a government office building in Africa as an example. Departments like education, agriculture, and environment are less likely to receive support from the construction projects. Furthermore, while business of the recipient country may be booming under stimulation from foreign investment, local companies are often at a competitive

[14] Wen, "Current Focus of Environmental Protection," *Xinhua Net*, April 18, 2006.

disadvantage, and they have to follow the rules set up by Chinese companies in some respects. Their imitation of the Chinese enterprises deprives them of the possibility to grow on their own. Also, for local African communities and society, the growth in economic indicators may not provide them with better healthcare, education, or employment. In some projects, the interests of the citizens are even sacrificed for those of the government and enterprises. In addition, all aid or investment projects face repayment pressures. If the operation fails to deliver the expected outcome, the entire society may confront a debt crisis.

From such projects, the Chinese government is fulfilling political interests while enterprises are obtaining profits (with risks of course). The whole Chinese society is gaining energy and resources, economic growth, and reduced environmental stress. However, if China's actions are not regulated in the aid and investment process, without a real concern for the well-being of the recipient country and responsibility for its own aid and investment, China will face international criticism and its long-term interests and international image will be hurt. It is a new challenge for Chinese society to accurately and comprehensively understand the impact of its aid and investment and take actions to supervise and improve China's investment overseas. In addition, one of the shareholders—the Chinese worker—is rarely noticed. The workers are in fact managed in a stringent way with a confined life circle, "quasi-military" management, frequent overtime, and delay in their salary payments. Moreover, they are faced with a poor working environment, threats to their personal safety, and a separation from their families. Concerns about these issues will also help us reflect on our own domestic labor issues.

B. *Does Chinese Overseas Aid and Investment Promote or Interfere with the Balance of Development in Recipient Countries?*

The Chinese have been showing their existence overseas more often and with more influence. This is a result of the Chinese government's improved support system for overseas aid and investment.

Chinese overseas investment is highly concentrated in areas supported by the "Going Global Strategy," like large engineering projects, mining, and communications technology. After some Chinese companies enter African markets via aid projects, they expand investment or break into other areas. To encourage small- and medium-sized enterprises and high-tech industries to go global, the Chinese government supports them through policies or funding like the Small and Medium Enterprise

International Marketing Fund (中小企业国际市场开拓资金). All of the above mentioned will add to the competitive disadvantages of local small- and medium-sized enterprises in developing countries, impeding the development of their local economies.

In Africa, China has set up the China-Africa Development Fund (中非发展基金). The fund was one of the eight major cooperative measures announced at the China-Africa Cooperation Forum Beijing Summit in 2006. The fund's focus of investment is on infrastructure, basic industries, agriculture, manufacturing, and other fields. However, in reality, a large part of the investment projects have been in mining industries.[15]

It is obvious that resources are the highlights of China's overseas investment. For recipient countries, they need to develop not only their resources sector, but also other sectors including manufacturing and agriculture. More importantly, education, health care, pensions, and social security system are the fundamental backbone for a society to develop in the long term. However these are not easily improved under foreign aid and investment as currently delivered.

All the above arguments do not aim to blame China for the imbalance of development of the recipient countries. On the contrary, the recipient countries should have development and coordination plans on their own. The key point here is we should acknowledge that imbalanced aid and investment are having an adverse effect on the recipients, and then we can start thinking of how to make improvements.

C. *Should Aid and Investment Projects Be Implemented through Information Transparency and Public Participation to Balance the Interests of All Parties?*

It can be inferred from the analysis of stakeholders that the impact of China's overseas investment is quite complex. For this reason, ensuring that every interest group has a chance to participate in the decision-making process is the prerequisite for balancing the interests of all parties.

How does an aid or investment project come into being? Do the projects address all the needs of different groups in regards to their distribution, design, and operation? Are there any strict pre-assessment procedures to reduce the negative impacts on the environment and society? Are there

[15] Zan Chunyan, "China-Africa Development Fund for the Past Six Months," *21st Century Business Herald*, June 12, 2010.

any relatively open channels for the public to participate in and supervise the projects?

No system is perfect unless it can receive feedback and execute self-correction. To guarantee this, it is necessary to ensure there is information transparency and that channels exist for public participation. Without such considerations, the potential negative impact of aid and investment may be worse.

D. *What Kind of Impact Do Donors or Investors Have during this Process?*

China's aid to foreign countries is usually delivered through companies, to strengthen their influence in recipient countries. These companies are more likely to analyze the process from an economic perspective, ignoring the environmental and social implications. They are also actively involved in the assistance plans, which may result in favor of their own interests instead of those of the recipient countries.

Chinese companies operate domestically in a way that is less than perfect. They tend to apply their "high efficiency" methods to other countries, which may create problems. When facing problems of environmental pollution or community conflict, Chinese companies usually turn to superior officials for solutions, ignoring the voices of civil society. This Chinese way of solving problems may exacerbate the conflicts instead.

Chinese culture and its management style affect the recipient countries as well, exemplified by labor issues (involving Chinese workers and local workers) and cultural conflicts. Among these, labor issues usually focus on the proportion of Chinese workers (Chinese companies hire more Chinese workers than the host country stipulates); salary (although Chinese companies mostly act in accordance with the minimum wage of the host country, it is often not enough for the workers to live); and physical conflicts due to lack of understanding, respect, and communication. At the same time, labor issues arise among Chinese workers, too.

It is the authors' hope that reflections and answers to the above questions can help us to view the impact of China's overseas investment from a neutral standpoint.

III. *New Internationalism: Transgressing Global and Local*

> China is a large developing socialist country, and its future development will not only directly affect the future of the Chinese people, but will also have an important impact on the development and progress of Asia and

> the world ... I hope that Chinese enterprises will learn from the advanced experiences of foreign enterprises. They must go out and temper themselves in the winds and storms of economic globalization, and build up their own competitiveness ...
>
> —Jiang Zemin

Most often, Western media depicts China's overseas investment negatively. China's footprints overseas are closely linked with stereotypes of geo-politics, the fight for resources, environmental damage, human rights violations, and social injustice. In such a context, it is risky for NGOs to pay attention to China's overseas investment. On the one hand, some criticisms from the West have a tendency to attack on ideological grounds, so we are easily trapped in a mentality of defending our homeland. We may strongly deny the accusations and try to refute the critics. On the other hand, when we focus on the injustices in individual incidents, it is likely to be interpreted as, intentionally or unintentionally, standing in the position of the West, or even trying to help it to contain China.

In this situation, Chinese civil society must say no to over-simplified responses based on ideology. It is important that we comprehensively and objectively assess China's overseas investment. On the one hand, we need to question and criticize a one-sided pursuit of economic growth and reflect on its impact on environment and society. On the other hand, we cannot measure China's interests only. Moreover, as we only have one earth, we should seek a common goal in a greener future, by sharing experiences and learning from each other around the globe. Under the current circumstances, Chinese environmental NGOs are struggling between the conflicting images of being nationalists or traitors. To avoid this, they must apply a new internationalist perspective, to break the boundary between global and local and deal with economic globalization.

In May 2007, while the African Development Bank held an annual conference in Shanghai, Sino-African NGOs jointly organized a public forum on investment issues to explore cooperation possibilities between the NGOs.[16] More and more NGOs concerned with global environmental and developmental issues, including NGOs from Asia, Africa, and Latin America, contacted Chinese NGOs in hope of strengthening communication and mutual cooperation. However, the Chinese environmental NGOs were relatively immature and lacked resources, and were busy struggling

[16] "China/Africa: Civil Society Meeting," *AfricaFocus Bulletin*, http://www.africa.upenn.edu/afrfocus/afrfocus081007.html, August 10, 2007.

to cope with numerous domestic environmental issues. Therefore, there were few NGOs paying attention to issues concerning the overseas investment of Chinese enterprises. However, as the situation is changing, Chinese NGOs should continue this trend. Chinese NGOs should analyze the environmental and social impact of China's investment overseas from a broader perspective, and exchange opinions with NGOs from other countries to enhance communication and understanding. They should provide advice and intervene to mitigate the adverse effects of China's overseas investment.

Back in the 1980s, the environmental movement in Europe and America proposed the slogan "Think Globally, Act Locally." This underlines the fact that human beings have only one earth, and environmentalists worldwide should deal with local problems based on a common goal. Facing the climate change crisis, global and local factors are already mixed in nature and are inseparable. We are no longer faced with a simple dichotomy, differentiating between global and local. Every local environmental incident is under the context of globalization, and every global trend spreads out by local carriers. Energy waste at home consumes coal and emits carbon dioxide, while global climate change accelerates the melting of glaciers on the Tibetan Plateau. Chinese cities and towns face overwhelming household garbage, which on the one hand represents a crazy consumerism, but on the other, developed countries are dumping outdated incinerator technology in China. The reason why Chinese enterprises go to Africa for logs and minerals is also to meet domestic needs for consumption, production, and re-export. We should ask ourselves how much of the material we are enjoying today comes from exploiting others. Today's environmental movement should transgress the boundary between global and local, combining reflection and action (think and act globally) more than ever before.

Pollution goes beyond borders. So we are concerned about the environmental problems brought about by the Chinese investment in Africa or other developing countries and we are dedicated to encouraging Chinese enterprises to follow the best environmental standards recognized by international criteria. By doing this, we are adding further impetus for China's transformation to a greener country and gaining more experience that will help improve domestic environmental standards and social responsibility. For example, the green credit policy is growing in popularity, and there are higher expectations for companies to make information public, welcome public participation, and share benefits in communities when financing. NGOs can introduce this new way of thinking, while urging

Chinese companies and the Chinese government to set up and follow the guidelines. We should have the insights that the China's overseas investment is a two-way process via which we can bring things to other people while learning from them.

Undoubtedly, more and more Chinese companies will invest overseas and the corresponding disputes concerning the environment and resources will also arise. Chinese environmental NGOs should shoulder the responsibilities as world citizens and follow up the process by communicating with international NGOs and local communities in the recipient countries. They should introduce the advanced experience from abroad, and take a perspective beyond narrow nationalistic interests, in order to pave a path for China, as well as for the world, towards sustainable development.

IMPROVING CHINA'S AID AND INVESTMENT TO AFRICA WITH A MORE OPEN ATTITUDE

Yi Yimin

Abstract: In the last decade, China's investment in Africa has increased by 46% annually, with an accumulated volume of 32.3 billion USD by August 2010. Investments in Africa account for more than 1/10 of China's total overseas investment. In 2010, China's investment there continued to speed up. It was reported that Ghana and Nigeria would receive over 10 billion USD in aid or investment and that other countries, including Zambia and the Republic of Congo, would receive over 5 billion USD. With China's aid and investment to Africa focusing on resource acquisition and market share expansion, the environmental and social impacts in Africa must not be overlooked.

Keywords: environmental impact, resources, overseas investment, Africa, hydropower

In 2010, three incidents involving Chinese investment in Africa sparked the interest of the international media and civil society. First, in June 2010, the Industrial and Commercial Bank of China (ICBC) (中国工商银行) agreed to invest 500 million USD in Ethiopia's 1.75 billion USD Gibe III Dam. The project will affect the Omo River, an international waterway, and the vulnerable ecosystems of Lake Turkana, upon which 300,000 residents of northern Kenya rely. Then, on October 15, supervisors at the Chinese-owned Collum mine in Zambia shot and wounded 11 workers during a dispute—showing that Chinese investment in mining in Africa can create serious labor relations issues, as well as environmental problems. Finally, in November, the China National Offshore Oil Corp (CNOOC) and Ghana National Petroleum Corp joined forces to attempt to buy the US firm Kosmos Energy's stake in the deepwater Jubilee offshore oil field in Ghana. This deal did not go through, but commentators linked it to earlier reports that China would provide the country with 13 billion USD in loans for oil and gas infrastructure, agricultural development, and other construction projects, reminding people of the "resources in exchange for projects and credit" model in China's overseas investment.[1]

[1] "IMF: Chinese Loans to Ghana are a 'Good Thing'," http://cn.reuters.com/article/currenciesNews/idCNnCN1387505201010226 (accessed October 26, 2010).

Fossil fuels, mining, and dams are the main areas of Chinese investment in Africa. These three types of projects reflect a number of characteristics and problems surrounding Chinese aid and investment on the continent.

According to international standards, aid and investment are different in the sense that preferential aid is provided at no cost to the beneficiary, or at a very low interest rate—aimed at humanitarian assistance and social and economic development rather than capital returns. China invests and provides aid in a wide range of fields in Africa. These include: infrastructure, government office building construction,[2] hydropower engineering,[3] agriculture, medical assistance, emergency relief, exemption of loans due, exchange and training programs, scholarships, and volunteer work. Although there are differences in emphasis across different African nations, the most funding is for the construction of government buildings, dams, and other infrastructure, the so-called "complete set of foreign aid projects" (对外援助成套项目). These are full-service aid projects advocated by the Chinese government, where the project implementer is a Chinese firm and project loans are transferred directly from the Export and Import Bank of China (China EXIM Bank) (中国进出口银行) to the firm rather than to the recipient nation, and equipment is required to be imported from China. These aid projects play an important role in helping Chinese firms expand overseas.

As for Chinese investment in Africa, the fields include: construction (including infrastructure, hydropower engineering, and real-estate development), energy, mining, agriculture, forestry, information and communication technology, industry, finance, and medicine. But most investment is concentrated in construction, energy, and mining.

In summary, Chinese aid and investment have found their ways into many fields closely related to the improvement of people's livelihood there, highlighting the good will of China. However, the capital configuration of these programs shows that priority has been given to resource and market development. It is also in these fields that the negative environmental and social impacts of Chinese aid and investment are most serious.

---

[2] This classification is different from the official one. Office building and dam construction projects are listed as different from infrastructure construction projects because these projects, while helping improve China's relationship with the recipient government, do not benefit the local economy or residents.

[3] Although dam projects may generate power and improve irrigation, they are listed as an independent category because these large-scale projects tend to have complicated environmental and social consequences.

I. *Imbalance in Aid and Investment*

Foreign aid and investment are by no means new in Africa. In the last few decades, Africa has received large amounts of Western (particularly European) aid and investment, which, instead of bringing development, leave the continent with many problems. These include heavy debt burdens, foreign aid dependency, and a lack of planned industrial development. Against this background, China comes on the stage as a new aid provider, whose aid and investment mechanisms, ideas, and practices have been keenly observed and studied. In spite of some criticism, many African development researchers and NGOs—disheartened by Western aid—are now hoping that the Chinese aid model will bring real development to Africa.

China has experimented with linking aid to economic development, both at home and abroad. In the controversial "Angola Model," the recipient country uses its oil reserves to secure loans for oil exploration and processing, and infrastructure projects. This means the recipient nation can turn resources into cash when funds are short and develop its extractive industries. It also often secures resource-development rights for China's state-owned oil companies in the recipient nation.

However, it is not enough to focus only on economic development. For the citizens of the recipient nation, justice, participation, and sustainability should be a part of the planning process right from the start. For example, although local companies can get business contracts from Chinese aid and investment projects, they also face competition from Chinese companies. Sometimes, the GDP growth does not bring local African communities more job opportunities or better working conditions and social welfare. On the contrary, in some regions, foreign investment has given rise to social conflicts, for example in the Niger Delta.[4] Moreover, projects like office building and stadium construction, while benefiting only a limited group of people, can easily lead to a debt crisis if there are difficulties repaying the loans. This can increase the burden on the whole local society.

---

[4] Since 2005, there have been many cases of vandalism and violence against oil pipelines, workers, and refinery facilities in the Niger Delta. In 2009, the Movement for the Emancipation of the Niger Delta opposed a CNOOC project in the region, calling it and other similar projects "a disaster to the oil resource rich communities" and "an immediate threat to farmland."

There is no denying that these Chinese projects have caused environment problems in Africa, including pollution, overdevelopment, habitat destruction, a decrease in biodiversity, land-use changes, water crises, and even climate change. These environmental problems affect both African and Chinese stakeholders and China's national image. Therefore, the supervision of Chinese overseas investment behavior poses a new challenge.

It seems that China regards Africa primarily in terms of its resources and potential market, and prioritizes those fields over aid, leading to some negative consequences for recipient nations. Prior research on the "resource curse" in countries such as Nigeria suggests that over-reliance on resource development for economic growth can produce some benefits for society, but can also gravely damage the local environment.

China hopes to contribute to the agricultural and infrastructural development of Africa and to use this commercial cooperation model to make aid projects more sustainable—but it risks putting Chinese firms in absolute control of projects, with potentially negative effects. Chinese companies may ultimately make the decisions about whether a project goes ahead, influencing policy and allowing the company risk-free entry to the recipient nation.

There is a risk that Chinese companies may apply "high-efficiency" methods overseas that leave farmers in a weak position, as has been the case in the "company plus farmer" (公司加农戶) agricultural investment models in China. The pursuit of profit has also led small private enterprises from China to become involved in illegal trading—in timber or ivory for example—to the detriment of the local environment.

II. *Extending China's Domestic Development Models to Africa*

In order to understand the environmental and social impacts of China's investments in Africa, one should probably understand the impacts of Chinese investment at home in China, since there are many similar problems. Or to put it in another way, a proper understanding of China's domestic investment and social models is a prerequisite to understanding China's investment behavior in Africa.

Thirty years of reform and opening up in China have led many Chinese—particularly those who have enjoyed the fruits of this process—to believe that economic growth will lead to development in other areas. That belief has been extended to China's overseas investment. But many others worry about the negative impacts of China's breakneck growth: the destruction of China's environment, the pillaging of resources, and rising social injustices.

Importantly, within China there are also examples of loans and engineering projects being exchanged for resources contracts. For example, in November 2010 it was reported that the local government of Ordos (鄂尔多斯), in Inner Mongolia, had decided manufacturing and high-tech projects of a certain size would be allocated corresponding amounts of coal.[5] This led a number of firms to rush to invest and claim their coal. This type of "resources for investment" deal has been accused of selling national resources on the cheap, as well as severely damaging the Ordos grasslands.

The Chinese government's Western Development strategy (西部大开发), first implemented in 2000, led to domestic investors favoring the western parts of the country, which is rich in resources and contains huge market potential, but is economically backward compared to the east. However, the single-minded focus on developing resources—much like in Africa—has created negative environmental and social consequences. An essay by Yang Yong (杨勇), titled "Facing Up to the Environmental Misery of Resource Development in Western China" points out that a series of large and sensitive projects are under construction in China's west, worsening environmental damage and creating social inequality and conflicts.[6] As development proceeds and capital is invested, the potential to prioritize greener industries or tourism is forgotten in favor of resource development.

III. *Environmental and Social Impacts of Chinese Investment in Africa*

China's investment in energy, mining, dam construction, forestry, and agriculture has had complicated environmental and social impacts.

Among Chinese energy investment programs, Chinese oil programs in such countries as Sudan, Angola, Gabon, Nigeria, and Niger have greatly

[5] Xiang Nan, "Ordos's 'Resource for Investment' Sells State Assets Cheap," *Securities Times*, November 2, 2010.

[6] Yang Yong, "Facing up to the Environmental Misery of Resource Development in Western China," *Green Book: Annual Report on Environment Development in China (2010)*, ed., Yang Dongping (Beijing: Social Sciences Academic Press, 2010).

increased the African supply of oil to China. In 2009, China imported 61.42 million tons of crude oil from Angola, Sudan, and Libya, accounting for 30.1% of China total oil imports that year.[7] In the first six months of 2010, Angola, exporting 21.7 million tons of oil to China, became Chinese biggest oil supplier.[8]

Mining has been prioritized in the last few years. According Zeng Shaojin (曾绍金), deputy director of the China Mining Association (中国矿业联合会), at the Tianjin International Mining Conference in November 2010, 280 Chinese mining enterprises had invested in 203 projects overseas in the previous 18 months. Zambia is the most favored recipient country for Chinese mining investment. Moreover, as in the mining industry in China, private Chinese investment is very active in Africa. However, as shown in the above-mentioned labor conflict in Zambia, private Chinese mining enterprises are troubled with labor problems, violence, and reliance on the local government in problem solving.[9] Furthermore, they have been blamed for bribery and ignoring the opinions and requests of local residents and NGOs.

The complicated environmental and social impacts of environmentally sensitive resource investment in the petroleum and mining industries also make Chinese investment in these fields more risky. The operating protocols of these large investments have attracted attention at home. According to a report in December 2010, central state-owned enterprises (SOEs) lost 4 trillion RMB in overseas investment as a result of the absence of an effective monitoring system of Chinese overseas investment and the inappropriate management and debt of overseas Chinese enterprises.[10] This report set the stage for a discussion about the safety and monitoring of Chinese overseas investment. It has been suggested that since central SOEs essentially invest overseas on behalf of the Chinese, we must consider the establishment of relevant management and accountability systems.[11] It is hoped that SOEs will, in the future, attach more importance

---

7 Tian Chunrong, "Analysis of China's Oil Import and Export in 2009," *International Petroleum Economics*, no. 3 (2010).

8 "A New Chapter in Sino-African Energy Cooperation," Xinhuanet, http://news.xinhuanet.com/world/2010-11/18/c_12791850.htm (accessed November 18, 2010).

9 Chen Zhu and Zhang Boling, "Chinese Entrepreneur Uses Gun against Africans," *New Century*, October 24, 2010.

10 "Central SOEs Lose 4 Trillion RMB Overseas," *21st Century Business Herald*, December 8, 2010.

11 Wu Jiang, "Overall Profit Volume Can't Cover Central SOEs' Huge Loss Overseas," *Shanghai Financial News*, December 14, 2010.

to the profit rate of state assets and more importantly the better use of the profits generated from overseas investment, so as to achieve social justice and minimize the negative environmental and social impacts on Africa.

Dam construction projects have also sparked people's interest. There has been much coverage of Chinese dam construction projects in Africa, including the Bui Dam in Ghana (which flooded a quarter of Bui National Park), the Kongou Dam in Gabon (which damaged the forest of Ivindo National Park), the Kafue Gorge Lower Dam in Zambia (which increased the ecological pressure on the Kafue Plain and other neighboring national parks), the Merowe Dam in Sudan (the water level fluctuation and sludge of which did harm to the local ecosystem, water quality, and public health),[12] the Mphanda Nkuwa Dam in Mozambique, and the Gibe III Dam in Ethiopia.

The Mphanda Nkuwa Dam is a typical case. This project, first proposed in April 2006, was not authorized until August 2010 by Mozambique's Minister of Energy, Salvador Namburete, with the Export-Import Bank of China as the most likely loan provider for 70% of the investment.

When the Mphanda Nkuwa Dam project was first proposed in 2006, Mozambique's environmental protection organizations actively lobbied the China EXIM Bank against the project for the following reasons: Firstly, the Cahora Bassa Dam constructed on the Zambezi River in the 1970s could generate more electricity than needed. With all the power sold to South Africa, Maputo has to buy electricity from South Africa. What Mozambique lacks is not electricity but the power grid (which only covers 20% of Mozambique). In fact, even this is not much needed, because most poor Mozambicans cannot afford electricity. The actual electricity use rate of the country is only 10%. Secondly, with the government working on resetting the eco-balance of the lower valley of the Zambezi River, the Mphanda Nkuwa Dam project will make this endeavor a mission impossible. And lastly, the power generated by the 2.3 billion USD new dam will still mainly go to South Africa, instead of Mozambique.

Many dam projects, either finished or under construction, have had similar impacts on the local environment and society. Residents in the neighboring communities are usually excluded from the benefits of electricity and economic development, only to find themselves relocated and having to give up fishing and farming to the projects. The lack of transparency

---

[12] Information on these dam projects comes from "China's Environmental Footprints in Africa" posted on the website of International Rivers, accessed May 18, 2008.

in these foreign-funded projects makes it difficult for relevant African NGO's efforts to protect the interests of local residents.

More caution should be taken in dam construction projects on international rivers. In October 2010, Kenyans protested against the ICBC's 500 million USD loan to the Gibe III Dam in Ethiopia. As the largest infrastructure project ever implemented in Ethiopia, the Gibe III Dam was expected to dramatically cut the runoff of the Omo River, endanger the survival of Lake Turkana and the livelihoods of up to 300,000 residents, and ultimately intensify tribal conflicts in the lake area and border conflicts between Ethiopia, Kenya, and Sudan. So far, the ICBC has not confirmed its loans to the project or responded to the inquiries of African environmental protection NGOs.

Excessive and illegal timber trading has led to heavy deforestation in many African countries, including Gabon, Cameron, Equatorial Guinea, the Republic of Congo, and Mozambique. According to local NGOs in forest protection and timber trade, the forest protection laws, regulations, and policies in these countries are not strictly enforced. Forest law enforcers turn a blind eye to illegal cutting, transporting, and trading, while local residents cut more trees than they are authorized to. For example, 94% of the timber from Tanzania in 2005 was the result of illegal cutting.[13] Deforestation and its impact on the local environment have seriously damaged the livelihood of local residents who live off the forest. Moreover, Chinese businessmen have been accused of hunting protected species in the process.[14]

Foreign investment in agriculture has also affected local farmers. Fertile land with good water supplies and convenient transportation is grabbed from local farmers by foreign investors, sometimes even with help from the local government. What's more, in competition with financially and technologically more capable foreign agricultural enterprises, local farming loses. Moreover, foreign invested agricultural projects generally focus on the growth of certain crops, which greatly decreases the diversity of crops. In fact, Africa's "zero input" cultivation techniques and local crops, more suitable to local culture, climate, and soil than China's "high input" agriculture model, can not only save labor and water resources, but also minimize the use of pesticides and fertilizer. Although there is no denying

[13] "NGOs and Forest Governance in East Africa," *Forest Trends*, http://www.forest-trends.org/~foresttr/documents/files/doc_844.pdf.

[14] Michel, Serge and Michel Bury, *Chinafrique*, Beijing, CITIC Press Corp., 2009, 35.

that the "high input" agriculture model can achieve the industrialization of agriculture, reduce costs, and increase harvest, this model poses a serious challenge to African farmers at present.

IV. *How to Improve China's Aid and Investment Behaviors in Africa*

A. *Open-Mindedness toward Discussions on the Impacts*

Some Chinese involved with the African projects have started considering the impacts of Chinese aid and investment. The Chinese government is doing the same. At the International Forum on Chinese Enterprises' Social Responsibility in October 2009, Director and Spokesman Yao Jian (姚坚) from the Ministry of Commerce said, "Chinese enterprises are now building stadiums and parliament buildings in other developing countries, while their Japanese counterparts are building schools… It is necessary for Chinese enterprises to adjust their problematic model of internationalization."[15]

While it may be true that there is no perfect way of providing aid and investment, the recognition of negative impacts is the precondition for solutions.

Therefore, we should be open-minded in the discussion about the impact of Chinese aid and investment in Africa, so as to identify our experiences and lessons and ultimately allow all interested parties to share their opinions on improvement.

B. *Better Information Disclosure and Enhanced Public Participation*

Up to now, much attention has been paid to the environmental and social impact assessment standards and procedures for Chinese overseas aid and investment projects. According to the *Full-Service Aid Project Management Measures* (对外援助成套项目管理办法), any full-service aid project must be based on a feasibility study and professional study completed by commissioned third party agencies. However, the *Measures* do not specify the details of these studies or the weight of the environmental and social assessment. Additionally, according to the *China EXIM Bank*

---

[15] Ye Yijian, "Spokesman Yao Jian from Ministry of Commerce Says: China Should Not Only Build Parliament Buildings," *21st Century Business Herald*, http://www.21cbh.com/HTML/2009-10-9/148930.html (accessed October 8, 2009).

Project Environmental and Social Assessment Outline (中国进出口银行贷款项目环境及社会影响评估大纲) issued in 2007 and the *Environmental Protection Policies Followed by China EXIM Bank* (中国进出口银行所遵守的环境保护政策), the environmental feasibility study of any project financed by China EXIM Bank must be approved by the environmental administration of the recipient country or region. In case of the absence of local environmental protection standards or standards lower than their Chinese counterparts, the Chinese national environmental protection standards shall apply.

While they have the potential to reduce environmental and social risks, these policies are quite general, without detailed implementation standards or necessary plans for information disclosure and public participation, making it difficult to assess the enforcement of these policies.

C. *Sharing and Re-examination of China's Development Model*

China, once the biggest receiver of international aid, needs to re-examine its own experience and development model so as to improve its aid and investment overseas.

Many questions now face China: How can international aid solve real problems in education, infrastructure, poverty elimination, and safe drinking water? How can we protect the local economy when opening to the outside world? How can we protect agriculture and the interests of local farmers? How can we correct the negative environmental and social impacts of economic development? We can learn from the discussions at home and the ensuing efforts to mend our ways.

CONCERNS FOR THE LANCANG-MEKONG[1] LIFELINE

Yu Yin

Abstract: In recent years, the rapid expansion of foreign trade and investment by Chinese enterprises in the Mekong Region, and a series of downstream impacts from the Lancang cascade hydropower development, have notably affected the image of China in the hearts of downstream people. The author believes that civil society can help to reduce the risks of overseas investment. Civil society organizations, through field investigations and objective analysis, are able to show investors the potential risks. They can promote stakeholders' dialogue and communication, ensure information transparency, and facilitate conflict resolution. Meanwhile, Chinese civil society organizations should, through people-to-people diplomacy, pass the kindness of the Chinese people to those living in the Mekong Region, and protect the national image.

Keywords: Mekong, overseas investment, river basin management, hydropower development, public participation, people-to-people diplomacy

The China-ASEAN Free Trade Area came into force in 2010. Statistics show that China-ASEAN bilateral trade volume reached 362.9 billion USD in 2011. China is now the largest trade partner of the members of ASEAN, and the ASEAN trade bloc has become China's third largest trade partner.[2] Chinese enterprises have also made substantial investments in this area.

Chinese enterprises are confronted with opportunities as well as risks in the international arena. At the center of global attention is whether Chinese enterprises can develop an attitude of "responsible power" in a globalized context. Over the past twenty years, the risks resulting from Chinese hydropower investment and development in the Mekong Region have notably affected China's image in the downstream nations. As someone who has lived in the Mekong Region for years, the author hopes to examine the causes of increasing concerns by analyzing cases of both Chinese overseas investment and watershed management. This will

[1] The river is known as the Lancang in the upper reaches of the river, and Mekong in the lower sections.

[2] Huang Yanqing, Liu Xiaoxiao, "China-ASEAN Trade Volume expanded 45 times in past 20 years," July 12, 2012, http://intl.ce.cn/specials/zxxx/201207/12/t20120712_23486604 .shtml.

help us better understand the concerns of downstream nations and risks of investment development and help build up mutual trust between China and the Mekong Region countries if we mutually seek sustainable development.

I. *The Cultural Significance of the Mekong River and Its Impact on the Livelihood of Mekong Region Nations*

The Mekong River is one of the few rivers in the world that flows longitudinally. Its estimated length is 4,880 kilometers and it originates in Zaduo County (杂多县) in China's Qinghai Province. After passing through glaciers in the Tibetan region and alpine meadows, mountains, and gorges in Yunnan Province, it flows through the wide valleys and alluvial plains in Myanmar, Thailand, Laos, and Cambodia. It finally discharges into the South China Sea via the Mekong Delta in Vietnam. This very important river in continental South East Asia nurtures millions of people and is deemed as the mother river and the lifeline in that region.

Why should civil society in the Mekong Region be so concerned about dam construction there? This is inseparable from the river's important role in the food and water supply of local people and its cultural significance to them. The drainage area of the Mekong River is 811,000 square kilometers, which covers approximately 90% of the land area of Laos and Cambodia, the northeastern part of Thailand (about 1/3 of its land area), and the fertile delta in southern Vietnam. The watershed is inhabited by 60 million people, the majority of them making their living by farming and fishing. For these people, the river and its ecological system are their source of life and food.[3]

The Mekong Region supports a biodiversity second only to the Amazon, with river fauna the most diverse in the world. The region is home to almost 1,300 species, including the Mekong Giant Catfish, the world's largest freshwater fish, and rare species of freshwater river dolphin. Fish and rice are the staple food for people in the region. Fish remains the major source of protein and an indicator of food security. Fishing is the source of income for the majority of riverside inhabitants. In Cambodia,

[3] Mekong River Commission, "State of Basin report 2003," http://www.mrcmekong.org/assets/Publications/basin-reports/state-basin-executive-sum2003.pdf.

at least half of the nation is engaged in fishing, with 12% of GDP coming from the freshwater fishing industry.[4]

The Mekong River has also shaped the culture and religious beliefs of the local inhabitants. The Songkran Festival and dragon boat racing are both connected with the Mekong River. People in Laos, Thailand, and Cambodia believe in Naga, the magic patron dragon of the Mekong River. They offer gifts to Naga to ask for blessings of their safety when they launch their boats. The Mekong River is intimately related to the local community, as fishing societies in every village have a managerial system in which the elderly pass down fishing skills and knowledge to young children.

II. *Chinese Investment in the Mekong Region*

The year 2010 was the tenth anniversary of the implementation of the "Going Global" (走出去) policy of the Chinese government. Southeast Asia, which is often viewed as China's "backyard," has become a focus of trade development and investments for the Chinese government and enterprises. China's overseas investment has concentrated on hydropower projects, mining, and agriculture, attracting both investments and international concern. For instance, according to official statistics, China's investment in Myanmar in 2010 surpassed 8.17 billion USD, of which 5 billion USD was invested in hydropower projects, 2.15 billion USD in oil and natural gas projects, and nearly 1 billion USD in mining.[5] The 900-odd kilometer natural gas pipeline that runs through Myanmar from east to west was launched in June 2010, and has drawn much concern from all over the world.

Meanwhile, other major hydropower stations have been constructed one after another. These projects are concentrated in regions where Myanmar's ethnic minority militias are active. Myanmar's government has increased its military deployment in the regions in order to keep the projects intact, thus exacerbating tensions between government militia and local minorities. The hydropower projects and natural gas pipelines

---

[4] An Introduction to Cambodia's Inland Fisheries, Mekong River Commission, November 2004. http://www.mrcmekong.org/assets/Publications/report-management-develop/Mek-Dev-No4-Mekong-Fisheries-Cambodia-Eng.pdf.

[5] Aung Hla Tun, "Chinese Investment in Myanmar Tops $8 Billion USD This Year," *Reuters*, August 16, 2010.

require large-scale land acquisition. However, the fact that regulations and policies of compensation go under the table and that government militias force locals to relocate has increased risks to Chinese enterprises in Myanmar.[6] In April 2010, a series of bomb blasts hit worksites and residential areas of three dam projects developed and built by China Power Investment Cooperation (CPI) (中国电力投资集团公司): N'Mai Kaichin, Mali Kaichin, and Irawaddy Myitsone Dams—all part of the Confluence Region Hydropower Project (CRHP). The bombs injured Chinese staff and damaged facilities.[7] According to statistics from the Burma Rivers Network, China has invested in at least 20 major hydropower projects—none of which were given comprehensive environmental and social impact assessments on the internationally recognized biodiversity hotspots and indigenous people inhabiting the region.

The Chinese enterprises have also aroused controversy in Cambodia. As reported, ZTE Energy Company (中兴能源有限公司) has been using washing methods for gold mining there, causing a series of problems such as pollution of water sources, impact on crop growth, and clashes with local farmers—which have so far remained unresolved.[8] The Chinese enterprises in Southeast Asia are thus confronted with political, ecological, and social risks, but have also been presented with valuable opportunities to learn to "be responsible" for their operations.

In order to learn about international standards for environment and society, Sinohydro Company (中国水利水电建设集团公司) has applied for political risk insurance from MIGA (Multilateral Insurance Guarantee Agency) for the Nam Ngum 5 (NN5) Dam, the first BOT (Build, Operate, Transfer) project in Laos. MIGA is a member of the World Bank Group that provides political risk insurance for foreign investors in developing countries. Before the board of trustees of MIGA approves the project guarantee, the applicant enterprise needs to make improvement plans using MIGA's guidelines for social and environmental safety, including those concerning dam safety, involuntary resettlement, and the rights of indigenous peoples. Sinohydro is the first state-owned enterprise that has applied for political risk insurance from international agencies under the World Bank, which implies that both Sinohydro and MIGA need to learn

---

6 Sarah Birke, "Ethnic tensions grow in Myanmar," *The National*, July 3, 2010.

7 Li Wei, Jin Xin, "Bomb Hits Myanmar Power Station Invested in by Chinese Enterprises," *Global Times*, April 19, 2010.

8 Gong Minghe, "Cambodians Come to learn Chinese Experiences on Environmental Impact Assessment, Environmental Conservation," *Southern Daily*, October, 09, 2010.

from each other and make adjustments in their collaboration. MIGA's regular on-spot investigation and management advice thus enable Sinohydro to expand their knowledge of the local environment and society, to disclose information to NGOs, and pay further attention to the impact of hydropower projects on local environments and people's livelihood, as well as to seek approaches to alleviate negative impacts.

In December 2010, the Global Environmental Institute, a Beijing-based Chinese non-governmental organization, signed a memorandum with the NN5 Power Company of Sinohydro in Laos, officially launching a biogas development project in the local community affected by the construction of NN5 Dam. The biogas project will serve to explore patterns of sustainable investment and development of Chinese enterprises overseas, raise support for the local community, improve their relationship with locals, and promote a positive interaction between the two parties, with a view to achieve a win-win situation between environment and economic growth.

In April 2010, the China-ASEAN Environmental Cooperation Center was established by the Ministry of Environmental Protection of China, aimed at promoting regional cooperation on the environment between ASEAN member states and China. In September 2010, with the support of NGOs, the Research Center for Environmental Impact Assessment under the Ministry of Environmental Protection of China went to Cambodia to train local officials. Meanwhile, Chinese enterprises are learning to cope with new situations, which can be perceived, for instance, in the effort by ZTE Energy to agree to a dialogue with a German NGO when it sent out a Chinese intern after their first request was declined by ZTE.[9]

Based on the above-mentioned cases, we can see that Chinese enterprises are confronted with many risks in their investment development in the Mekong Region. Professor Qin Hui (秦晖) from the Department of History at Tsinghua University recently pointed out that Chinese construction of hydropower stations in the upper reaches of the Lancang River were confronted with a complex interest relationship between the downstream nations, and therefore received both support and opposition. Only oppositional voices were heard because "Bad news travels far, while good news stays put."[10] This pattern also applies to China's investment

[9] Lü Minghe, "Chinese Teachers, Cambodian Students," *Southern Weekend*, October 7, 2010.

[10] Qin Hui, "On the Mekong, A Better Way (2)," *China Dialogue*, December 29, 2010.

in the Mekong Region. On the other hand, the Chinese enterprises' public relations crisis provides an opportunity to learn lessons and make improvements. Only with an open mind can we achieve smooth communication with host country governments and NGOs, learn about the local situation, and seek solutions for local issues.

III. *China's Lancang Hydropower Development and Responses from Mekong Region Countries*

In March 2010, the Mekong Region countries suffered varying degrees of environmental and economical losses that were a result of falling water levels in the Mekong River. The affected villagers blamed China's dam construction on the Lancang River, which, as they claimed, intensified the drought downstream.[11] Governments of the river basin countries set about negotiating with China, hoping to get more detailed information. In February 2010, the water volume of the Mekong was reduced to half of its previous year's flow, leading to the paralysis of transportation along the river section from Jinghong (景洪) in China to Laos, with 26 cargo barges stranded.[12] The reduced water volume also posed a food security threat to the locals. Despite the fact that both Cambodian Prime Minister Hun Sen and the CEO of the Mekong River Commission Jeremy Bird announced that the hydropower stations on the Lancang River did not cause the drought in the Mekong Region, the connection between hydroelectrical projects and the disaster remained a focus of controversy.

In early February, civil society groups in Thailand pressed the Thai government to negotiate with Beijing and request the Chinese government to release hydrological data for the Lancang River section, including the water volume in the dam reservoir.[13] On April 3, the Thai People's Network for the Mekong jointly signed an open letter and sent it to the Chinese Embassy in Thailand, petitioning the Chinese government to stop building dams on the upper Mekong River's mainstream so as to free the water flow. The open letter held that the 2010 drought and the flood in 2008 seriously affected the livelihood and food security of the people in downstream countries. They also maintained that dams not only do not

11 Cao Haili, "China Mekong Crisis in Public Relations."
12 Dai Zhenhua, "Lancang River Transportation Halted for Draught."
13 "China 'Will Heed River Concerns'," *Bangkok Post*, March 7, 2010.

prevent floods in the wet season, but worsen the drought in the dry season. Whether in the wet season or dry season, the dams lead to unnatural rapid rising and falling of water levels, posing threats to the river ecosystem, food safety, culture, social stability, local economy, and trade, as well as tourism in downstream countries.[14]

Nevertheless, Chinese diplomatic officials rebutted the accusations. On March 11, they held a press conference at the Chinese Embassy in Thailand, refuting the accusations concerning the impact of dams in China on the Mekong drought and maintaining that the drought was closely connected with global climate change.[15] In early April, China sent a delegation led by Deputy Minister of Foreign Affairs Song Tao (宋涛) to attend the first Mekong Summit. China promised to provide relevant hydrological data and invite the officials of four Mekong countries to visit hydroelectric stations on the upper stretches of the Lancang River.

IV. *The Controversy over Dams among the People in Lower Mekong Countries*

After WWII, development projects flooded into the Mekong Region. Major hydropower stations and irrigation reservoirs were built, forests were lumbered for export, monoculture forest and crops planted, and high pollution industrial zones opened. Behind the economic growth was the gradual destruction and deprivation of the sources of livelihood that the locals relied on, and they were forced to leave home to make a living in other towns. The gap between the rich and the poor, the urban and the rural, intensified the conflict between governments, enterprises, and local communities over the use of resources and power. At the same time, more and more people set out to investigate and research the impacts of "development projects," to disclose the final beneficiaries and victims, and to encourage the locals to reflect on an alternative development pattern.

The Pak Mun Hydropower station is located on the Mun River, a major tributary of the Mekong, only 5.5 kilometers from the confluence of the

---

[14] Mekong Community Media Project, "Two Important Lessons from Mekong Mainstream Dams in China," *Mekong-Lanna Natural Resources and Cultural Conservation Network*, April 1, 2010.

[15] Zhu Li, "Press Conference of Chinese Embassy in Thailand: the Drought in the Mekong Watershed," Xinhuanet, http://news.xinhuanet.com/world/2010-03/11/content_13150928.htm, March 11, 2010.

two rivers. The project was started in 1991 and completed in 1994. It was constructed by the Electricity Generating Authority of Thailand (EGAT) with support from the World Bank at a total cost of 260 million USD. Unfortunately, the dam blocked the migration of fish species between the Mun and the Mekong, having a disastrous impact on the fish resources in the Mun River, and affecting the livelihood of more than 20,000 people in the area, among who most were fishermen. The Anti-Pak Mun Movement, having lasted more than twenty years, is one of the important landmarks of civil society movements in Thailand. Since the completion of the dam, the participants in the movement have held demonstrations and sit-ins at the site of the dam many times. They also gathered together in front of Bangkok Parliament and Government House, petitioning to meet the Prime Minister or the President and to solve the problems of their lost livelihood. Under the pressure of the activists' continuous effort, in 2000 the government appointed Ubon Ratchathani University to make an environmental and social impact assessment. The report advised the government to open the gates of the Pak Mun Dam for a long period of time. The government ignored the advice, which caused thousands of angry villagers to gather in front of the Government House in Bangkok again to petition for the implementation of the solution. Some villagers clashed with police during their demonstration.[16]

The Anti-Pak Mun Movement gave birth to an influential civil network, Assembly of the Poor (AOP), which is a coordinating network for the poor to strive for sustainable rights. Since the 1990s, it has organized many protests for the poor to defend their means of livelihood and protect natural resources. These protests not only raised the issue of Pak Mun Dam, but also other issues related to dam construction, land acquisition, and industrial pollution of the local environment in other parts of the country.

Another movement is taking place in Cambodia. Downstream villagers in Ratanakiri Province have been affected by the construction of the Yali Falls Dam in Vietnam for years, and they have continuously striven for their rights of subsistence, management of natural resources, compensation for their livelihood, and installation of an alert system from the governments of the two countries.

Based on the above mentioned lessons of the past development process in the Mekong Region, two important observations can be made:

---

[16] Supara Janchitfah, "Catalogue of Woes Goes Ever On," *Bangkok Post*, August 13, 2000, http://www.livingriversiam.org/thai/pm/pm_n/PMDnE99.htm, (accessed April. 10, 2009).

1. *Reconsideration of development models.* People have started to reevaluate the social and environmental problems caused by GDP-oriented development, such as the forming of interest groups and the distribution of social and economic benefits. Civil society organizations resort to protests to arouse social concerns about the development, and meanwhile search for more means of sustainable development.

2. *The importance of information transparency, civil participation, and conversations with stake-holders.* Local citizens have usually been excluded from the development process, which makes it difficult for them to enjoy benefits and therefore increases the undesirable side effects of economic development. The affected citizens usually have no access to comprehensive information as well as participation in decision-making and conversation. This leads to social disharmony and instability and increases the tensions between stake-holders. Only when the citizens have the right to know and participate can there be a possibility to alleviate the tensions and promote harmonious social development.

V. *The Complex Relationship between Countries and Their People in the Mekong Region*

In theory, the governments of the Mekong Region countries and the Mekong River Commission are spokespersons of their people. However, there is room for improvement for them in terms of information transparency for local people and communication between countries about water resource usage. For instance, the Mekong River Commission has regularly issued and updated hydrological data on its website, but the technically-loaded data makes it nearly inaccessible for a non-professional audience. Yet if one starts tracking this data, many errors can be found. When the water level suddenly rises and falls, the data only shows the average flow rate of the day, which has caused suspicion in many people.

The differences between the political systems of the Mekong Region countries have also directly affected the relations between governments and citizens.

Moreover, when governments manage watershed affairs, they often strive to maximize the use of resources and attract foreign investment rather than emphasize overall sustainable development. Therefore, the conflict of interests between governments and their people will also be reflected in foreign investment projects in these regions, causing certain risks for investors.

It has also been observed that despite the sometimes divided interests of Chinese government and enterprises, people in downstream countries have stereotyped images of both, which has to do with investment operations. For a long time the Chinese government has taken the lead in pushing the policy of "Going Global." Many investment projects in the Mekong Region have been promoted by the government, while the enterprises are the actual operators and the two are inseparable stakeholders.

VI. *Proposition: Civil Society Assumes Responsibility for Chinese Investment in the Mekong Region*

In the process of pursuing overseas investment, enterprises and government departments should be aware of the role of academics, media, and civil society—which conduct field research and clarify the risks of investment projects. Civil society serves to ensure information transparency and promote the stakeholders' dialogue and communication, and facilitates conflict resolution.

The civil society in the Mekong Region countries has done much field research on natural resources management in many places of the Mekong watershed. This can help Chinese researchers and potential investors to comprehensively learn about the region. In addition, civil society hopes to build up links for communication with Chinese investors in order to effectively deliver the opinions of locals to investors and other departments and get spontaneous feedback. More importantly, they encourage the local community to participate in the process of decision-making about development.

In recent years, China has been extremely cautious about encouraging public diplomacy or people's diplomacy. However, as Han Fangming (韩方明), vice chairman of the Foreign Affairs Committee of the Chinese People's Political Consultative Conference (CPPCC) indicated, developing public diplomacy as a way to boost China's "soft power" is inevitable.[17] China needs to demonstrate its confidence as a country with increased national strength. The Chinese government often views opposing voices from other countries with suspicion. This has not played a positive role in the integration of the Chinese people into the international community

[17] Han Fangming, "Strengthening Public Diplomacy and Showing the Real China in Different Ways," *China Cultural Newspaper*, December 3, 2010, http://hanfangming.blog.sohu.com/163721427.html.

and proper treatment of opposition from international civil organizations. The Chinese government would benefit from allowing Chinese civil society more freedom. This would improve the national image, and pass the kindness of the Chinese people to people in the Mekong Region via people-to-people diplomacy. By cooperating with international organizations, Chinese civil society can also learn about the international standards of environmental management and social development, including standards of environmental and hydrological information transparency and corporate social responsibility. In addition, Chinese civil society can deal with real cases and promote the progress of Chinese enterprises in this field, as well as advising the government on managing and regulating overseas investment of the Chinese enterprises.

PART EIGHT

APPENDIX

ANNUAL INDEXES: 2010 & 2011 ENVIRONMENTAL TRENDS

The developing trends of China's major environmental problems will be presented with tables and figures to generate a direct image and a general picture. Based on the key environmental issues of 2010 and 2011, data selected here include the emission and control of air pollutants, discharge and control of water pollutants, energy consumption, and the water quality of rivers, lakes, and seawater. Several of the tables and figures compare data from 2009, the mid-point of the *Planning Objectives of Environmental Protection for the Eleventh Five-Year Plan Period* (十一五环境保护规划目标), with the year 2005, the year selected as the benchmark. With the changing trends of environmental problems during the mid-term of the Eleventh Five-Year Plan period, both the effects of governmental efforts and the possibility of reaching the objectives for the five years can, to some extent, be observed.

I. *Air Quality*

Sulfur dioxide emission in 2009 totaled 22.144 million tons, soot emission was 8.472 million tons, and industrial dust was 5.236 million tons, down 4.6%, 6%, and 11.7% respectively.

The total emission of sulfur dioxide in 2010 was 21.851 million tons, down 1.32% from 2009. Soot emission was 8.291 million tons, down 2.2% from 2009. Industrial dust emission totaled 4.487 million tons, down 14.3% from 2009.

Table A.1: Annual changes in the emission of major air pollutants.

Year	Emission of sulfur dioxide (million tons)			Emission of soot (million tons)			Emission of industrial dust (million tons)
	Total	Industrial	Domestic	Total	Industrial	Domestic	
2006	25.888	22.348	3.540	10.888	8.645	2.243	8.084
2007	24.681	21.400	3.281	9.866	7.711	2.155	6.987
2008	23.212	19.913	3.299	9.016	6.707	2.309	5.849
2009	22.144	18.661	3.483	8.472	6.039	2.433	5.236
2010	21.851	18.644	3.207	8.291	6.032	2.259	4.487

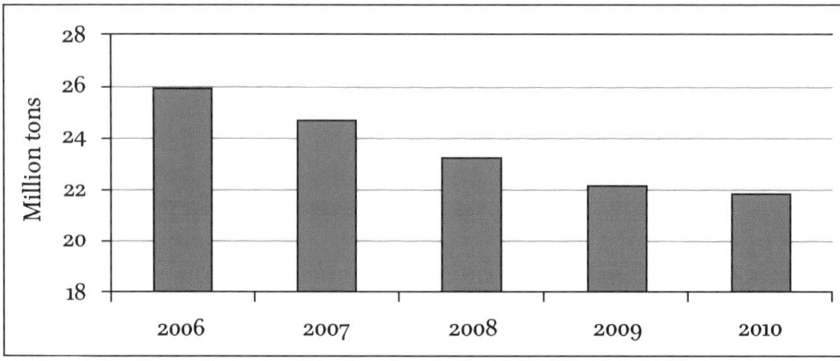

Figure A.1: Emission of sulfur dioxide, 2006–2010.

Compared with the figures from the benchmark year 2005, the total emission of sulfur dioxide was reduced by 14.29%, exceeding the original emission reduction target by 10%.

In 2009, the State Council convened the third meeting of the leading group on energy conservation and emissions reduction, while its general office issued its working arrangement. The results of evaluation for all the provinces, autonomous regions, municipalities directly under the administration of the central government, the State Grid, and the five largest power companies were released on the total emission reductions of major pollutants. Also released was the bulletin on the emission of major pollutants in each region during the first half of 2009. Regions and companies with pronounced problems were made public, deadlines were specified for improvement, and monetary punishment was imposed in some cases. Written warnings were issued to eight provinces (or regions) that were slow in emissions reduction in the first half of 2009, while the top officials of the local governments were summoned to give them guidance, supervision, and examination.

Table A.2: Emission of SO_2 (2004–2009) compared to the target of the eleventh five-year plan.

Emission \ Year	2004	2005	2006	2007	2008	2009	Target of Eleventh Five-Year Plan
Total (million tons)	22.549	25.493	25.888	24.681	23.212	22.144	22.950

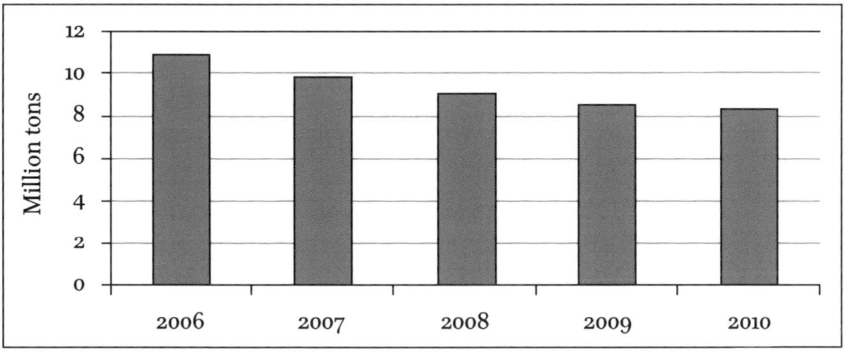

Figure Figure A.2: Emission of industrial soot, 2006–2010.

Table A.3: Air quality ranking of China's province-level municipalities and capital cities of provinces, 2011.

Rank	City	Percentage of days with air quality of Level I and II (%)	Days with excess concentration of sulfur dioxide	Days with excess concentration of sulfur dioxide	Days with excess concentration of inhalable particles	Percentage of days with air quality of Level I and II in previous year (%)	Rank in previous year
1	Haikou	100	0	0	98	100	1
2	Kunming	100	18	0	222	100	2
3	Lhasa	100	0	0	67	99	3
4	Guangzhou	99	0	4	228	98	4
5	Fuzhou	99	0	0	239	96	6
6	Nanning	96	0	0	216	96	5
7	Guiyang	95	30	0	277	94	8
8	Hohhot	95	61	0	174	96	7
9	Nanchang	95	35	0	269	94	9
10	Changchun	94	3	0	308	93	10
11	Changsha	93	2	0	275	93	11
12	Shanghai	93	0	0	220	92	12
13	Yinchuan	92	14	0	308	91	13
14	Hangzhou	91	0	1	289	87	15
15	Shenyang	91	42	0	275	90	14
16	Chongqing	89	2	0	281	86	20
17	Tianjin	88	42	0	254	84	25
18	Shijiazhuang	88	22	0	294	87	17
19	Chengdu	88	0	0	298	87	16
20	Nanjing	87	1	0	291	83	26
21	Ji'nan	87	5	0	333	85	23
22	Xining	87	0	0	326	87	18
23	Harbin	87	0	0	324	85	21

Table A.3 (*cont.*)

Rank	City	Percentage of days with air quality of Level I and II (%)	Days with excess concentration of sulfur dioxide	Days with excess concentration of sulfur dioxide	Days with excess concentration of inhalable particles	Percentage of days with air quality of Level I and II in previous year (%)	Rank in previous year
24	Zhengzhou	87	23	0	298	87	19
25	Taiyuan	85	71	0	192	84	24
26	Wuhan	84	0	0	285	79	28
27	Xi'an	83	2	0	265	83	27
28	Hefei	83	0	0	303	85	22
29	Beijing	78	1	0	277	78	29
30	Urumqi	75	50	0	284	74	30
31	Lanzhou	67	0	0	331	62	31

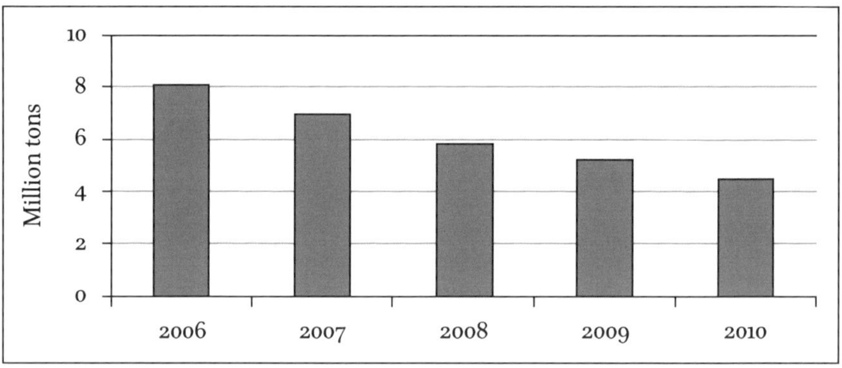

Figure A.3: Emission of industrial dust, 2006–2010.

II. *Water Quality*

A. *Water Systems*

In 2009, the water quality of the seven major water systems was slightly polluted in general, mostly unchanged compared with the previous year. Among the 203 rivers and the 408 nationally monitored cross-sections of surface water, 57.3% was at Level I, II, or III; 24.3% was at Level IV and V; while 18.4% was worse than Level V. Major pollution indexes were hyper-manganate, 5–day biochemical oxygen demand (COD5), and ammonia

Table A.4: Water quality of China's 7 major water systems, 2009.

Index \ Water system	Yangtze River	Yellow River	Pearl River	Songhua River	Huaihe River	Haihe River	Liaohe River
Level I, II, III	87.4%	68.2%	84.9%	40.5%	37.3%	34.4%	41.7%
Level IV	5.8%	4.5%	12.1%	47.6%	33.7%	10.9%	13.9%
Level V	2.9%	2.3%	3.0%	2.4%	11.6%	12.5%	8.3%
Worse than Level V	3.9%	25.0%	0	9.5%	17.4%	42.2%	36.1%

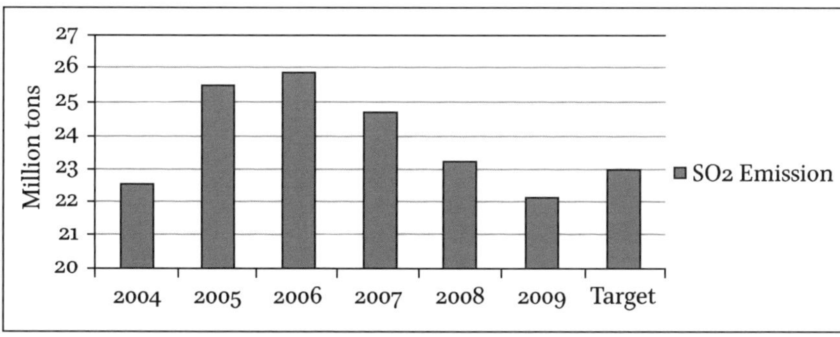

Figure A.4: Emission of SO_2 (2004–2009) compared to the target of the eleventh five-year plan.

nitrogen. Water in the Pearl and Yangtze Rivers was in good quality, the Songhua and Huaihe Rivers were lightly polluted, the Yellow and Liaohe Rivers moderately polluted, and Haihe River was severely contaminated.

In 2010, the water quality of the seven major water systems (Yangtze, Yellow, Pearl, Songhua, Huaihe, Haihe, and Liaohe Rivers) was slightly polluted in general. Among the 204 rivers and the 409 nationally monitored cross-sections of surface water, 59.9% were in Category I, II, or III, 23.7% in Category IV and V, and 16.4% were worse than Category V. The major pollution indexes were hypermanganate, COD5, and ammonia nitrogen. Water in the Pearl and Yangtze Rivers had good quality, the Songhua and Huaihe Rivers were slightly polluted, the Yellow and Liaohe Rivers were moderately polluted, and Haihe River remained heavily polluted.

Table A.5: Water quality of China's 7 major water systems.

	Yangtze River	Yellow River	Pearl River	Songhua River	Huaihe River	Haihe River	Liaohe River	Total
Category I, II, III	88.6%	68.2%	84.9%	47.6%	41.9%	37.1%	40.5%	58.4%
Category IV	6.6%	4.5%	12.1%	35.7%	32.5%	11.3%	16.3%	17.0%
Category V	1.0%	6.8%	0.0%	4.8%	9.3%	11.3%	18.9%	7.4%
Worse than Category V	3.8%	20.5%	3.0%	11.9%	16.3%	40.3%	24.3%	17.2%

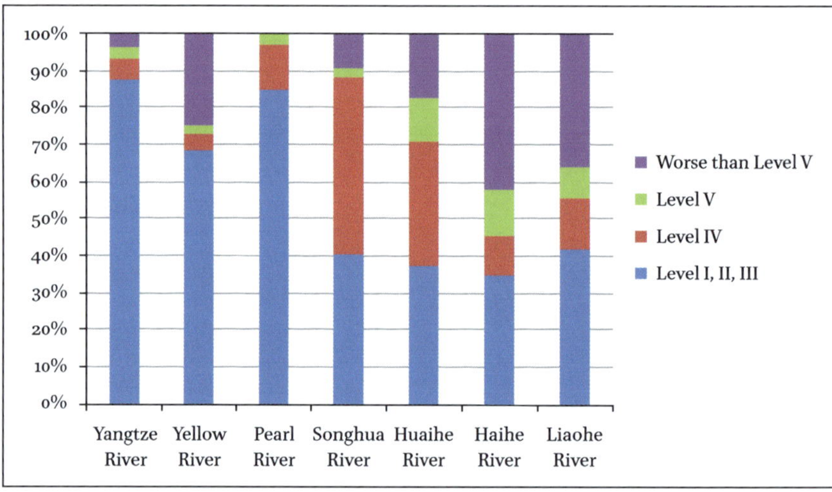

Figure A.5: Water quality of China's 7 major water systems, 2009.

B. *Lakes and Reservoirs*

In 2009, among the 26 lakes and reservoirs with close governmental monitoring, the water quality of one lake was Level II, accounting for 3.8 %; five were Level III, accounting for 19.2%; six were Level IV, accounting for 23.1%; five were Level V, accounting for 19.2%; and nine were worse than Level V, accounting for 34.6%. The major pollution indexes were total nitrogen and total phosphorus. As to the level of eutrophication, one lake (3.8%) was high, two (7.7%) were intermediate, and eight (30.8%) were low. The rest (57.7%) of the lakes were mesotrophic.

In 2010, the water quality of one lake was in Category II, accounting for 3.8%; five were in Category III, accounting for 19.2%; four were in

Category IV, accounting for 15.4%; six were in Category V, accounting for 23.1%; and ten were worse than Category V, accounting for 38.5%. The water quality of large reservoirs was better than large freshwater and urban lakes.

As to the level of eutrophication of these 26 lakes and reservoirs, one lake (3.8%) was high, two (7.7%) were intermediate, and 11 (42.3%) were low. The rest (46.2%) of the lakes and reservoirs were mesotrophic.

Table A.6: Water quality of lakes (reservoirs) with close governmental monitoring, 2004–2010.

	Category I, II, III	Category IV, V	Worse than Category V
2004	26.0%	37.0%	37.0%
2005	28.0%	29.0%	43.0%
2006	29.0%	23.0%	48.0%
2007	49.9%	26.5%	23.6%
2008	21.4%	39.3%	39.3%
2009	23.1%	42.3%	34.6%
2010	23.0%	38.5%	38.5%

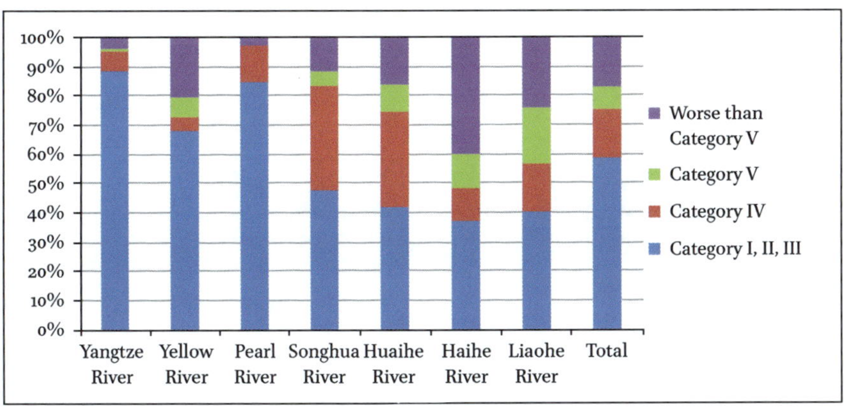

Figure A.6: Water quality of China's 7 major water systems, 2010.

Table A.7: Water quality of key lake types, 2010.

Types of lakes	Amount	Category I	Category II	Category III	Category IV	Category V	Worse than Category V	Major pollutant
Three lakes*	3	0	0	0	0	1	2	Total nitrogen and total phosphorus
Large freshwater lakes	9	0	0	3	0	3	3	
Urban lakes	5	0	0	0	2	1	2	
Large reservoirs	9	0	1	2	2	1	3	
Total	26	0	1	5	4	6	10	
Percentage (%)		0	3.8	19.2	15.4	23.1	38.5	

* Three lakes refer to Taihu Lake, Dianchi Lake, and Chaohu Lake

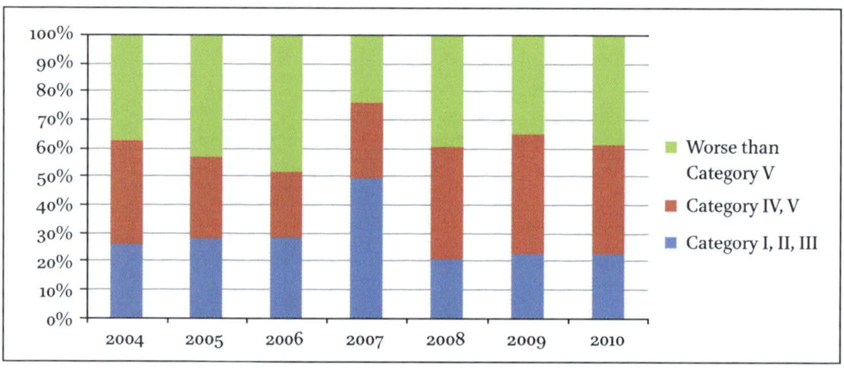

Figure A.7: Water quality of lakes (reservoirs) with close governmental monitoring, 2004–2010.

1. *Taihu Lake* (太湖)—In 2010, the general water quality was worse than Category V. Major pollutants include total nitrogen and total phosphorus. The lake body was slightly eutrophicated. Compared with 2009, the water quality was mostly unchanged.

2. *Dianchi Lake* (滇池)—In 2009, the water quality of Dianchi Lake in general was worse than Level V, with major pollution indexes being total phosphorus and total nitrogen, and the water quality mostly unchanged from the previous year. In 2010, the general water quality of Dianchi Lake was again worse than Category V, with major pollutants being total phos-

phorus, total nitrogen, and hypermanganate. Both Caohai Lake (northern part of Dianchi) and Waihai Lake (the southern part) were highly eutrophicated, which was unchanged from 2009.

3. *Chaohu Lake* (巢湖)—In both 2009 and 2010, the general water quality in Chaohu Lake was Category V. The major pollutants included total nitrogen, total phosphorus, and oil. The western half of the lake had intermediate eutrophication, while the other half was slightly eutrophicated. Compared with previous years, the water quality was mostly unchanged.

4. *Other large freshwater lakes*—Among the nine freshwater lakes under close governmental monitoring, the lakes of Jingpo (镜泊湖), Erhai (洱海), and Bosten (博斯腾湖) had a water quality of Category III in 2010; Hongze (洪泽湖), Poyang (鄱阳湖), and Nansi (南四湖) had Category V; while Dalai (达赉湖), Baiyangdian (白洋淀), and Dongting (洞庭湖) were worse than Category V. Major pollutants for the large freshwater lakes were total nitrogen, total phosphorus, and hypermanganate. Compared with 2009, the water quality of Hongze Lake improved; the lakes of Poyang, Nansi, and Dongting deteriorated; and the other large lakes were mostly unchanged.

The lakes of Jingpo, Erhai and Bosten were mesotrophic in 2010; Hongze, Poyang, Nansi, and Dongting were slightly eutrophicated, while Dalai and Baiyangdian had an intermediate level of eutrophication.

Table A.8: Water quality of 9 largest freshwater lakes, 2004–2009.

Lakes	2004	2005	2006	2007	2008	2009	Major pollution indexes	Eutrophication
Taihe	< V	< V	< V	< V	< V	< V	TN, TP	Low
Dianchi	< V	< V	< V	< V	< V	< V	TP, TN	High
Chaohu	< V	< V	V	V	V	V	TP, TN, oil	Intermediate (western part); low (eastern part)
Erhai	III	III	III	III	II	III	—	Intermediate
Dongting	V	V	V	IV	V	V	TP, TN	Intermediate
Hongze	< V	< V	< V	< V	< V	V	TN, TP	Low
Nansi	< V	V	< V	V	IV	IV	Oil, TP, TN	Intermediate
Poyang	< V	IV	V	IV	IV	IV	Oil, TP, TN	Low
Jingbo	IV	IV	IV	IV	IV	III	Hypermanganate	Intermediate

Note: TN = Total Nitrogen; TP = Total Phosphorus

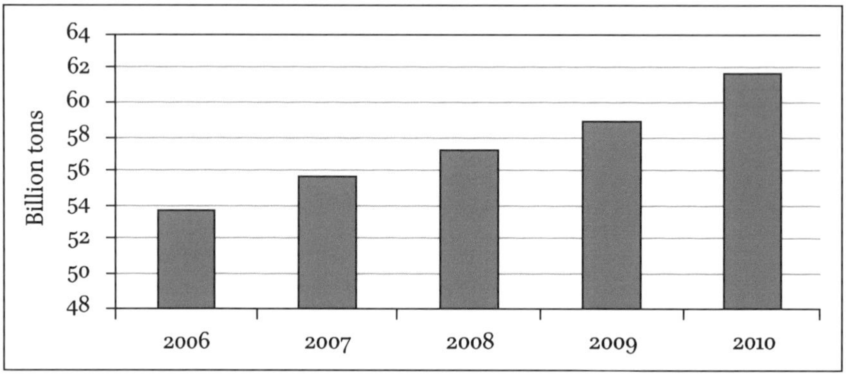

Figure A.8: Total wastewater discharge, 2006–2010.

Table A.9: Water quality of 9 largest freshwater lakes, 2010.

Name of Lake	General eutrophication index	Level of eutrophication	Water quality	Major pollutants
Dalai	65.2	Intermediate	Worse than Category V	Hypermanganate, total phosphorus, total nitrogen
Baiyangdian	60.3	Intermediate	Worse than Category V	Ammonia nitrogen, total phosphorus, total nitrogen
Hongze	58.2	High	Category V	Total phosphorus, total nitrogen
Boyang	51.5	Low	Category V	Total phosphorus, total nitrogen
Nansi	50.7	Low	Category V	Total phosphorus
Dongting	50.4	Low	Worse than Category V	Total nitrogen, total phosphorus
Jingpo	43.3	Mesotrophic	Category III	—
Erhai	40.6	Mesotrophic	Category III	—
Bosten	38.1	Mesotrophic	Category III	—

6. *Urban lakes*—In 2010, among the five urban lakes under monitoring, Kunming Lake (昆明湖) in Beijing and East Lake (东湖) in Wuhan had a water quality of Category IV, Xuanwu Lake (玄武湖) in Nanjing a Category V, and both West Lake (西湖) in Hangzhou and Daming Lake

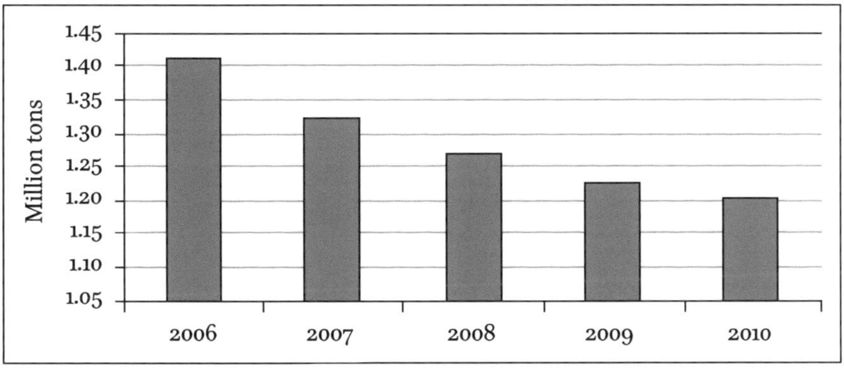

Figure A.9: Total ammonia nitrogen discharge, 2006–2010.

(大明湖) in Ji'nan were worse than Category V. The major pollutants of these lakes were total nitrogen and total phosphorus. Compared with the previous year, the water quality of these five urban lakes was mostly unchanged. In 2010, Kunming Lake was mesotrophic, while the other four were slightly eutrophicated.

7. *Large reservoirs*—In 2010, among the nine large monitored reservoirs, the water quality of Miyun Reservoir (密云水库) in Beijing was in Category II; Qiandao Lake (千岛湖) in Zhejiang Province and Dongpu Reservoir (董铺水库) in Anhui Province was in Category III; Danjiangkou Reservoir (丹江口水库) in Hubei and Henan Province as well as Yuqiao Reservoir (于桥水库) in Tianjin were in Category IV; Songhua Lake (松花湖) in Jilin Province was in Category V; while Menlou Reservoir (门楼水库) in Shandong Province, Dahuofang Reservoir (大伙房水库) in Liaoning Province, and Laoshan Reservoir (崂山水库) in Shandong Province were all worse than Category V. The major pollutant for these reservoirs is total nitrogen. Compared with the previous year, the water quality in Yuqiao Reservoir improved, Songhua Lake and Dahuofang Reservoir deteriorated, and the rest were mostly unchanged. Laoshan Reservoir is slightly eutrophicated, while the rest are in a mesotrophic condition.

Table A.10: Water quality of 5 urban lakes, 2010.

Name of lake	Eutrophication Index	Level of eutrophication	Water quality	Major pollutants
East	57.4	Low	Category IV	Total phosphorus, total nitrogen
Xuanwu	56.2	Low	Category V	Total nitrogen, total phosphorus
Daming	51.7	Low	Worse than Category V	Total Nitrogen
West	51.0	Low	Worse than Category V	Total Nitrogen
Kunming	46.4	Mesotrophic	Category IV	Total Nitrogen

Table A.11: Water quality of large reservoirs, 2010.

Name of reservoir	Eutrophication index	Level of eutrophication	Water quality	Major pollutant
Laoshan	52.1	Low	Worse than Category V	Total nitrogen
Songhua Lake	49.8	Mesotrophic	Category V	Total nitrogen, total phosphorus
Yuqiao	46.1	Mesotrophic	Category IV	Total nitrogen
Dongpu	45.6	Mesotrophic	Category III	—
Dahuofang	45.5	Mesotrophic	Worse than Category V	Total nitrogen
Menlou	37.8	Mesotrophic	Worse than Category V	Total nitrogen
Miyun	35.5	Mesotrophic	Category II	—
Danjiangkou	35.0	Mesotrophic	Category IV	Total nitrogen
Qiandao Lake	33.1	Mesotrophic	Category III	—

C. *Key Water and Hydropower Projects*

1. *The area of the Three Gorges Reservoir*—In 2010, the water quality was generally good. Among the six cross-sections under national monitoring, two had their water quality in Category I while the rest were all in Category II.

2. *The east route of the South-to-North Water Diversion Project*—In 2010, the water was slightly polluted in general. Among the 10 cross-sections under national monitoring, 60% were in Categories I–III, 30% were in

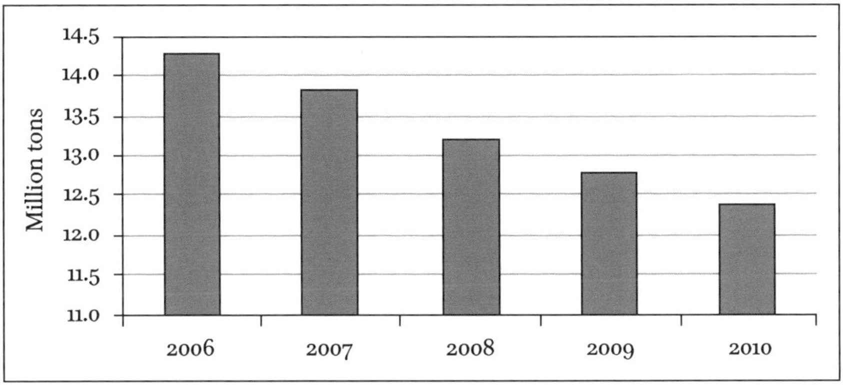

Figure A.10: Total COD discharge, 2006–2010.

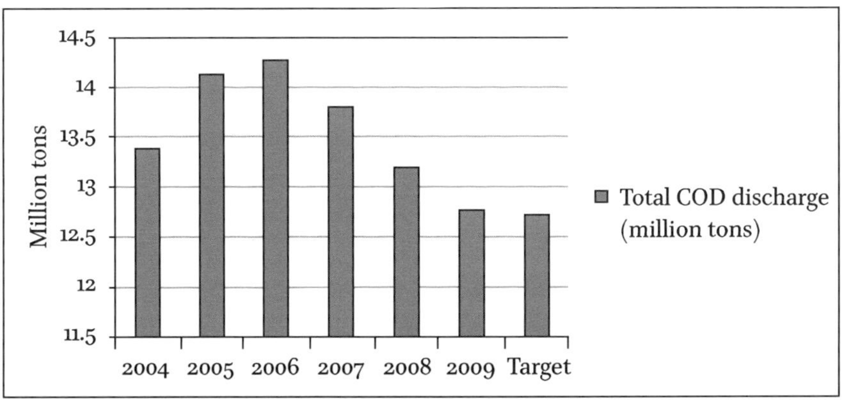

Figure A.11: Total COD discharge (2004–2009) compared to the target of eleventh five-year plan.

Category IV, and 10% were worse than Category V. Major pollutants included hypermanganate, COD5, and oil. Compared with 2009, the water quality improved.

D. *Groundwater Quality*

With 4,110 monitoring stations, groundwater quality in 182 cities nationwide was monitored in 2010. The analytical results show that water quality at 418 stations was high, accounting for 10.2% of all the stations. Water quality at 1,135 stations (27.6%) was good, 206 stations (5.0%) moderate, 1,662 stations (40.4%) poor, and 689 stations (16.8%) bad.

Groundwater quality nationwide allows for no optimism. In total 1,759 monitoring stations (42.8%) fall into the categories of high, good, and moderate water quality, while 2,351 stations (57.2%) have poor and bad water quality.

Groundwater quality of China's major cities was mostly unchanged. Cities with an improving groundwater situation are concentrated in eastern China, with only a few scattered in the northern, northeastern, and northwestern parts of China. Cities with a deteriorating groundwater quality are mostly in north, northeast, and northwest China, and very few are in the eastern, central-southern, and southern regions.

E. *Water Quality in Key Centralized Drinking Water Source Areas in China's Major Cities*

In total 395 centralized drinking water source areas in 113 key environmental protection cities nationwide were monitored, among which 245 were surface water source areas and 150 were groundwater. The monitoring results showed that the total water intake in these key cities was 22.03 billion tons, 76.5% (16.85 billion tons) of which is up to the standards, while 23.5% (5.18 billion tons) were of substandard quality.

F. *Discharge of Wastewater and Major Pollutants*

In 2009, wastewater discharge nationwide was 58.92 billion tons, up 3.0% from the previous year. Chemical oxygen demand (COD) discharge was 12.775 million tons, down 3.3% from the previous year. Ammonia nitrogen discharge was 1.226 million tons, down 3.5% from the previous year.

Discharge of wastewater nationwide in 2010 was 61.73 billion tons, up 4.7% from the previous year. Discharge of COD was 12.381 million tons, down 3.1%, while discharge of ammonia nitrogen was 1.203 million tons, down 1.9%.

G. *Seawater Quality*

In general the water quality of offshore sea areas in China was slightly polluted in 2010 and had not changed much in the previous two years.

The monitored area of offshore sea in 2010 was 279,225 square kilometers in 2010, of which the water quality of 177,825 square kilometers was in Category I and II, 44,614 square kilometers in Category III, while the remaining 56,786 square kilometers was in or worse than Category IV.

Table A.12: Annual change of China's discharge of wastewater and major pollutants.

Item Year	Discharge of wastewater (billion tons)			Discharge of COD (million tons)			Discharge of ammonia nitrogen (million tons)		
	Total	Industrial	Domestic	Total	Industrial	Domestic	Total	Industrial	Domestic
2006	53.68	24.02	29.66	14.282	5.415	8.867	1.413	0.425	0.988
2007	55.68	24.66	31.02	13.818	5.111	8.708	1.323	0.341	0.983
2008	57.20	24.19	33.01	13.207	4.576	8.631	1.270	0.297	0.973
2009	58.92	23.44	35.48	12.775	4.397	8.378	1.226	0.273	0.953
2010	61.73	23.75	37.98	12.381	4.348	8.033	1.203	0.273	0.930

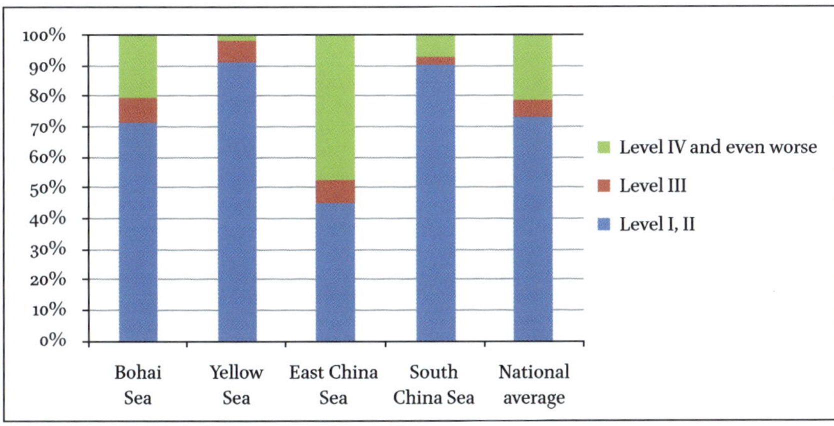

Figure A.12: Water quality in offshore areas, 2009.

According to the calculations from monitoring stations, 62.7% of seawater is in Category I and II, down 10.2 percentage points from the previous year. Seawater of Category III has a share of 14.1%, up 8.1 percentage points, while 23.2% is in or worse than Category IV, up 2.1 percentage points.

In 2010, among the four largest sea areas, the offshore seawater of the South China Sea and Yellow Sea has a good quality. Water quality in the Bohai Sea is poor, and the East China Sea bad. Water quality of Beibu Gulf (北部湾) and the estuary of Yellow River was high, Jiaozhou Gulf (胶州湾) moderate, Liaodong Gulf poor, while Bohai Gulf, Hangzhou Gulf, and the estuaries of the Yangtze, Minjiang, and Pearl Rivers was bad. Compared with the previous year, monitoring stations with seawater quality of Category I and II increased by 25 percentage points in Jiaozhou Gulf, while the same indicator decreased by more than 20 percentage points in Bohai Gulf as well as the estuaries of the Yangtze and Pearl Rivers. Water quality in other gulfs is mostly unchanged from the previous year.

Table A.13: Water quality in offshore areas, 2009.

Quality \ Sea	Bohai Sea	Yellow Sea	East China Sea	South China Sea	National average
Level I, II	71.4%	90.7%	45.2%	90.1%	72.9%
Level III	8.2%	7.4%	7.4%	3.0%	6.0%
Level IV and even worse	20.4%	1.9%	47.4%	6.9%	21.1%

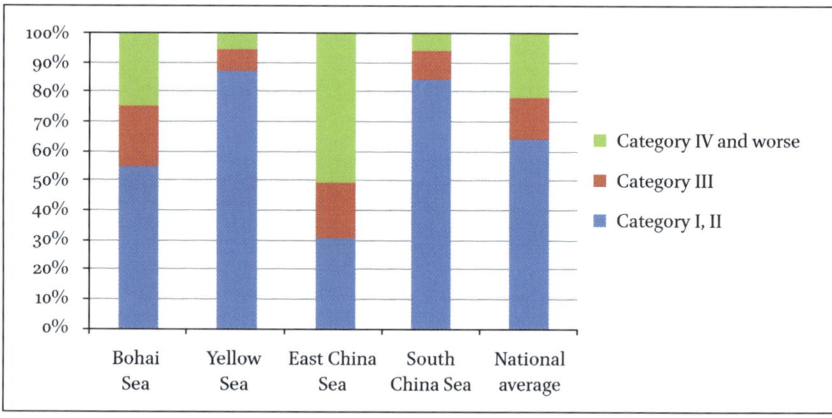

Figure A.13: Water quality in offshore areas, 2010.

Table A.14: Water quality of offshore sea areas, 2010.

Seawater quality	Bohai Sea	Yellow Sea	East China Sea	South China Sea	National average
Category I, II	55.1%	87%	30.6%	84%	64.2%
Category III	20.4%	7.4%	18.9%	10%	14.2%
Category IV and worse	24.5%	5.6%	50.5%	6%	21.7%

H. *Discharge of Land-Sourced Pollutants into Sea Areas*

Water quality of the 192 monitoring cross-sections in the rivers that drain into the sea was in general poor for 2010. The quantity of pollutants discharged from rivers into the sea was larger than that directly discharged into the sea. The quantity of pollutants discharged from rivers into the East China Sea was larger than that into the other sea areas.

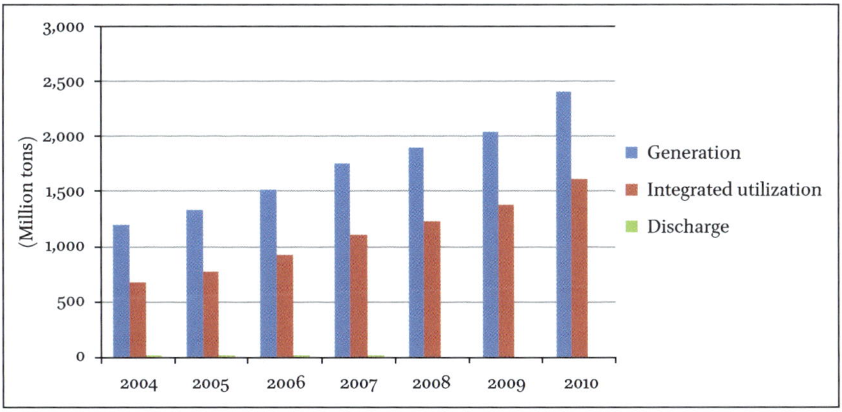

Figure A.14: Generation and discharge of industrial waste, 2004–2010.

Table A.15: Water quality of monitoring cross-sections in rivers draining into the sea, 2010.

Sea area	Water quality						
	Category I	Category II	Category III	Category IV	Category V	Worse than Category V	Total
Bohai Sea	0	0	9	4	7	28	48
Yellow Sea	1	3	15	16	5	13	53
East China Sea	0	0	5	7	6	7	25
South China Sea	0	10	21	20	8	7	66
Total	1	13	50	47	26	55	192

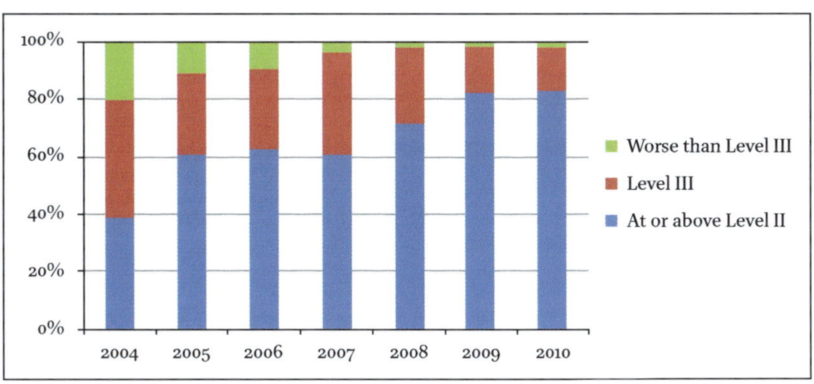

Figure A.15: Urban air quality, 2004–2010.

Table A.16: Total quantity of pollutants discharged from rivers into the 4 largest seas, 2010.

Sea area	Hypermanganate (tons)	Ammonia nitrogen (tons)	Oil (tons)	Total phosphorus (tons)
Bohai Sea	156,000	35,000	2,300	3,600
Yellow Sea	290,000	45,000	4,000	8,400
East China Sea	2,427,000	394,000	28,800	189,000
South China Sea	1,091,000	183,000	17,000	35,600
Total	3,964,000	657,000	52,100	236,700

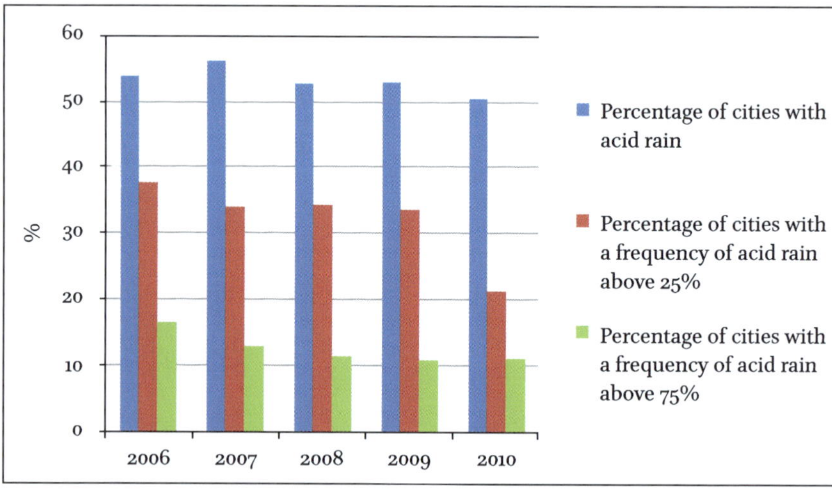

Figure A.16: Occurrence of acid rain in China, 2006–2010.

In 2010, the total quantity of major pollutants discharged from the 192 monitoring cross-sections into the sea consisted of: 3,964 million tons of hypermanganate, 657,000 tons of ammonia nitrogen, 52,100 tons of oil, and 236,700 tons of total phosphorus.

Total wastewater discharge from 461 industrial, domestic and mixed sources (each with a daily discharge capacity of more than 100 tons) directly into the seas reached 5.092 billion tons in 2010, among which were 219,400 tons of COD, 1,215 tons of oil, 22,870 tons of ammonia nitrogen, 2,091 tons of total phosphorus, 244.6 kilograms of mercury, 1,076 kilograms of hexavalent chrome, 1,064 kilograms of lead, and 392 kilograms of cadmium.

Table A.17: Direct discharge of pollutants into the seas, 2010.

Source of Pollution	Wastewater (million tons)	COD (tons)	Oil (tons)	Ammonia Nitrogen (ton)	Total phosphorus (ton)	Mercury (kg)	Hexavalent Chrome (kg)	Lead (kg)	Cadmium (kg)
Industrial	1,498	28,200	91	1,182	67	3.49	603	553	157
Domestic	845	44,800	373	5,248	796	22.11	359	130	180
Mixed	2,748	146,400	751	16,440	2,038	219	114	381	55
Total	5,092	219,400	1,215	22,870	2,901	244.6	1,076	1,064	392

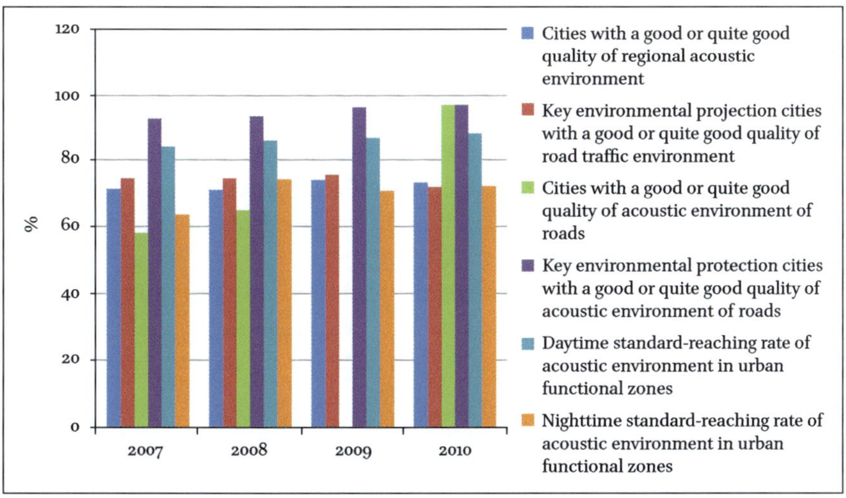

Figure A.17: Urban Acoustic Environment in China, 2007–2010

Table A.18: Pollutants directly discharged into the 4 largest seas, 2010.

Sea area	Wastewater (million tons)	COD (tons)	Ammonia nitrogen (tons)	Oil (tons)	Total phosphorus (tons)
Bohai Sea	181	7,900	1,100	74.7	66.9
Yellow Sea	878	45,000	5,300	81.5	774.3
East China Sea	2,978	118,700	11,200	598	1068
South China Sea	1,055	47,700	5,300	460.8	991.4

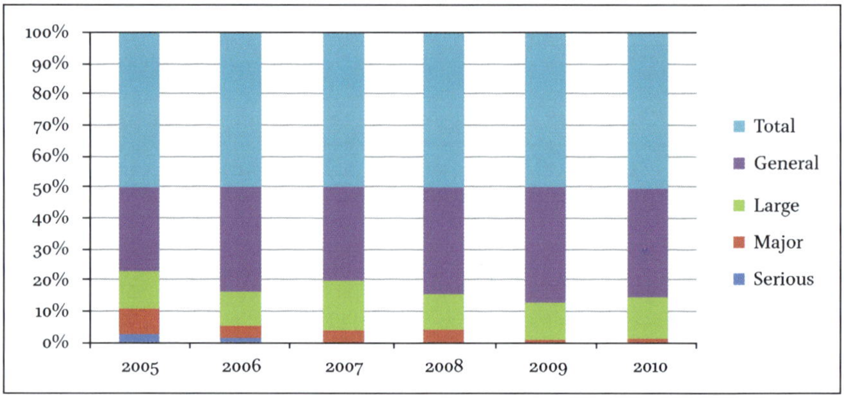

Figure A.18: Emergency environmental events, 2005–2010.

III. *Solid Waste*

In 2009, the generation of industrial solid wastes nationwide was up 7.3% from the previous year, the amount of discharge was down 9.1%, while the recycled amount (including recycling the storage from the previous years) increased by 12%.

In 2010, the total generation of industrial solid waste in China was 2.409 billion tons, up 18.1% from the previous year. The quantity of solid waste discharge was 4.982 million tons, down 29.9% from the previous year. The quantity of integrated solid waste utilization, storage, and treatment was 1.617 billion tons, 239.183 million tons, and 572.638 million tons respectively, accounting for 67.1%, 9.9%, and 23.8% of the quantity. The total quantity of hazardous wastes generated was 15.868 million tons, while the quantity of integrated hazardous waste utilization, storage, and treatment was 9.768 million tons, 1.663 million tons, and 5.127 million tons respectively.

Table A.19: Generation and handling of industrial hazardous solid waste, 2010

Generation (million tons)		Integrated utilization (million tons)		Storage (million tons)		Treatment (million tons)	
Total	Hazardous waste	Total	Hazardous waste	Total	Hazardous waste	Total	Hazardous waste
2,409.435	15,868	1,617.72	9.768	239.183	1.663	572.638	5.127

Table A.20: Generation and discharge of industrial waste, 2004–2010.

	2004	2005	2006	2007	2008	2009	2010
Generation (million tons)	1,200	1,340	1,515	1,757	1,901	2,041	2,409.435
Integrated utilization (million tons)	678	770	926	1,104	1,235	1,383	1,617.720
Discharge (million tons)	17.92	16.54	13.02	11.97	7.82	7.1	4.982

Table A.21: Urban air quality, 2004–2010.

	2004	2005	2006	2007	2008	2009	2010
At or above Level II	38.6	60.3	62.4	60.5	71.6	82.5	82.8
Level III	41.2	29.1	28.5	36.1	26.9	16.2	15.5
Worse than Level III	20.2	10.6	9.1	3.4	1.5	1.3	1.7

IV. *Urban Environment*

A. *Air Quality*

Air quality nationwide in 2009 and 2010 was generally good and better than previous years, though pollution remained severe in certain cities.

In 2010, air quality was monitored in 471 cities at and above the county-level nationwide in 2010, and the monitored pollutants included sulfur dioxide, nitrogen dioxide, and inhalable particles. Air quality in 3.6% of the monitored cities reached the standard of Level I, 79.2% of Level II, 15.5% of Level III, and 1.7% were worse than Level III. Altogether 85.5% of the county-level cities are up to the standards, slightly higher than that of the cities at and above the prefecture-level.

Among the cities at the prefecture-level (including prefectures, autonomous prefectures, and leagues in Inner Mongolia), 3.3% reached the national air quality standard of Level I, 78.4% of Level II, 16.5% of Level III, and 1.8% worse than Level III.

In terms of annual average density of inhalable particles, 85% of cities were up to or above Level II, while 1.2% worse than Level III in 2010.

B. *Acid Rain*

Distribution of acid rain nationwide barely changed over the last few years, while the pollution remained severe. Acid rain mainly concentrated in the area along and south of the Yangtze River and east of the Tibetan

Table A.22: Frequency of occurrence of acid rain, 2010.

Frequency of acid rain	0	0–25%	25–50%	50–75%	≥75%
Number of cities	245	89	57	49	54
Percentage (%)	49.6	18.0	11.5	9.9	11.0

Table A.23: Occurrence of acid rain in China, 2006–2010.

	2006	2007	2008	2009	2010
Percentage of cities with acid rain	54.0	56.2	52.8	52.9	50.4
Percentage of cities with a frequency of acid rain above 25%	37.8	34.2	34.4	33.6	21.4
Percentage of cities with a frequency of acid rain above 75%	16.6	13.0	11.5	10.9	11.0

Plateau, including the majority of Zhejiang, Jiangxi, Hunan, and Fujian Provinces, the Yangtze River Delta, southern Anhui Province, western Hubei Province, southern Chongqing Municipality, southeastern Sichuan Province, northeastern Guizhou Province, northeastern Guanxi Zhuang Autonomous Region, and central Guangdong Province.

Among the 494 monitored cities and counties in 2010, 249 had the occurrence of acid rain, accounting for 50.4%. In total, 160 cities and counties had a frequency of occurrence above 25%, accounting for 32.4%, while 54 had a frequency above 75%, accounting for 11.0%.

C. *Measures and Actions*

On November 4, 2010, the MEP released the *Annual Report on the Prevention and Control of Vehicle Pollution in China (2010)* (中国机动车污染防治年报(2010年度)) to make public, for the first time in China, the emission of vehicle pollutants and to systematically introduce the progress of prevention and control of vehicle pollution. According to the report, vehicle pollution in China is increasingly serious, with vehicle exhaust gas being one of the major sources of air pollution in large and medium-size cities. Cars are the major emitters of vehicle pollutants, contributing more than 70% of the emitted carbon monoxide and hydrocarbon, and more than 90% of nitrogen oxide and particulates. With almost three decades of development, China has made great progress in environmental management of vehicles. Pollution from light cars manufactured in 2009 was cut by over 90% from that in 2000. Due to the fast upgrading of emission standards, vehicle emissions have not increased proportionally with the fast growing number of vehicles. Compared with the statistics in 1980, China's cars in use have expanded by 25 times, but the total emission has increased only 12 times, effectively easing the huge pressure of vehicle growth upon the environment.

D. *Acoustic Environment*

On October 29, 2009, the MEP submitted a report to the Environmental Protection and Resources Conservation Committee of the National People's Congress on the progress, problems, and suggestions for future work of the prevention and control of noise pollution in China. In concert with the committee, the Ministry also sent officers to Beijing and Zhejiang Province to investigate noise pollution, and completed an evaluative report on the work on the prevention and control of noise pollution.

In 2010, the acoustic environment in 73.7% of urban areas nationwide was good or quite good, while 72.5% of key environmental protection cities were in the same condition. The acoustic environment of road traffic in 97.3% of cities nationwide was good or quite good, while the figure for key environmental protection cities was also 97.3%. The acoustic environment in 88.4% of all urban functional zones was up to the standards in the daytime, and the figure was 72.8% in the evening.

Among the 331 monitored cities in 2010, 6% were of a good quality regarding regional acoustic environment, 67.7% quite good, 25.4% slightly polluted, and 0.9% moderately polluted. Compared with the previous year, cities with a good quality of regional acoustic environment increased by 0.1 of a percentage point, cities with a quite good quality went down 1.0 percentage point, slightly polluted cities went up 1.1 percentage points, and moderately polluted cities went down 0.2 percentage points.

The average equivalent sound level range of regional environmental noise in key environmental protection cities was 43.4–60.1 dB (A). About 72.5% of cities were of a good or quite good quality of regional acoustic environment, 26.6% slightly polluted, and 0.9% moderately polluted.

Among the 331 monitored cities, 68% were of a good quality of acoustic environment of road traffic, 29.3% were quite good, 1.2% were slightly polluted, 1.2% were moderately polluted, and 0.3% were heavily polluted. Compared with the previous year, cities with a good quality of acoustic environment of road traffic went up by 0.9 percentage points, cities with a quite good quality went up 1.8 percentage points, slightly polluted went down 3.0 percentage points, moderately polluted went up 0.3 percentage points, and the figure was unchanged for the heavily polluted cities.

The average equivalent sound level range of road traffic noise in key environmental protection cities in 2010 was 63.6–73.3 dB (A). About 56.6% of cities were of a good quality of acoustic environment of road traffic, 40.7% quite good, 0.9% slightly polluted, and 1.8% moderately polluted.

Noise in urban functional zones: In the 248 monitored cities, 7,621 point-times were up to the standard in the daytime annually, accounting for 88.4% of the daytime monitoring point-times. The figure was 6,276 point-times in the evening, accounting for 72.8% of the nighttime monitoring point-times. About 87.8% of the urban functional zones in key environmental protection cities were up to the standard in the daytime, while the figure for the evening was 68.8%.

The daytime percentage of functional zones reaching the standards was higher than the nighttime figure, while functional zones of Category 3 had a higher rate than the other types of zones.

Table A.24: Monitoring points reaching the standards in urban functional zones nationwide, 2010.

Types of functional zones	Category 0		Category 1		Category 2		Category 3		Category 4	
	Day	Night	Day	Night	Day	Night	Day	Night	Day	Night
Point-times up to the standards	116	105	1,687	1,521	2,150	2,024	1,579	1,503	2,089	1,123
Monitoring point-times	193	193	1,969	1,969	2,456	2,456	1,673	1,673	2,334	2,334
Standard-reaching rate (%)	60.1	54.4	85.7	77.2	87.5	82.4	94.4	89.8	89.5	48.1

Table A.25: Urban acoustic environment in China, 2007–2010.

	2007	2008	2009	2010
Cities with a good or quite good quality of regional acoustic environment	72.0	71.7	74.6	73.7
Key environmental protection cities with a good or quite good quality of road traffic environment	75.2	75.2	76.1	72.5
Cities with a good or quite good quality of acoustic environment of roads	58.6	65.3		97.3
Key environmental protection cities with a good or quite good quality of acoustic environment of roads	92.9	93.8	96.5	97.3
Daytime standard-reaching rate of acoustic environment in urban functional zones	84.7	86.4	87.1	88.4
Nighttime standard-reaching rate of acoustic environment in urban functional zones	64.1	74.7	71.3	72.8

V. *Serious and Major Environmental Events in China*

In 2009, the MEP received reports of and properly handled 171 emergency environmental events, an increase of 26.7% from the previous year.

Categorized by the causes, there were 63 events caused by work safety accidents, accounting for 36.84% of the total with an increase of 6 events from the previous year. There were 52 traffic accidents, accounting for 30.41%, 27 more than the previous year. Pollution discharge from enterprises caused 23 events, or 13.45% of the total. The figure was unchanged from the previous year. There were 33 environmental events caused by other factors, accounting for 19.3% of the total, with an increase of 3 events.

Categorized by the types of pollution, there were 80 water pollution events: 61 of air pollution, 3 of solid waste pollution, 16 of land pollution, 2 of oceanic pollution, and 9 other types of environmental pollution.

In 2010, the MEP received reports of and properly handled 156 emergency environmental events, down 8.78% from the previous year. Among these, there were 5 major environmental events, 41 large events, 109 general events, and 1 event whose severity has yet to be decided.

Categorized by the causes, 69 events were caused by work safety accidents, accounting for 44.2% of the total with an increase of 6 events from the previous year; 28, or 18%, were traffic, 24 less than in 2009; 17, or 10.9%, were pollution incidents from enterprises, 6 less than in 2009; and 42, or 26.9%, were natural disasters and other factors, 9 more than the previous year.

Categorized by the types of pollutions, there were 65 water pollution events, 66 air pollution events, 4 land pollution events, 10 oceanic pollution events, 1 noise pollution event, and 10 other events without pollutants.

Table A.26: Emergency environmental events, 2005–2010.

	2005	2006	2007	2008	2009	2010
Serious	4	3	1	0	2	0
Major	13	15	8	12	2	5
Large	18	35	35	31	41	41
General	41	108	66	92	126	109
Total	76	161	110	135	171	156

VI. *Public Survey on the Degree of Satisfaction with Environmental Conditions*

To find out how the public feels about the environmental conditions and the changing environmental trends in China, the MEP launched a public survey nationwide in 2009 on the degree of satisfaction with environmental conditions.

The survey questionnaire was conducted on about 6,000 urban and rural citizens in 30 provinces (autonomous regions and municipalities) excluding Hong Kong, Macau, Taiwan, and Xinjiang Uygur Autonomous Region. The major results are as follows:

About 55.6% of the people surveyed said they were "satisfied" or "mostly satisfied" with the current environmental conditions.

The surveyed had the highest degree of satisfaction with the quality of drinking water, and lowest satisfaction with noise pollution and refuse disposal.

In general, rural residents had a lower degree of satisfaction than their urban fellows, particularly in terms of refuse disposal and the quality of drinking water.

People from northeast China had the highest degree of satisfaction with the environmental conditions, while the lowest degree was found in north China.

In terms of the improvement of environmental conditions, 65.8% of those surveyed from urban areas and 57.9% from rural areas said they were "satisfied" or "mostly satisfied." People from northeast China had the highest degree of satisfaction with the environmental conditions, southwest China ranked second, while the lowest satisfaction was found in north China.

Public opinion and suggestions concentrate on the following aspects: while promoting economic and social development, more attentions should be paid to environmental protection; monitoring and supervision should be enhanced on the environmental conditions; and more efforts should be made in the management of the environment.

TERMINOLOGICAL GLOSSARY

Beijing Consensus (北京共识): The Washington Consensus is an expression used to define the liberal measures promoted by the IMF, World Bank, and U.S. Treasury. Although the Beijing Consensus has no precise definition, it is often used to describe an alternative economic development model for developing countries.

Black box operations (黑箱操作): This refers to efforts by the government to escape the supervision of society, the public, and the media by blocking information. Their actions are said to take place within a black box.

Cascade development (梯级开发): Strategy of building several smaller hydropower dams as opposed to one very large dam.

China Model (中国模式): This term is often used to vaguely describe the economic and political policies of China (state-guided capitalism and authoritarianism) which have contributed to its economic rise after the Reform and Opening Up period.

Circular economy (循环经济): The China Daily describes the circular economy as a model for economic growth which aims at environmental protection, pollution prevention, and sustainable development.

Under this model, resources are used with higher efficiency and reused and recycled when possible, so that pollution is minimized and waste is reduced as much as possible. It also involves the transformation of industrial organization and allocation, urban infrastructure, environmental protection, technological paradigms, and social welfare distribution.

An important part in China's effort toward sustainable development, the circular economy was officially raised as a target for future growth in 2004. The State Council issued a file about promoting it in July 2005, making it a key guideline in the Eleventh Five-Year Plan and to achieve the specific goals in energy conservation and pollution reduction before 2020.

City mineral resources (城市矿产): This refers to recyclable iron and steel, nonferrous metals, rare metals, plastics, rubber that is generated and stored in discarded mechanical and electronic equipment, electric wires and cables, communication devices, automobiles, household electric appliances, electronics, metal and plastic packaging materials, and other discarded materials produced during the process of industrialization and urbanization. In 2010, the National Development and Reform Commission (NDRC) and Ministry of Finance organized and developed a model base for "city mineral resources" with a plan to establish 30 model bases within five years to reutilize minerals that are stored in urban discarded waste.

"Company Plus Farmer" model (公司加农戶): This is a model of agricultural investment that involves a group of farm households that work in combination with a large company. This mode of operation began in the 1980s and is aimed at providing a win-win situation for all involved by enabling the produce from farms to be brought to a larger market. But it has been criticized as being unbalanced in favor the company.

"Development of a New Socialist Countryside" policy (社会主义新农村建设): This policy is aimed at bringing the economic standing of China's 800 million rural residents up to the level of those living in cities. The policy focuses more on the redistribution of resources and rebalancing of incomes, rather than economic development, through rural investments, agricultural subsidies, and improved social services.

Farmer guerrillas (农民游击队): This refers to rural people who collect garbage and recyclables in the city to sell.

Going Global Strategy (走出去战略): This is a government policy that encourages Chinese enterprises to invest abroad. Since 1999, the government has introduced several schemes to assist domestic companies in investing overseas.

Green credit policy (绿色信贷): The policy of not issuing credit to heavily polluting companies until they implement environmental protection measures in order to comply with relevant laws. Also, companies that are found to violate environmental standards after receiving loans could have their loans recalled. This policy was implemented in 2007 and strengthened in 2008.

Low-carbon economy (低碳经济) (LCE): A low-carbon economy is characterized by a certain level of carbon productivity (economic output per unit of carbon emission) aimed at high economic growth with low-carbon emissions. Low-carbon development is the transition towards LCE and it will be achieved through technological breakthroughs and institutional regulation. The results of low-carbon development are more efficient energy use, an optimized energy structure, and more rational consumer behavior. In December 2006, for the first time, the Chinese government put forth the idea of developing a LCE in the *National Assessment Report on Climate Change* (气候变化国家评估报告).

Man-made forest (人工纯林): A forest that is mostly populated by a single species of plant. Since 1980, a campaign of "Hilly Land Allotted for Private Use" has pushed people to replace natural forests with man-made commercial forests composed of fruit trees, bamboos, pines, Chinese firs, rubber trees, and eucalyptus. China's man-made forests contribute to ecological deterioration and rank first in the world in terms of size.

Three high industries (三高产业): These are industries with high energy consumption, high emissions, and high pollution.

Three rivers and three lakes (三河三湖): This refers to Huaihe River, Haihe River, Liaohe River, Lake Taihu, Lake Dianchi, and Lake Chaohu.

Urban diseases (城市病): This term refers to problems caused by urbanization in China's big cities. These problems arise from a major lack of resources and shortfall in infrastructure capacity coupled with a rapidly growing population. The problems, which include traffic congestion and pollution, are thought to pose a hidden danger to the stability of China's large cities and are a result of a lack of proper city planning.

"Western Development" strategy (西部大开发): This is a central government policy to develop China's western regions, which includes the municipality of Chongqing, the provinces of Gansu, Guizhou, Qinghai, Shaanxi, Sichuan, and Yunnan, and the autonomous regions of Guangxi, Inner Mongolia, Ningxia, Tibet, and Xinjiang. The policy aims to promote economic growth through encouraging both foreign and domestic investments into oil and other natural resource exploitation, poverty alleviation, ecological restoration, and large infrastructure projects.

INDEX TERMS